# FIELD GUIDE TO
# Whales, Dolphins and Porpoises
# NORTH AMERICA

Mark Carwardine

Illustrated by
Martin Camm

With additional illustrations by
Rebecca Robinson
Toni Llobet

BLOOMSBURY WILDLIFE
LONDON · OXFORD · NEW YORK · NEW DELHI · SYDNEY

*For Mum and Dad*

BLOOMSBURY WILDLIFE
Bloomsbury Publishing Plc
50 Bedford Square, London, WC1B 3DP, UK
Bloomsbury Publishing Ireland Limited,
29 Earlsfort Terrace, Dublin 2, D02 AY28, Ireland

BLOOMSBURY, BLOOMSBURY WILDLIFE and the Diana logo are trademarks of
Bloomsbury Publishing Plc

First published in the United Kingdom 2026

Copyright © Mark Carwardine, 2026
Illustrations © as credited on p. 203, 2026

Mark Carwardine has asserted his right under the Copyright, Designs and Patents Act, 1988,
to be identified as Author of this work

For legal purposes the Acknowledgements on p. 204
constitute an extension of this copyright page

All rights reserved. No part of this publication may be: i) reproduced or transmitted in any form, electronic or mechanical, including photocopying, recording or by means of any information storage or retrieval system without prior permission in writing from the publishers; or ii) used or reproduced in any way for the training, development or operation of artificial intelligence (AI) technologies, including generative AI technologies. The rights holders expressly reserve this publication from the text and data mining exception as per Article 4(3) of the Digital Single Market Directive (EU) 2019/790

Bloomsbury Publishing Plc does not have any control over, or responsibility for, any third-party websites referred to or in this book. All internet addresses given in this book were correct at the time of going to press. The author and publisher regret any inconvenience caused if addresses have changed or sites have ceased to exist, but can accept no responsibility for any such changes

A catalogue record for this book is available from the British Library

Library of Congress Cataloguing-in-Publication data has been applied for

ISBN: PB: 978-1-3994-2441-7 ePub: 978-1-3994-2440-0; ePDF: 978-1-3994-2443-1

2 4 6 8 10 9 7 5 3 1

Design by Julie Dando
Maps and scale drawings by Julie Dando
Printed and bound in Dubai by Oriental Press.

To find out more about our authors and books visit www.bloomsbury.com and sign up for our newsletters
For product safety related questions contact productsafety@bloomsbury.com

# CONTENTS

| | |
|---|---|
| How to use this book | 4 |
| The challenges of identification | 6 |
| Cetacean topography | 10 |
| Barnacles to whale lice | 12 |
| Quick ID guides | |
|     Bow-riding dolphins and porpoises | 18 |
|     Identifying whales by their flukes | 20 |
|     Identifying whales by their blows | 22 |
|     Identifying male beaked whales by their lower jaws | 24 |
|     Cetaceans of the North American Atlantic Ocean | 26 |
|     Cetaceans of the North American Pacific Ocean | 28 |
|     Cetaceans of the North American Arctic Ocean | 30 |
| Information for working at sea | 31 |
| **Right and bowhead whales** (family Balaenidae) | 32 |
| **Gray whale** (family Eschrichtiidae) | 44 |
| **Rorquals** (family Balaenopteridae) | 48 |
| **Sperm whales** (families Physeteridae, Kogiidae) | 78 |
| **Narwhal and beluga** (family Monodontidae) | 86 |
| **Beaked whales** (family Ziphiidae) | 92 |
| **Blackfish** (family Delphinidae) | 124 |
| **Shorter-beaked oceanic dolphins** (family Delphinidae) | 150 |
| **Longer-beaked oceanic dolphins** (family Delphinidae) | 164 |
| **Porpoises** (family Phocoenidae) | 190 |
| Caring for whales, dolphins and porpoises | 196 |
| Glossary | 197 |
| North American species checklist | 200 |
| Sources and resources | 202 |
| Artists' biographies | 203 |
| Image credits | 203 |
| Acknowledgements | 204 |
| Index | 206 |

# HOW TO USE THIS BOOK

## SPECIES NAMES AND AUTHORITY
The currently accepted common name of the species is given in English. Many have alternative names – some in common usage, many rare or regional and others historical – and these are all provided as well. Names in other languages are not given, purely for reasons of space (each species has lots of different names in almost every language, so the number of possibilities is huge). Common names can vary but each species has only one scientific name, which is given in italics. The authority (the person credited with first publishing the name of the species) is also provided. If the original name has changed – e.g. the species has been transferred to a new genus (not the one in which it was originally described) – the original author's name and date are put in parentheses, for example: (Linnaeus, 1758). If it is not in parentheses, it is still in the original genus.

## IUCN STATUS
This is given for each species (together with the year of assessment). This is the official status on the International Union for Conservation of Nature Red List of Threatened Species, which is the most authoritative, objective and comprehensive list of species that have been rigorously evaluated for their risk of extinction. There are eight categories: Extinct (no reasonable doubt that the last individual has died); Extinct in the Wild (known only to survive in captivity or as a naturalized population well outside its historic range); Critically Endangered (extremely high risk of extinction in the wild); Endangered (very high risk of extinction in the wild); Vulnerable (high risk of extinction in the wild); Near Threatened (likely to qualify for a threatened category in the future); Least Concern (does not qualify for a more at-risk category); and Data Deficient (not enough data to make an assessment). A ninth category – Not Evaluated – is for species not yet assessed.

## POPULATION
Counting cetaceans is notoriously difficult and, inevitably, estimates are of variable accuracy (and, of course, some are more recent than others). However, while most population figures should be viewed cautiously, this section provides what is known and gives a guide to the estimated abundance of the species.

## CLASSIFICATION AND TAXONOMY
The guide follows the taxonomic arrangement, and the scientific names of species and subspecies, recommended by the Society for Marine Mammalogy's Committee on Taxonomy. This authoritative list is updated annually. It is in a constant state of flux, with new species being discovered and named and other species being combined or split as more information comes to light. Notes about the taxonomy of each species are given as background information and to point out any possible future changes.

## SIZE
The species silhouette shows the typical size of an adult to scale against a human diver (all human silhouettes represent 1.8m). L = length, WT = weight. Size ranges for adult males, adult females and calves are given, together with the maximum lengths and weights recorded to date. All measurements are in metric.

## AT A GLANCE
An abbreviated list of key features is provided for quick reference – outlining the main features to concentrate on to achieve a correct identification – including distribution, size and key physical and behavioral features. Size is based on a simple scale: small (up to 3m), medium (4–10m), large (11–15m), and extra large (more than 15m).

## ILLUSTRATIONS
The main illustration for each species shows a typical side view (if males and females look different they are illustrated separately). Other images show the opposite side view (if it is different), the upperside and underside, any subspecies, regional variations or age variations, and any relevant close-ups of dorsal fins, beaks, flippers or other features (sometimes including comparisons with similar species). Each illustration has its own annotations, pointing out the main distinguishing features useful for identification and any other interesting characteristics. See Cetacean Topography on p. 10 for a breakdown of the names of the various parts of a cetacean. There are additional illustrations showing the shape and size of the blow or spout from behind (for the larger species) and a typical dive sequence.

## DISTRIBUTION
The distribution maps show the known and/or presumed range of each species. For many, there is very little information and putting together a distribution map is like putting together a jigsaw puzzle with a few or many parts missing. The accompanying text explains in more detail what is known about the distribution and where there are gaps in our knowledge, as well as providing more information on habitat and depth preferences (within the mapped range, each species

occurs only in appropriate habitats and depths), migrations and other movements, and extralimital records.

## BEHAVIOR
Behavior such as breaching, spyhopping, lobtailing and reaction to boats can provide valuable clues in cetacean identification, and details are given in this section. Bear in mind, though, that behavior can vary enormously between individuals, from one region to another, with the seasons and according to many other factors.

## TEETH/BALEEN PLATES
The number of teeth is normally given as a count for each row (i.e. two rows in the upper jaw and two in the lower jaw). However, for simplicity, the figures in this guide show the range in total number of teeth in the upper and the same for the lower jaw. Baleen plates are present only in the upper jaw and the figure gives the total range for each side.

## FOOD AND FEEDING
This section includes short notes on the main prey species and foraging behavior.

## DIVING AND BLOW
The way a cetacean surfaces – the dive sequence – can be very distinctive in many species and is often a useful identification feature. This is illustrated, together with a short description of the main features to look for. This section also includes information on dive depth and dive times (with the maximum recorded to date) and blow characteristics (if a distinctive blow is visible).

## GROUP SIZE AND STRUCTURE
There is a great deal of variation in cetacean groupings, between individuals, regions and seasons, but the most commonly observed group sizes and structures are given, together with further information on variability.

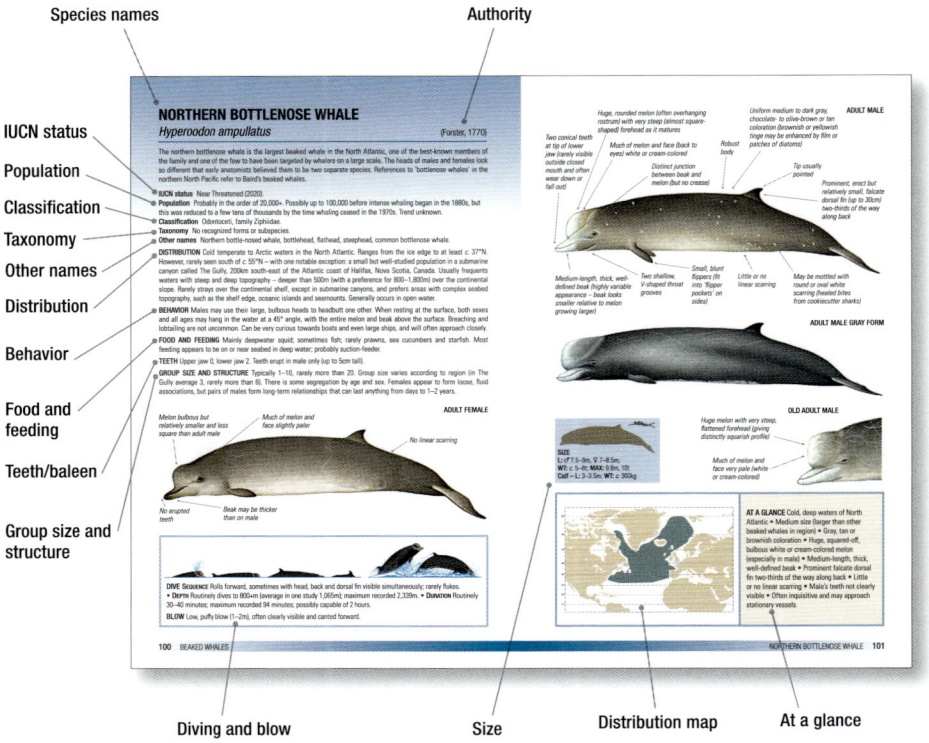

# THE CHALLENGES OF IDENTIFICATION

Identifying whales, dolphins and porpoises at sea can be enormously satisfying, but also quite challenging. In fact, it can be so difficult that even the world's experts are unable to identify every species they encounter – on most official surveys, at least some sightings have to be logged as 'unidentified'.

The trick is to use a relatively simple process of elimination, running through a mental checklist of 14 key features every time a new animal is encountered at sea. It is not often possible to use all of these features together and one alone is rarely enough for a positive identification. The best approach is to gather information on as many as possible before drawing any firm conclusions.

**1. Geographical location** There is not a single place in the world where every cetacean species has been recorded. In fact, there are not many places with records of more than a few dozen, so this immediately helps to cut down on the number of possibilities.

**2. Habitat** Just as cheetahs live on open plains rather than in jungles, and snow leopards prefer mountains to wetlands, most whales, dolphins and porpoises are adapted to specific marine or freshwater habitats. In this respect, marine charts can be surprisingly useful identification aids. Knowing the underwater topography could help to tell the difference between a minke whale (normally found over the continental shelf) and a superficially similar northern bottlenose whale (more likely to be seen over submarine canyons or in deep waters offshore).

**3. Size** It is difficult to estimate size accurately at sea, unless a direct comparison can be made with the length of the boat, a passing bird or an object in the water.

Remember that only a small portion of the animal (the top of the head and back, for example) may be visible at any one time. Larger species don't necessarily show more of themselves than smaller species, so size can be quite deceptive. It is therefore better to use four simple categories: small (up to 3m), medium (4–10m), large (11–15m), and extra large (more than 15m).

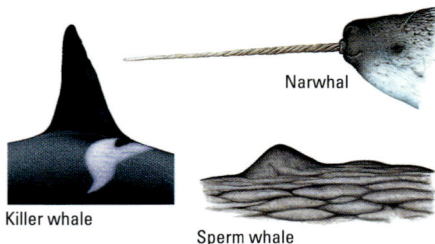

Killer whale

Narwhal

Sperm whale

**4. Unusual features** Some cetaceans have very unusual features, which can be used for a quick identification. These include the extraordinary long tusk of the male narwhal, the enormous dorsal fin of the male killer whale and the wrinkly skin of the sperm whale.

**5. Dorsal fin** The size, shape and position of the dorsal fin varies greatly between species and is a particularly useful aid to identification. Don't forget to look for any distinctive colors or markings on the fin.

Spinner dolphin

Pacific white-sided dolphin

Pygmy beaked whale

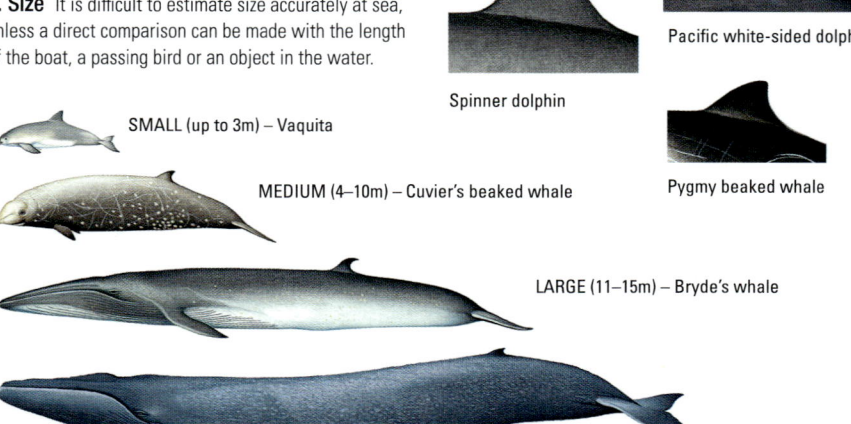

SMALL (up to 3m) – Vaquita

MEDIUM (4–10m) – Cuvier's beaked whale

LARGE (11–15m) – Bryde's whale

EXTRA LARGE (more than 15m) – Blue whale

6  INTRODUCTION

**6. Flippers** The length, color and shape of the flippers, as well as their position on the animal's body, vary greatly from one species to another. It is not always possible to see them, but flippers can be useful for identification in some species – in the humpback whale, for example, they are unmistakable.

Risso's dolphin

Bowhead whale

Rough-toothed dolphin

Humpback whale

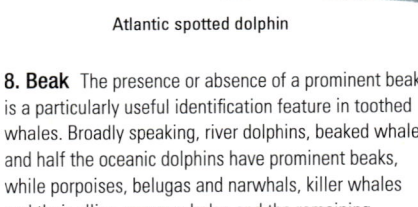
Atlantic spotted dolphin

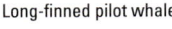
Long-finned pilot whale

**7. Body shape** Much of the time, whales, dolphins and porpoises do not show enough of themselves to provide an overall impression of their shape. Sometimes, however, this can be a useful feature. Is the animal stocky or slim, for example? The shape of the melon (forehead) can also be distinctive.

**8. Beak** The presence or absence of a prominent beak is a particularly useful identification feature in toothed whales. Broadly speaking, river dolphins, beaked whales and half the oceanic dolphins have prominent beaks, while porpoises, belugas and narwhals, killer whales and their allies, sperm whales and the remaining oceanic dolphins do not. There is also great variation in the beak length from one species to another. And try to see if there is a smooth transition from the top of the head to the end of the snout (as in rough-toothed dolphins, for example) or a distinct crease (as in Atlantic spotted dolphins).

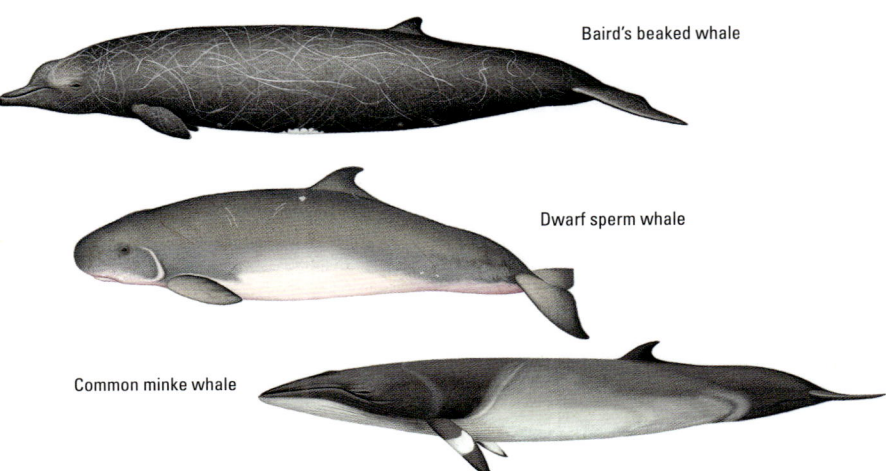

Baird's beaked whale

Dwarf sperm whale

Common minke whale

THE CHALLENGES OF IDENTIFICATION

**9. Color and markings** Many cetaceans are surprisingly colorful and have distinctive markings such as body stripes or eye patches. Bear in mind that colors at sea vary according to water clarity and light conditions, and the animal can appear much darker than normal if viewed against the sun.

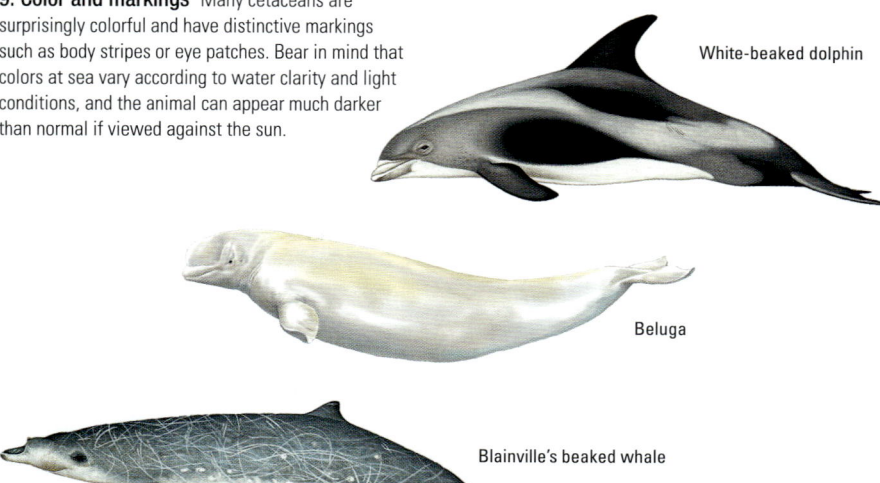

White-beaked dolphin

Beluga

Blainville's beaked whale

**10. Flukes** The flukes can be important in identifying larger whales. Some species lift their flukes high into the air before they dive, while others do not, and that alone can help to tell one from another. It is also worth checking the shape of the flukes, looking for any distinctive markings and noticing whether or not there is a deep notch between the trailing edges.

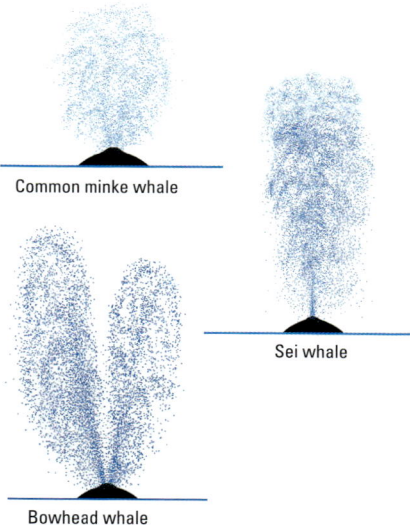

Common minke whale

Sei whale

Bowhead whale

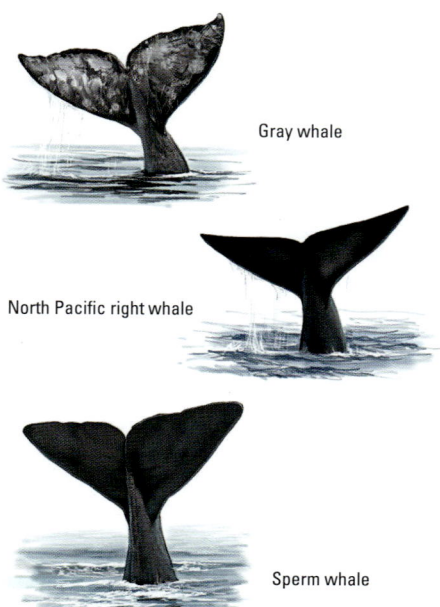

Gray whale

North Pacific right whale

Sperm whale

**11. Blow or spout** The blow is particularly distinctive in larger whales. It varies in height, shape, density and visibility between species and can be extremely useful for identification, especially on calmer days. But identifying a blow is not easy – if it is raining or windy, the blow can be bent out of shape, and there are variations between individuals – and the first blow after a deep dive tends to be stronger than the rest. However, experienced observers can often tell one species from another just by the blow, even from a considerable distance.

Fin whale

Longman's beaked whale

Dwarf sperm whale

**12. Dive sequence** The dive sequence can be surprisingly distinctive in many species. Variations include: the angle at which the head breaks the surface; how much of the head and beak (if present) are visible; whether or not the dorsal fin and blowhole are visible at the same time; whether the animal arches its back to dive (and how much it arches) or whether it merely sinks below the surface; the time interval between breaths; and the number of breaths before a deep dive.

**13. Behavior** Some species are more active at the surface than others, so any unusual behavior can sometimes be useful for identification purposes. Did it leap out of the water, for example, or was its behavior quite cryptic? The reaction to boats can also be helpful: common bottlenose dolphins may race over and bow-ride, while the similar-looking Atlantic humpback dolphins tend to be more shy and will not bow-ride.

Spinner dolphin

**14. Group size** Since some species are highly gregarious, while others tend to live alone or in small groups, it is worth noting the number of animals seen together. Estimating group size is notoriously difficult, because the animals are mobile and frequently change direction and, at the time of counting, any number of them can be hidden beneath the surface. Estimating the size of a large school of active dolphins is especially challenging – the tendency is usually to underestimate.

It is often tempting to guess the identification of an unusual whale, dolphin or porpoise that you have not seen very clearly. However, working hard at identification – and then enjoying the satisfaction of knowing that an animal has been identified correctly – is what makes a real expert in the long term. It is perfectly acceptable to record simply 'unidentified dolphin', 'unidentified whale' or 'unidentified beaked whale' if a more accurate identification is not possible. If you write detailed notes at the time, and then see the same species again in the future, it may be possible to turn a sighting previously recorded as 'unidentified' into a positive identification, days, weeks, months or even years later.

It does get easier with practice. After a while, you look at a whale and it triggers a switch in your brain. You know what it is likely to be – a single species or, perhaps, several possibilities – from its 'jizz' (the overall impression). At a glance, you get an overall impression that is more instinctive than something that can easily be put into words.

Part of the fun is that cetaceans are unpredictable. Never say 'never' on a whale-watching trip: just because the distribution maps suggest you're unlikely to see a particular species in a particular area, it doesn't mean to say it couldn't pop up anywhere at any time; and just because the guidebooks say that one species of whale doesn't fluke, it doesn't mean to say the particular individual next to your boat won't prove everyone wrong by lifting its tail high into the air.

# CETACEAN TOPOGRAPHY

## PARTS OF A BALEEN WHALE

## PARTS OF A BALEEN WHALE

## PARTS OF A TOOTHED WHALE

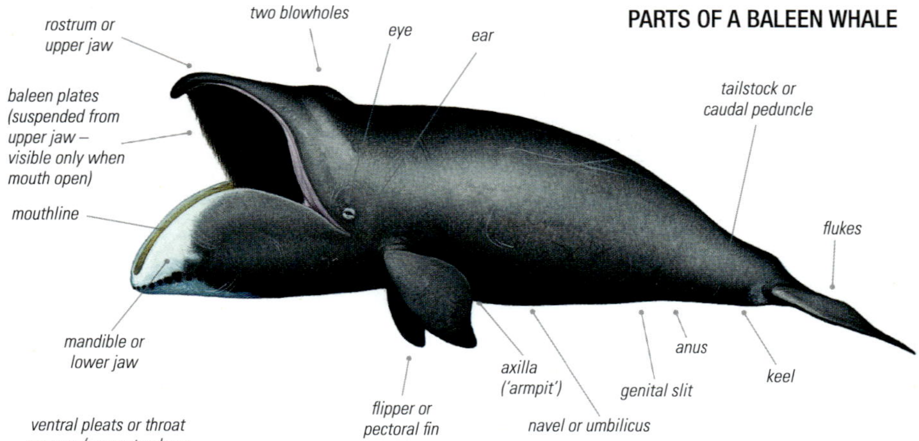

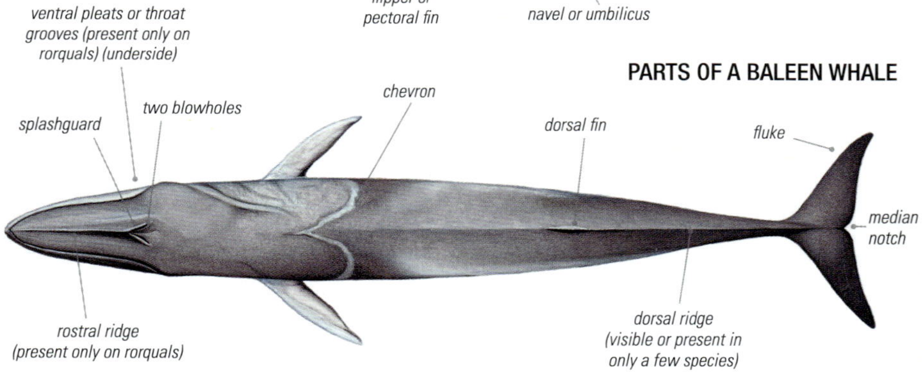

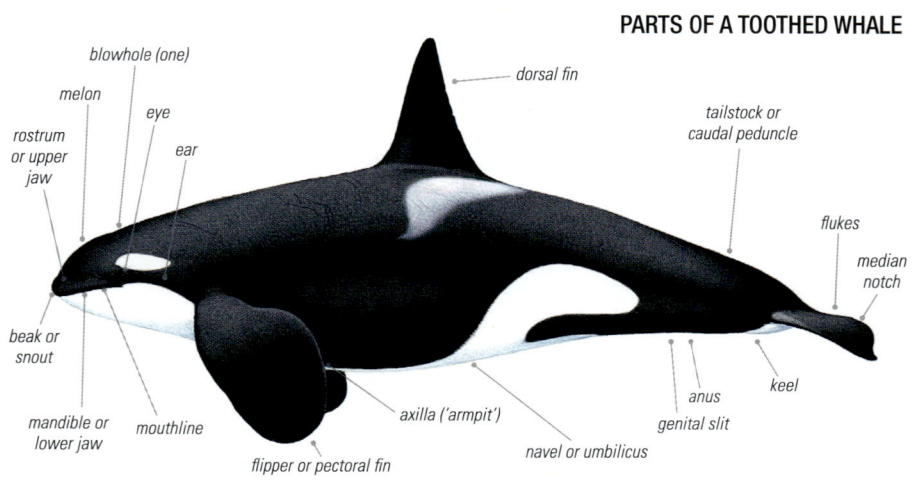

## PARTS OF A TOOTHED WHALE

## DORSAL FIN

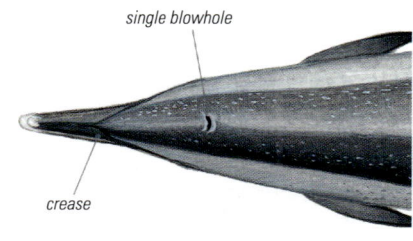
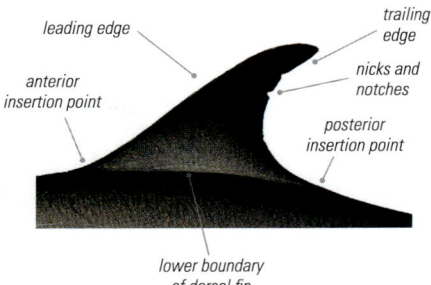

## HOW TO SEX A CETACEAN

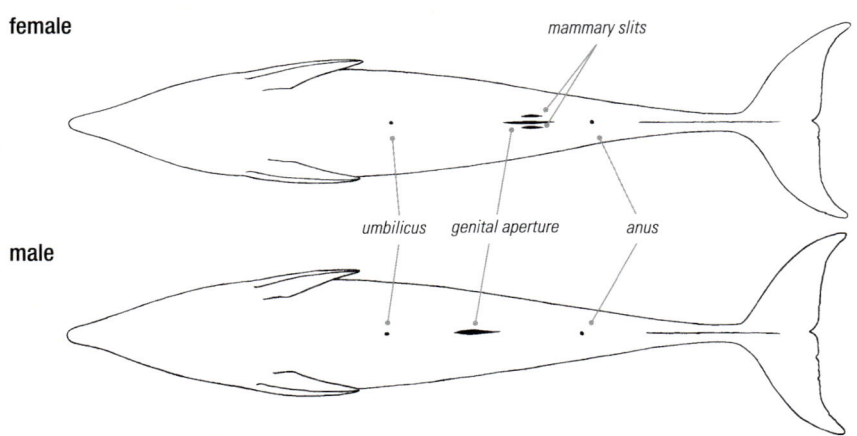

## PARTS OF A CETACEAN SKELETON

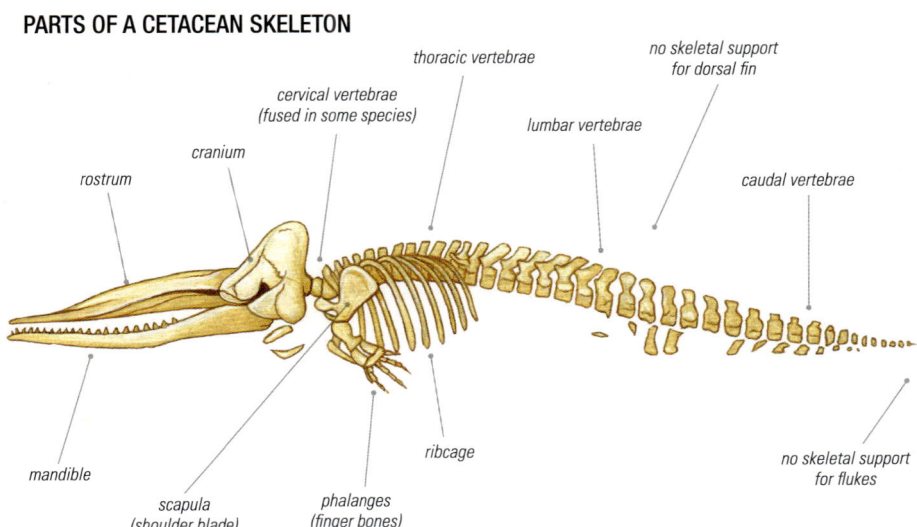

# BARNACLES TO WHALE LICE

An introduction to the ectoparasites and other creatures living on cetaceans.

## BARNACLES

Barnacles are crustaceans that as adults permanently attach themselves to a variety of inanimate and animate objects. They occur everywhere from the tropics to the poles and from the seashore to the ocean depths. There are more than 1,000 species altogether, but only eight live on cetaceans. They settle in the greatest numbers on large baleen whales, but also occur on some toothed whales.

Barnacles are not true parasites, because they do not obtain nourishment from their hosts and do not appear to cause infection or inflammation. They are hitch-hikers, getting a free ride as they filter their planktonic food out of the water that passes over them. However, heavy infestations of barnacles can create drag, reducing swimming efficiency, and may cause irritations. In some cases, they may actually be beneficial to the whales – humpbacks, for example, use their barnacle-encrusted flippers as weapons (like 'knuckledusters') in competition with other males and to fight off attacks by tiger sharks, false killer whales and other species. Most barnacles are hermaphroditic (individuals possess the reproductive structures of both sexes) and their life cycle usually includes six free-swimming stages; the final larval stage searches for a place to settle and anchors itself with a specially secreted cement. The breeding season of barnacles that cling to whales is probably synchronous with the breeding season of their hosts.

There are three main types of whale barnacle:

ACORN BARNACLE *Cryptolepas rhachianecti*

**Acorn barnacles** Four species in three genera live on cetaceans: *Coronula diadema* (found in large numbers on most humpback whales – one individual had 450kg of this species attached to its body – and in very small numbers on some other baleen whales and sperm whales); *Coronula reginae* (common on humpback whales and found in very small numbers on blue, fin, sei, right and sperm whales); *Cryptolepas rhachianecti* (found in large numbers on most gray whales); and *Cetopirus complanatus* (found on right whales). Named for their superficial resemblance to the acorns of oak trees, acorn barnacles are mound-shaped, and most often occur on the head, flippers and flukes.

RABBIT-EAR BARNACLE (*Conchoderma auritum*) attached to an acorn barnacle on a humpback whale's flipper

**Stalked, goose or gooseneck barnacles** Two species in one genus live on cetaceans (though there are isolated records of *Lepas* and *Pollicepes* species on whales). Stalked barnacles require a hard surface for attachment and frequently connect to other barnacles rather than directly onto a cetacean's skin. The rabbit-ear barnacle *Conchoderma auritum* is common worldwide on humpback whales – it usually attaches to the acorn barnacle *Coronula diadema* rather than the whale itself – and is found in very small numbers on blue, fin and sperm whales. Up to 7cm across, it also sometimes attaches to the teeth of adult male beaked whales and occasionally occurs on baleen plates. The much smaller *Conchoderma virgatum* – no more than 3.5mm across – is found mainly in the tropics and sub-tropics. It usually occurs on inanimate objects, such as driftwood or ships' hulls, but also attaches to sea snakes, ocean sunfish and some marine mammals (including, rarely, baleen whales). It occasionally settles directly onto a cetacean's skin, but normally attaches to a parasitic copepod or whale louse.

**Pseudo-stalked barnacles** Two species in two genera live on cetaceans. *Xenobalanus globicipitis* is unusual because it superficially resembles a stalked barnacle (it has developed an aberrant pseudo stalk). This curious dark, worm-like animal – which can be up to 5cm long – hangs from the trailing edges of the tails, dorsal fins and flippers of at least 34 cetacean species in tropical, sub-tropical and temperate waters worldwide. It is sometimes on the rostrum, and even on baleen plates and teeth. Pseudo-stalked barnacles

**PSEUDO-STALKED BARNACLES (*Xenobalanus globicipitis*)**

settle in the greatest numbers on the larger baleen whales, but are also found on killer whales, common bottlenose dolphins, Indo-Pacific finless porpoises and many other species. There can be just one or as many as 100 in a cluster. They burrow into the skin (and blubber) to various depths and, once attached with the shell base embedded in the host, do not move. *Tubicinella major* is found among the callosities on southern right whales. *Xenobalanus* still protrudes significantly, but *Tubicinella* is so deeply embedded that only the tip is exposed for feeding.

## REMORAS

Remoras (otherwise known as suckerfish or diskfish) occur mostly in tropical to warm temperate waters worldwide. Elongate and round in cross section, they use a flat, oval-shaped suction disc on the top of their head — which makes them look as if they are upside down — to stick onto cetaceans, sirenians, sharks, sea turtles and any other large marine object (including ships, submarines and even occasionally human divers). A modified dorsal fin, the disc resembles venetian-blind slats and, when the slats are lifted, they create a strong vacuum, enabling the fish to suck onto its host. There are also tooth-like projections, called spinules, that help to prevent slippage. The disc is so effective that, with fine control, a remora can slide quickly around the host's body without falling off (although they are capable of free-swimming).

Benefits to the remoras include hitching a free ride, protection from predators, a surface for meeting of males and females, and a swift passage of water over their gills (they cannot survive in still water). They opportunistically feed on parasitic copepods (which account for most of their diet), zooplankton and smaller nekton in the passing water, scraps from their hosts' meals, sloughing whale skin and whale faeces. Cetaceans are rarely harmed by remoras, which do not normally hurt or leave scars (although they may leave temporary marks). However, at least in the case of some spinner, pantropical spotted and bottlenose dolphins, they can cause persistent damage — usually large raw patches just below the dorsal fin — that potentially could become infected. They can also cause hydrodynamic drag (they have been dubbed hydrodynamic parasites for reducing swimming efficiency) and may be irritating. It is unknown why cetaceans generally tolerate them — there may be as yet unknown benefits — but some dolphin species have been seen biting off remoras from each other, and may leap and spin to dislodge them from uncomfortable positions.

**WHALESUCKER (*Remora australis*) adult on dolphin**

There are eight species altogether in the family Echeneidae, and only one of them is known to attach regularly to cetaceans. The whalesucker (*Remora australis* — formerly *Remilegia australis*) occurs only on cetaceans (although, if it falls off, it may attach to any passing animal or object until a preferred host passes nearby); pale sky blue, it is found in warm pelagic waters worldwide and grows to a length of 62cm. Different

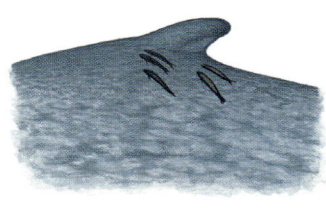

**WHALESUCKER (*Remora australis*) young on blue whale**

**WHALESUCKER (*Remora australis*)**

SHARKSUCKER (*Echeneis naucrates*)

sizes probably represent different life-history stages of this species (possibly with different diets – e.g. small individuals on blue whales could be young remora feeding on sloughed whale skin, larger individuals on common dolphins could be older remora feeding on larger food items). The sharksucker (*Echeneis naucrates*), which reaches about 90cm, has been found only on common bottlenose dolphins, but may occur on other cetaceans (identification in the field is difficult).

Remora larvae do not appear to be free-living in the plankton layer, but may hang onto the baleen plates of whales until they develop a disc.

## LAMPREYS

Lampreys belong to a primitive class of cartilaginous jawless fish, the Agnatha. Eel-shaped and lacking the scales, paired fins and jaws of true fish, they have a disc-shaped suction-cup around their mouth – which is wider than the mouth itself – ringed with sharp, horny teeth. They latch onto their unfortunate host and use their rough tongue to rasp away the animal's flesh, in order to feed on the blood and body fluids. Rather like leeches, they produce anti-coagulants to prevent the blood from clotting and increase the flow. After spending several years at sea, lampreys stop feeding and migrate to fresh water to spawn.

There are 43 species altogether, ranging from 15cm to 1.2m in length and living in coastal, cool temperate waters worldwide (except Africa). Thirty-two of these are almost always confined to fresh water and 18 are parasitic. Few lamprey–cetacean interactions have been described in detail, but two species in particular are known to attack cetaceans: the Pacific lamprey (*Lampetra tridentata* – formerly *Entosphenus tridentatus*), which is found in the North Pacific; and the sea lamprey (*Petromyzon marinus*), found in the North Atlantic, which is the largest lamprey species.

## COOKIECUTTER SHARK

The cigar-shaped cookiecutter shark is a strange-looking member of the family Dalatiidae. Reaching a maximum length of only about 50cm, but armed with large, serrated teeth on the lower jaw and tiny, spike-like teeth on the upper jaw, it is a menace to other marine animals. Its name comes from its nasty habit of biting neat, round, cookiecutter-shaped chunks of flesh from a variety of marine megafauna – especially cetaceans, but probably any other marine megafauna, including seals, dugongs, tuna and sharks (and a human on one verified occasion in Hawai'i in 2003). An ambush predator, with very large eyes for better vision in the dark depths, it attaches itself to its prey with its lips, then inserts its hook-like upper teeth and proportionately massive lower teeth, and spins its body to remove a plug of flesh. It leaves behind an oval or round hole or 'pit' up to *c.* 10cm across and *c.* 4cm deep (though usually narrower and shallower).

Cookiecutter sharks usually spend the day in deep

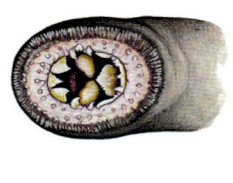

PACIFIC LAMPREY (*Lampetra tridentata*)

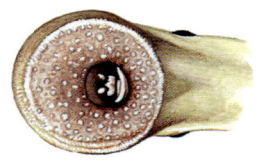

NORTH ATLANTIC LAMPREY (*Petromyzon marinus*)

SMALLTOOTH COOKIECUTTER SHARK (*Isistius brasiliensis*)

waters – sometimes down to 3.5km – then migrate to surface waters at night to feed. Light-emitting photophores, scattered on the belly and other parts of the body, lure would-be predators close enough to attack and then the predators become the prey.

There are three known species: smalltooth cookiecutter (*Isistius brasiliensis*), which is probably responsible for most cetacean attacks; the apparently rare bigtooth cookiecutter (*I. plutodus*); and the poorly known South China cookiecutter (*I. labialis*). The smalltooth cookiecutter occurs at water depths of at least 1,000m during the day, and is believed to migrate towards surface waters at night (when most attacks on cetaceans presumably occur). It lives in temperate to tropical seas, most commonly between 20°N and 20°S (though on occasion to 35°), which suggests that cetaceans pockmarked with the tell-tale oval or round scars of their bites have been in warmer waters at some point (species that live year-round in colder waters generally do not have cookiecutter shark bites).

Cookiecutter shark bites – which are much deeper than lamprey bites – have been recorded on at least 49 species of cetacean. They are harmful and painful, but do not normally cause death (except, possibly, when the sharks attack young calves or bite through the stomach wall). It can take several months for the bites to heal, though the scars may remain for many years (or even for life). In addition to their unique 'hit-and-run' feeding behaviour, cookiecutter sharks also eat free-living squid, small fish and crustaceans.

## WHALE LICE

Whale lice belong to an order of crustaceans called amphipods and are all in the family Cyamidae (their correct name is 'cyamid amphipods'). They are not 'lice' (which are insects) and were named incorrectly by whalers in the 1800s, who thought they looked and moved like human lice.

Seven genera and 28 species have been positively identified living on cetaceans: *Cyamus* (14 species), *Isocyamus* (5), *Neocyamus* (1), *Platycyamus* (2), *Orcinocyamus* (1), *Scutocyamus* (2) and *Syncyamus* (3). They tend to be host-specific on mysticetes and generalists on odontocetes. Some species of whale louse live exclusively on a single species of whale: e.g. *Cyamus boopis* lives on humpback whales (although there is one case of this species on a southern right whale) and *Cyamus scammoni*, *C. kessleri* and *C. eschrichtii* live on gray whales; *C. catodontis* lives exclusively on medium-sized and large male sperm whales; and *Neocyamus physeteris* lives exclusively on female and small-sized male sperm whales. They usually spend their entire lives on their whale of birth, but some do risk transferring to other whales when they come into direct physical contact with one another. There can be as many as 7,500 whale lice on a single whale (in some species, at least, exceptionally large numbers are a diagnostic indicator of poor health).

Measuring 3–30mm long (females are generally broader but shorter than males), they have no free-swimming larval stage and spend their entire lives clinging to their hosts with stout, grasping appendages tipped with exceedingly sharp, recurved claws. They have small heads and flattened bodies and require shelter to avoid being swept into the sea – if they fall off they are doomed – and usually aggregate in areas of reduced water flow, such as the deep ventral grooves of many baleen whales, the callosities of right whales

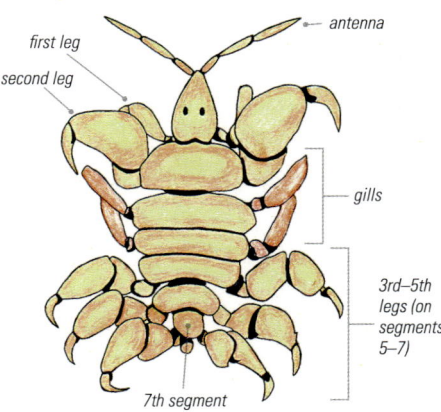

LOUSE anatomy

## Whale lice and their known hosts

| | | |
|---|---|---|
| *Cyamus balaenopterae* | Common minke whale, blue whale, fin whale | |
| *Cyamus boopis* | Humpback whale; also recorded on single southern right whale in Brazil | |
| *Cyamus ceti* | Bowhead whale, gray whale | |
| *Cyamus erraticus* | Southern right whale, North Atlantic right whale, North Pacific right whale | |
| *Cyamus gracilis* | Southern right whale, North Atlantic right whale, North Pacific right whale | |
| *Cyamus ovalis* | Southern right whale, North Atlantic right whale, North Pacific right whale, sperm whale | |
| *Cyamus eschrichtii* | Gray whale | |
| *Cyamus kessleri* | Gray whale | |
| *Cyamus scammoni* | Gray whale | |
| *Cyamus catodontis* | Sperm whale (medium-sized and large males only) | |
| *Cyamus mesorubraedon* | Sperm whale | |
| *Cyamus nodosus* | Narwhal, beluga | |
| *Cyamus monodontis* | Narwhal, beluga | |
| *Cyamus orubraedon* | Baird's beaked whale | |
| *Isocyamus antarcticensis* | Killer whale | |
| *Isocyamus deltobranchium* | Killer whale, long-finned pilot whale, short-finned pilot whale | |
| *Isocyamus delphinii* | False killer whale, melon-headed whale, short-finned pilot whale, long-finned pilot whale, Risso's dolphin, common dolphin, rough-toothed dolphin, Gervais' beaked whale, white-beaked dolphin, harbor porpoise | |

## Whale lice and their known hosts

| | | |
|---|---|---|
| *Isocyamus indopacetus* | Longman's beaked whale | |
| *Isocyamus kogiae* | Pygmy sperm whale | |
| *Neocyamus physeteris* | Sperm whale (females and small males), Dall's porpoise | |
| *Orcinocyamus orcini* | Killer whale | |
| *Platycyamus flaviscutatus* | Baird's beaked whale | |
| *Platycyamus thompsoni* | Northern bottlenose whale, southern bottlenose whale, Gray's beaked whale | |
| *Scutocyamus antipodensis* | Hector's dolphin, dusky dolphin | |
| *Scutocyamus parvus* | White-beaked dolphin | |
| *Syncyamus aequus* | Striped dolphin, spinner dolphin, common dolphin, common bottlenose dolphin | |
| *Syncyamus ilheusensis* | Short-finned pilot whale, melon-headed whale, Clymene dolphin | |
| *Syncyamus pseudorcae* | False killer whale, Clymene dolphin | |

or among the barnacles on the heads of gray whales. Females have a pouch, or marsupium, in which they protect their eggs, embryos and juveniles until they are old enough to cling onto the skin themselves.

Cyamids eat sloughed whale skin (and possibly other foods that adhere to the skin, such as bacteria and algae) and feed on damaged tissue. Though usually considered parasites, they might be more accurately described as cleaning symbionts. They are eaten by some fish (such as topsmelt silversides which often accompany gray whales in their breeding lagoons).

## DIATOMS

Many cetacean species often have a thin yellowish, brownish, greenish or orangish film – or irregular patches – of microscopic single-celled algae called diatoms over their skin. There are countless tens or even hundreds of thousands of species of diatoms – they are the key primary producers in the ocean – but only four genera and a small number of species have been found on cetacean skin. The cold-water *Bennettella* (formerly *Cocconeis*) *ceticola* is the most common species on baleen whales and killer whales in particular, and can cover their bodies after an extended stay in polar waters; it has never been found free-living.

In the Antarctic, the diatom layer takes about a month to develop, so its extent can be used to judge the length of time an animal has been in the region. Normally, whales slough and regenerate their skin continually (it needs to be regenerated to repair scars, sunburn, etc.), but the build-up of diatoms strongly indicates that this is not happening in cold waters (probably to limit heat loss). Indeed, it is believed that Antarctic killer whales make rapid migrations into tropical waters, where they incidentally shed the diatoms when they regenerate their skin tissue. Round trips typically last 5–7 weeks, and they return to the cold waters of the Antarctic looking much 'cleaner'. Otherwise, the diatom layer can get so thick that it may cause significant drag, slowing the animals down (diatoms are a big problem for ships – reducing speed by up to 5 per cent – hence the use of anti-fouling paints).

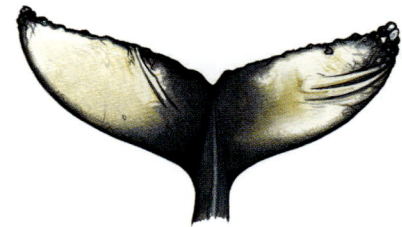

DIATOMS on the flukes of a humpback whale

# QUICK ID GUIDES

**BOW-RIDING DOLPHINS AND PORPOISES**

Killer whale

Pygmy killer whale

Short-finned pilot whale

False killer whale

Risso's dolphin

Long-finned pilot whale

Melon-headed whale

Northern right whale dolphin

BOW-RIDING DOLPHINS AND PORPOISES 19

## IDENTIFYING WHALES BY THEIR FLUKES

North Atlantic right whale

North Pacific right whale

Bowhead whale

Blue whale

Gray whale

Humpback whale

Sperm whale

Killer whale

**IDENTIFYING WHALES BY THEIR FLUKES**

## IDENTIFYING WHALES BY THEIR BLOWS

The height and intensity of a whale's blow, or spout, depends on many factors, including behavior, the size of the individual, when it occurs during the surfacing sequence, air temperature, the quality of the light and wind conditions. It is therefore important to bear in mind that a single whale's blow can vary from virtually invisible to tall and dramatic. Indeed, blow heights have been seriously underestimated in the past, not least because they largely disappear against a pale sea or sky background.

These illustrations show picture-perfect blows (from behind the whales) on the first surfacing after a long dive – in ideal conditions – and represent the maximum heights. Not all cetaceans have clearly visible blows, but these are the ones that are most distinctive and most useful for identification purposes.

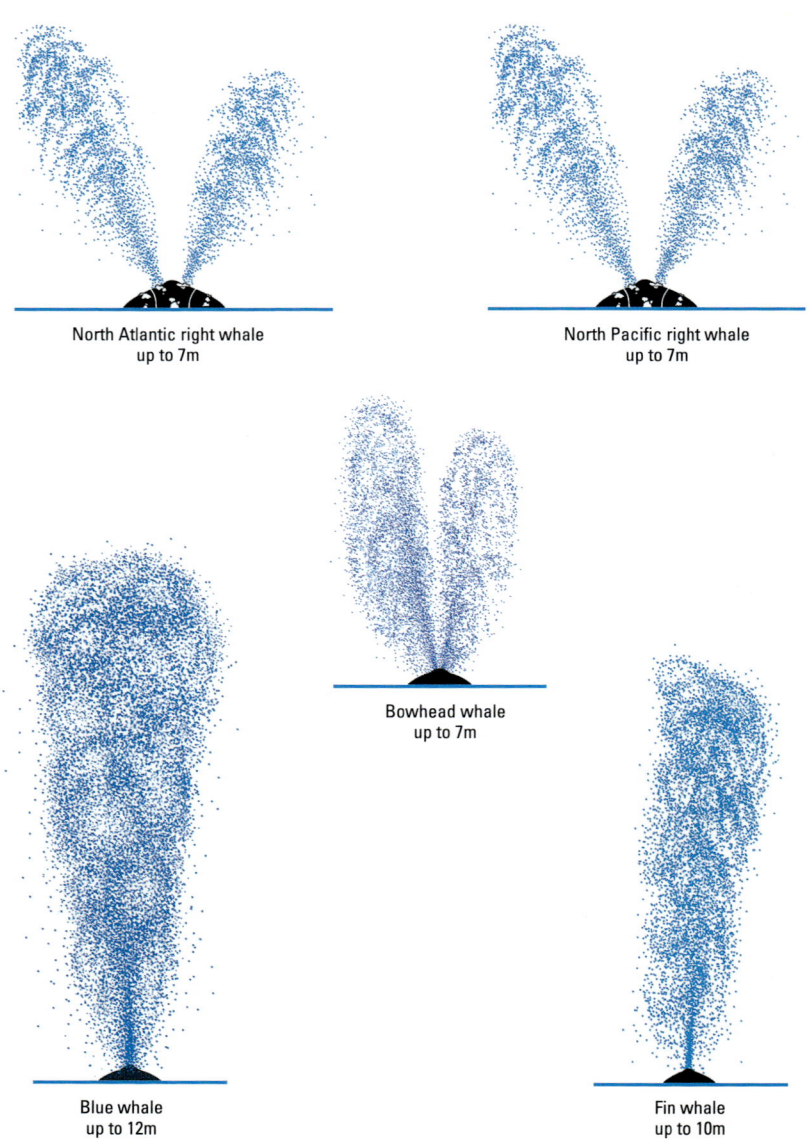

North Atlantic right whale
up to 7m

North Pacific right whale
up to 7m

Bowhead whale
up to 7m

Blue whale
up to 12m

Fin whale
up to 10m

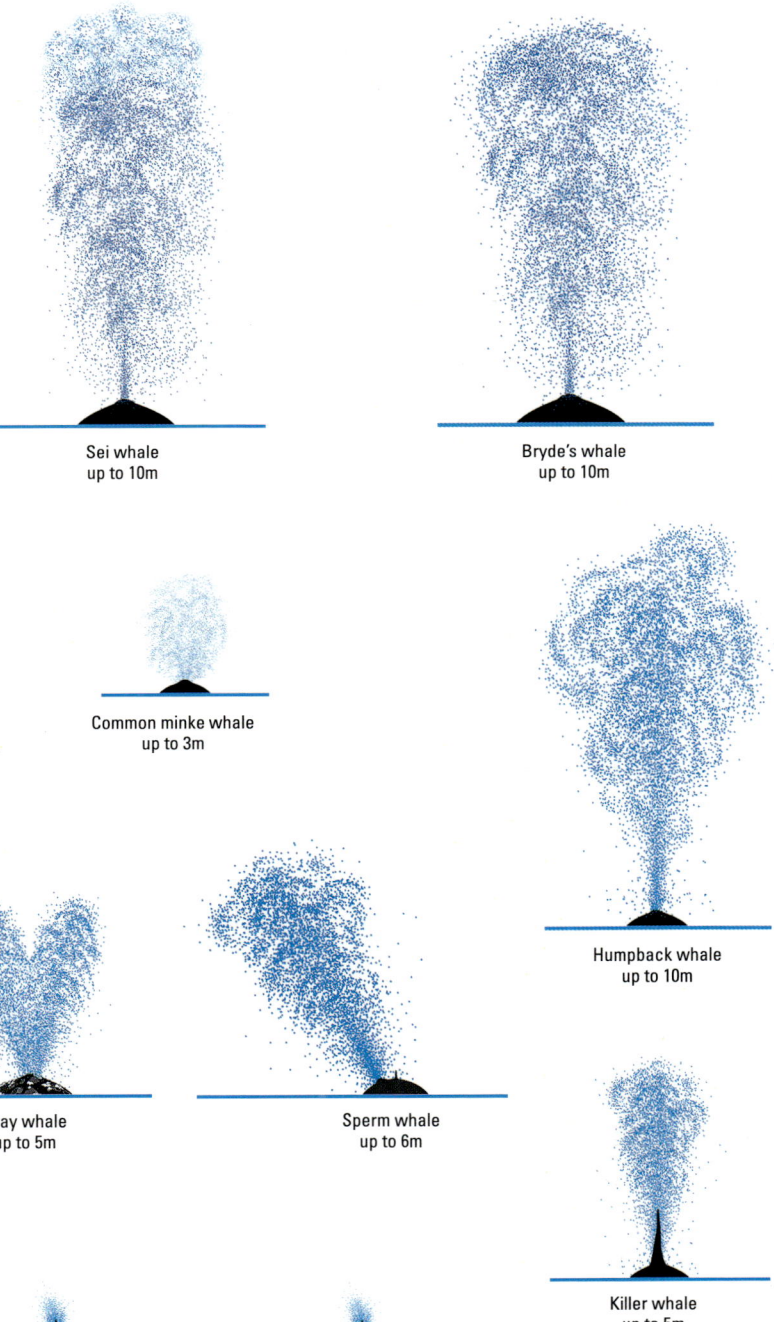

IDENTIFYING WHALES BY THEIR BLOWS

# IDENTIFYING MALE BEAKED WHALES BY THEIR LOWER JAWS

With a few exceptions, most adult male beaked whales have dramatically reduced dentition, with only one or two pairs of erupted teeth in the lower jaw and none in the upper jaw (most females have no erupted teeth at all). The number, position, size and shape of the males' teeth are usually the best clues to their identification.

*Flattened, triangular, forward-pointing teeth*

*Posterior teeth hidden when mouth closed*

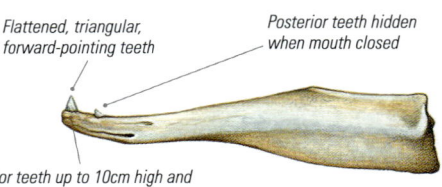

*Anterior teeth up to 10cm high and 10cm at longest point (twice as large as peg-like posterior teeth)*

**Baird's beaked whale**

*Two forward-pointing, conical teeth c. 8cm long (including portion buried in jawbone)*

*Rarely second pair erupts behind main pair*

**Cuvier's beaked whale**

*Two pairs of teeth near tip of lower jaw (anterior tooth much larger than posterior)*

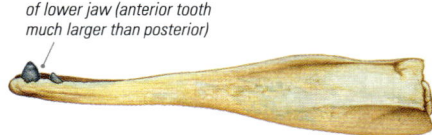

**Sato's beaked whale**

*Pair of pear-shaped teeth at tip of lower jaw*

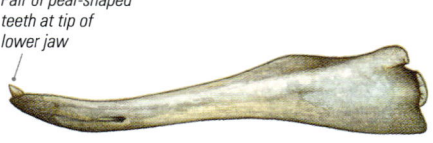

**Longman's beaked whale**

*Two teeth erupt at tip of lower jaw (lean forward)*

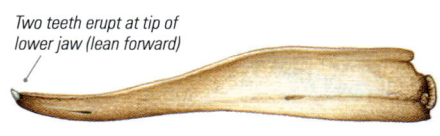

**Northern bottlenose whale**

*Two small, conical teeth at apex of arch*

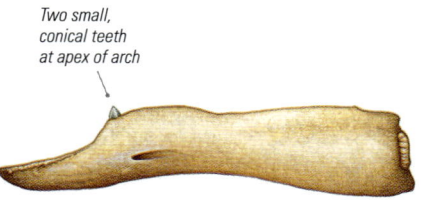

**Pygmy beaked whale**

*Two relatively large, laterally compressed triangular teeth c. 1–2cm behind tip of lower jaw*

**Perrin's beaked whale**

*Tooth shape resembles leaf of ginkgo tree*

*Tooth compressed laterally*

*Tooth as wide or wider than it is tall (6.5cm by 11.5cm) cf. Deraniyagala's beaked whale (taller than it is wide)*

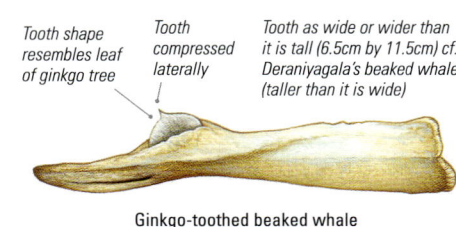

Ginkgo-toothed beaked whale

*Flattened tooth erupts from top of bony arch (angled forward at c. 45°)*

*Out of socket, tooth measures c. 15–18cm tall, 8–9cm wide, 4.5cm deep (but typically less than 2cm extends above gums)*

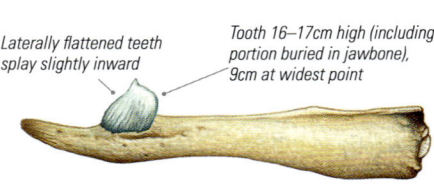

Blainville's beaked whale

*Laterally flattened teeth splay slightly inward*

*Tooth 16–17cm high (including portion buried in jawbone), 9cm at widest point*

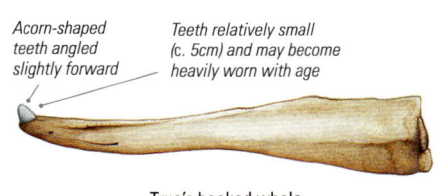

Hubbs' beaked whale

*Acorn-shaped teeth angled slightly forward*

*Teeth relatively small (c. 5cm) and may become heavily worn with age*

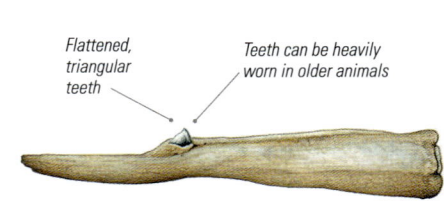

True's beaked whale

*Flattened, triangular teeth*

*Teeth can be heavily worn in older animals*

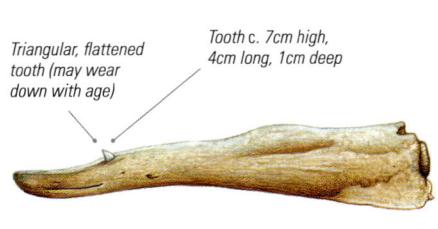

Sowerby's beaked whale

*Triangular, flattened tooth (may wear down with age)*

*Tooth c. 7cm high, 4cm long, 1cm deep*

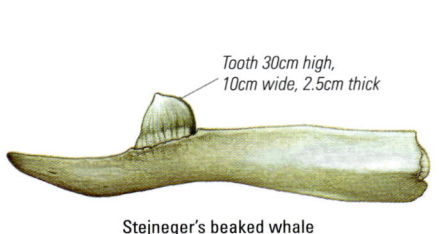

Gervais' beaked whale

*Tooth 30cm high, 10cm wide, 2.5cm thick*

Stejneger's beaked whale

# CETACEANS OF THE NORTH AMERICAN ATLANTIC OCEAN
(including the Caribbean Sea and Gulf of Mexico)

Please note: many species on this spread have very restricted ranges within this broad region, and it is not impossible to see species from other parts of the world outside their normal range. Relative sizes (for average-length males) are correct for the region.

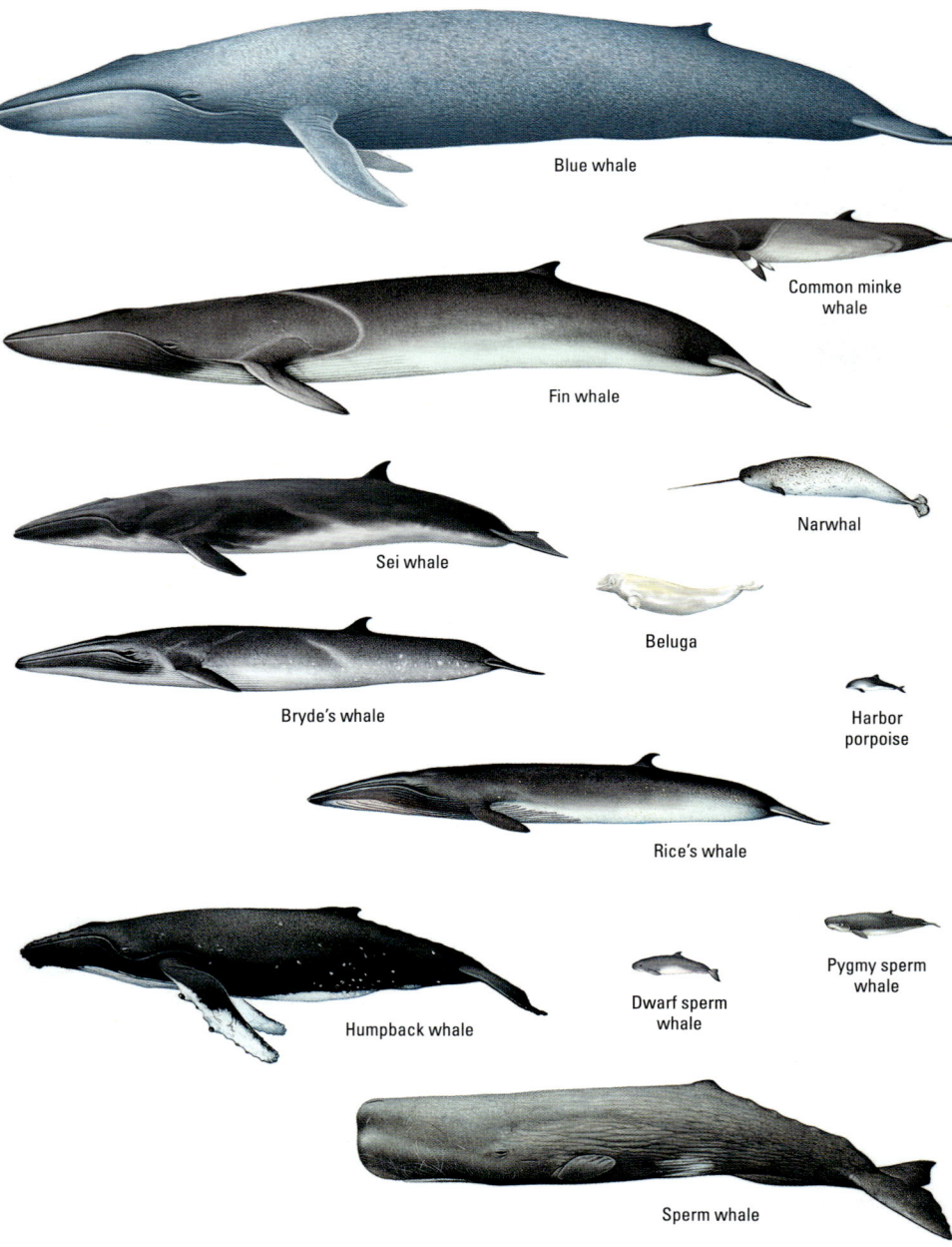

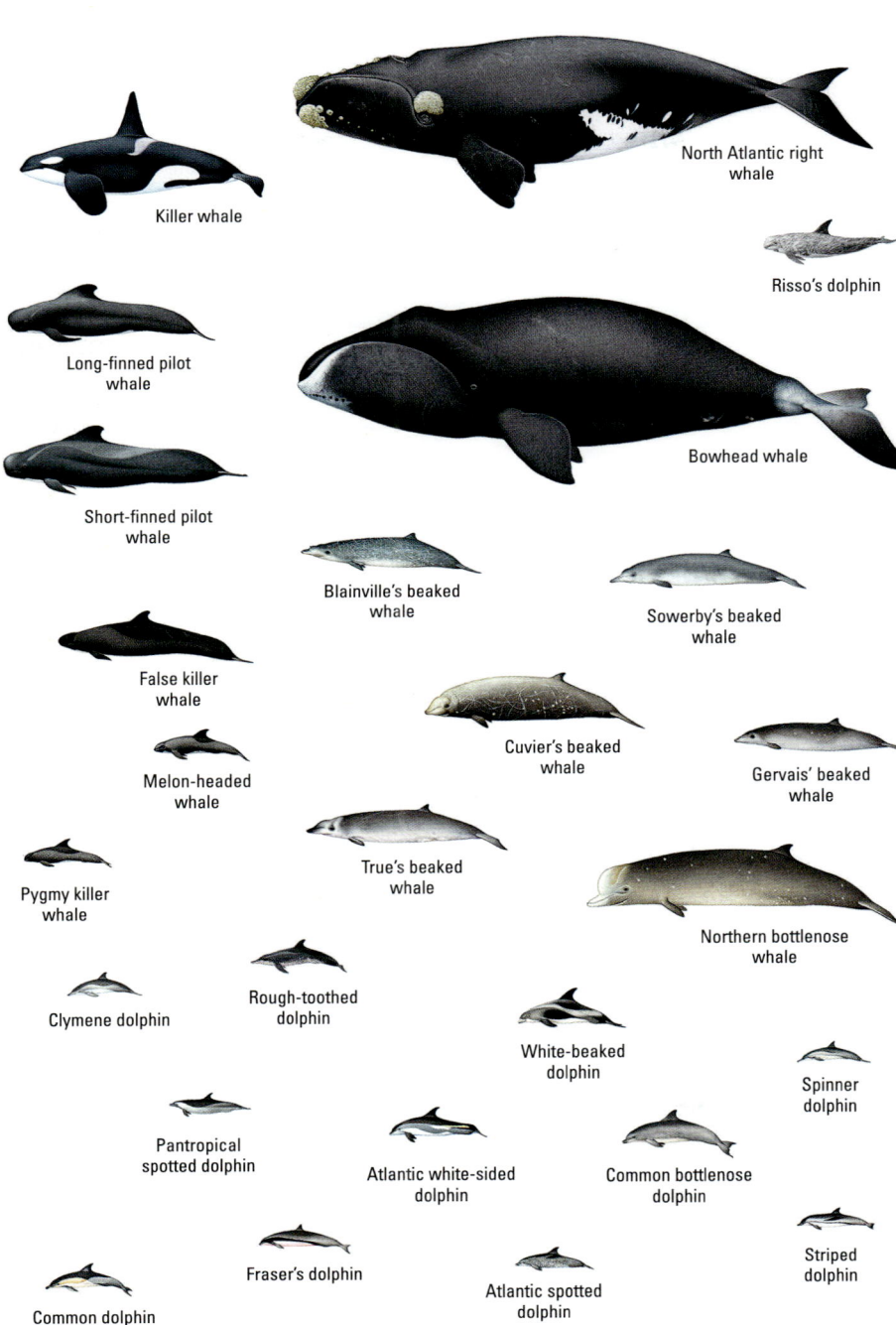

CETACEANS OF THE NORTH AMERICAN ATLANTIC OCEAN

# CETACEANS OF THE NORTH AMERICAN PACIFIC OCEAN
(including the Gulf of California, Gulf of Alaska and Bering Sea)

Please note: many species on this spread have very restricted ranges within this broad region, and it is not impossible to see species from other parts of the world outside their normal range. Relative sizes (for average-length males) are correct for the region.

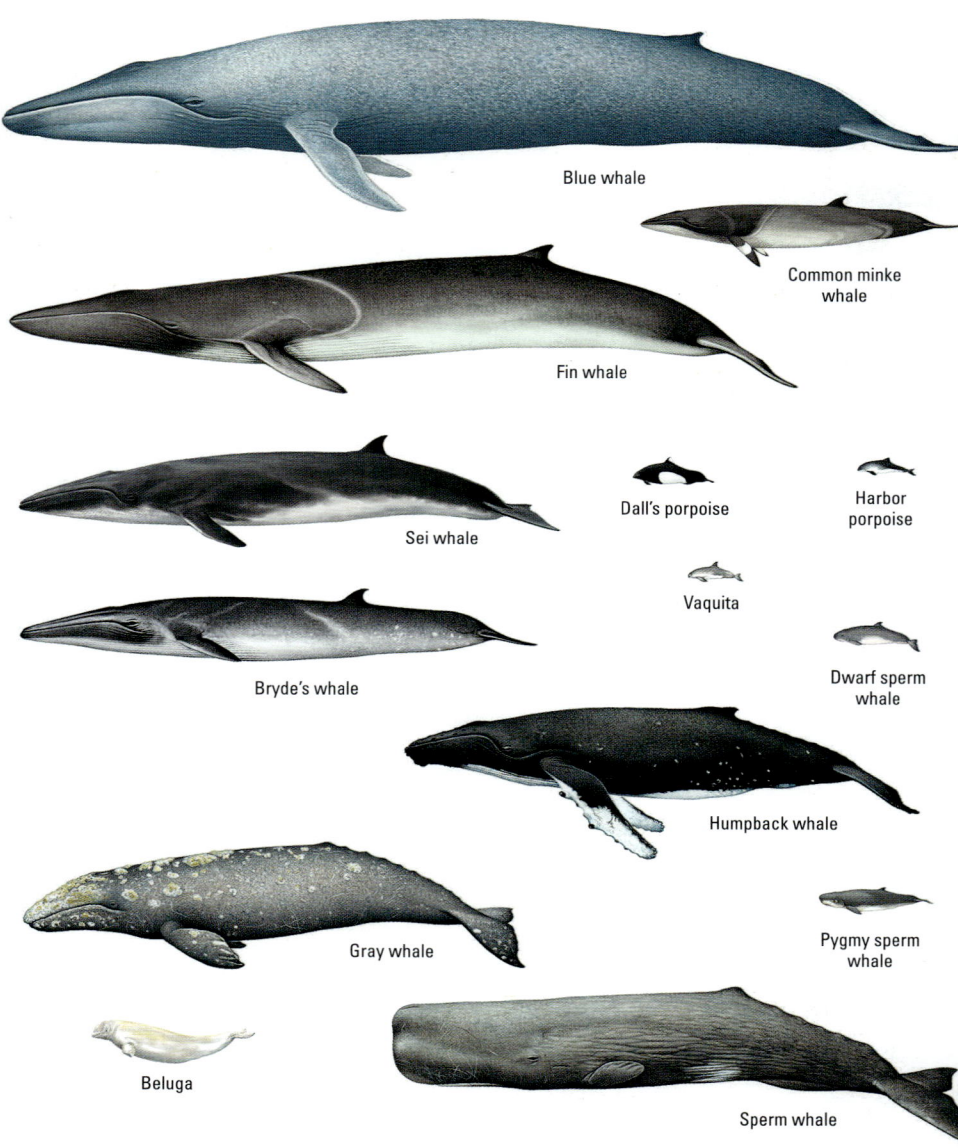

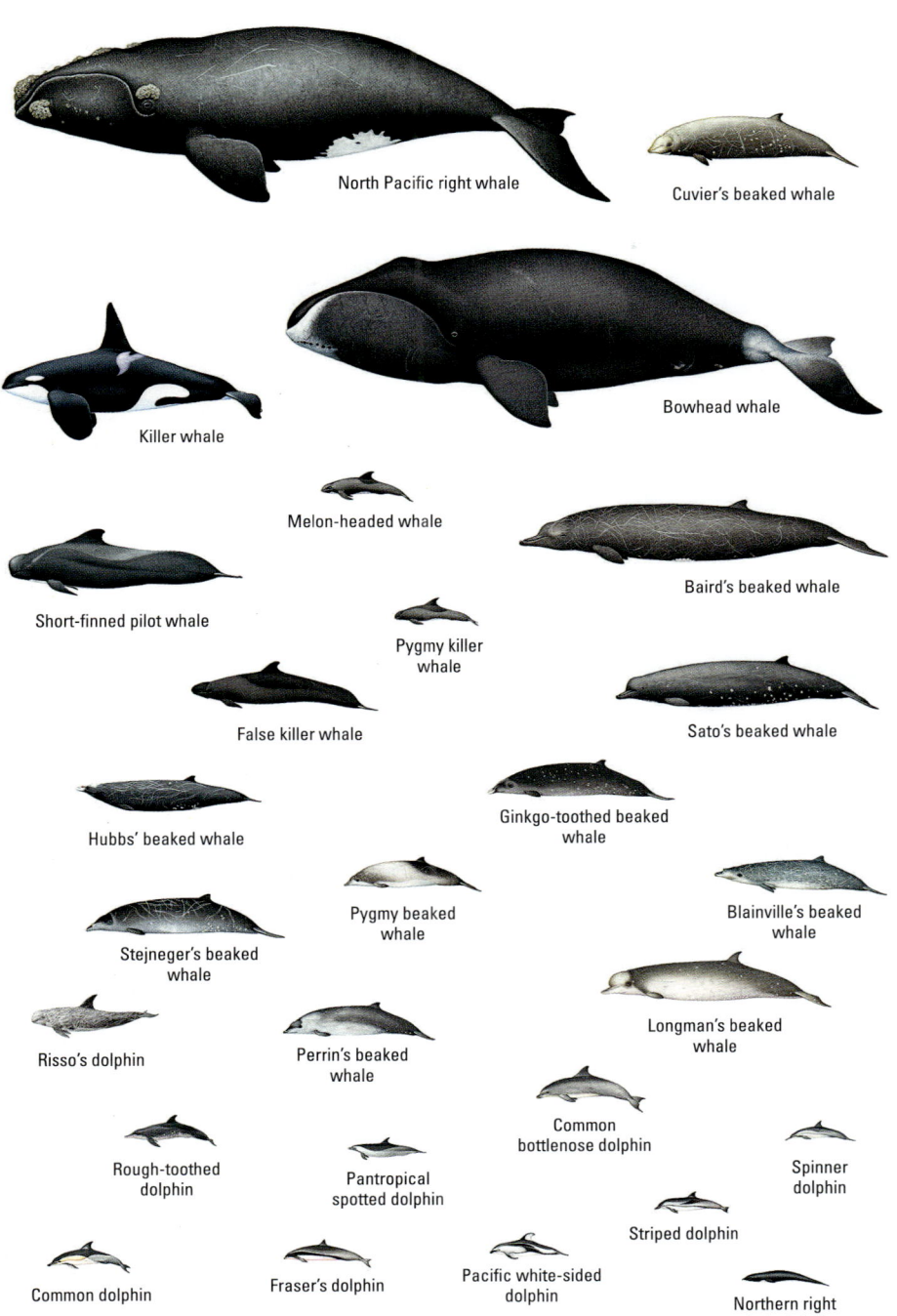

CETACEANS OF THE NORTH AMERICAN PACIFIC OCEAN

# CETACEANS OF THE NORTH AMERICAN ARCTIC OCEAN
(including the Beaufort Sea, Davis Strait, Baffin Bay and Hudson Bay)

Please note: many species on this spread have very restricted ranges within this broad region, and it is not impossible to see species from other parts of the world outside their normal range. Relative sizes (for average-length males) are correct for the region.

# INFORMATION FOR WORKING AT SEA

## RECOGNIZING THE SEA STATE
'Sea state' is the term used to describe sea conditions. A sea state of three or less, when there are no or few whitecaps, is best for whale watching. As the sea state increases, it becomes increasingly difficult to spot anything among the waves and spray.

| Sea state | Official term | Forecast description | Specification |
|---|---|---|---|
| 0 | Calm | Calm | Sea like a mirror |
| 1 | Calm | Light air | Small ripples; no crests or whitecaps |
| 2 | Smooth | Light breeze | Small wavelets; glassy crests; no whitecaps |
| 3 | Smooth | Gentle breeze | Large wavelets; crests begin to break; a few scattered whitecaps |
| 4 | Slight | Moderate breeze | Small waves; fairly frequent whitecaps |
| 5 | Moderate | Fresh breeze | Moderate; longer waves; many whitecaps; some spray |
| 6 | Rough | Strong breeze | Large waves; many whitecaps; frequent spray |
| 7 | Very rough | Near gale | Sea heaps up; white foam from breaking waves blows in streaks |
| 8 | High | Gale | Long; moderately high waves; edges of crests breaking; foam blows in streaks |
| 9 | Very high | Severe gale | High waves; dense streaks of foam; crests of waves topple; tumble and roll over; sea begins to roll; spray may affect visibility |
| 10 | Very high | Storm | Very high waves with long, overhanging crests; dense streaks of foam make sea appear mostly white |
| 11 | Phenomenal | Violent storm | Exceptionally high waves; sea covered in patches of foam; crests of waves blown into froth; visibility affected |
| 12 | Phenomenal | Hurricane | Air filled with foam and spray; sea completely white with driving spray; visibility seriously affected |

## READING A WIND CHART
Wind is illustrated on a weather chart using 'wind barbs', which are a convenient way to represent both wind speed and direction. The barbs point in the direction from which the wind is blowing (north is always 'up'). They also show the wind speed in knots: each short barb represents 5 knots, and each long barb 10 knots. Simply add the value of all the barbs to find the wind speed — a long barb and a short barb, for example, is 15 knots. If there are no barbs (or simply a dot) the wind speed is less than 2 knots, and a flag or triangle is used to show winds of 50 knots.

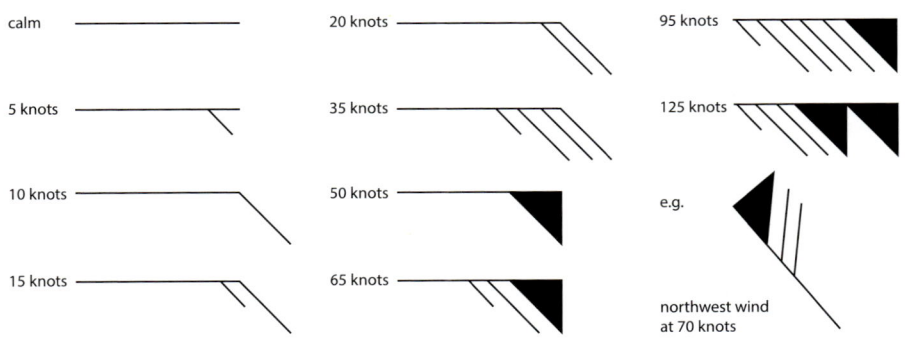

# NORTH ATLANTIC RIGHT WHALE
*Eubalaena glacialis* (Müller, 1776)

The North Atlantic right whale is one of the most closely studied – and most endangered – large whales in the world. The few animals around today are the survivors of nearly 1,000 years of commercial exploitation and, although hunting has stopped, they face new human-induced threats and are widely considered to be in very real danger of extinction.

**IUCN status** Critically Endangered (2020).
**Population** *c.* 372 (2023). Pre-whaling population unknown (one estimate of 9,000–21,000 based on ecological carrying capacity). Reduced to fewer than 100 by 1935. Decreasing (declined dramatically since a post-whaling peak of 483 in 2011), mainly due to vessel strikes, entanglements in fishing gear and climate change.
**Classification** Mysticeti, family Balaenidae.
**Taxonomy** No recognized forms or subspecies; split from the North Pacific right whale in 2000 due to genetic differences between both the whales and their lice (the two species were previously lumped together as 'northern right whale', *Eubalaena glacialis*).
**Other names** Atlantic right whale, northern right whale.

**DISTRIBUTION** Historically, there were two largely isolated populations on either side of the North Atlantic. However, the eastern population is considered functionally extinct. Western North Atlantic animals are highly mobile, but only pregnant females and a few other animals undertake predictable seasonal migrations. Occurs mainly in temperate and sub-polar coastal waters, including shallow basins, and in relatively deeper areas over the continental shelf.
**Eastern population** Historically, North Atlantic right whales probably ranged from the only known breeding ground, in Cintra Bay, off Western Sahara, to feeding grounds in the Bay of Biscay, off western Britain, around Iceland, across the Norwegian Sea to the North Cape (northern Norway) and possibly in the Mediterranean. Rarely seen in the north-east Atlantic nowadays, all identified individuals since 1960 have been migrants from the west, so a remnant eastern population seems unlikely (although a calf with foetal folds – suggesting that it had been born nearby – was observed in the Canary Islands, Spain, in December 2020).
**Western population** There has been a broad-scale distribution shift since 2011. Due to warming waters, and associated changes in abundance of preferred prey, sightings have decreased dramatically in the Gulf of Maine and the Bay of Fundy, which used to be critical spring and summer foraging grounds. There has been a general northeastward shift and, since 2105, most of the population has favored the Gulf of St Lawrence. There has also been increased use of the western Gulf of Maine and novel use of southern New England in recent years. Two were seen in the Bahamas in 2025.

In November and December, pregnant females (sometimes accompanied by a small number of juveniles and non-calving females) migrate south along the eastern seaboard of North America to the only known calving ground, in the relatively sheltered, shallow coastal waters of northern Florida and southern Georgia (mainly between Savannah and St Augustine); for some individuals, the calving grounds may extend as far north as Cape Fear, North Carolina, and occasionally further west into the Gulf of Mexico. They return to northern feeding grounds in March and April. Most animals (including most juveniles and adult males) do not migrate to these calving grounds; their wintering range is unknown.

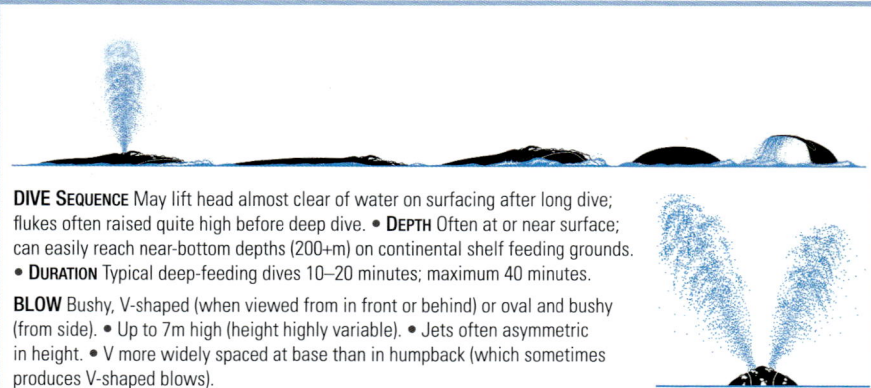

**DIVE SEQUENCE** May lift head almost clear of water on surfacing after long dive; flukes often raised quite high before deep dive. • **DEPTH** Often at or near surface; can easily reach near-bottom depths (200+m) on continental shelf feeding grounds. • **DURATION** Typical deep-feeding dives 10–20 minutes; maximum 40 minutes.

**BLOW** Bushy, V-shaped (when viewed from in front or behind) or oval and bushy (from side). • Up to 7m high (height highly variable). • Jets often asymmetric in height. • V more widely spaced at base than in humpback (which sometimes produces V-shaped blows).

**ADULT**

- Top of head sprinkled with callosities (naturally light to dark gray, but appear white, cream or yellowish due to presence of whale lice)
- Very strongly arched jawline
- Extremely stocky body (bordering on rotund)
- Massive head (up to one-third of body length)
- Maximum girth can exceed c. 60 per cent of total length
- Individuals sometimes appear mottled (caused by uneven sloughing of patches of skin)
- Smooth, broad back with no dorsal fin or ridge
- Predominantly black
- Upper margin of lower 'lip' serrated
- Largest callosity (the 'bonnet') on tip of rostrum
- Pattern of callosities varies between individuals but distributed in generally consistent locations
- No pleats or grooves on throat
- Eyes just above corners of mouth
- Large, broad flippers up to 1.7m long
- Many individuals have irregular white patch around navel on underside (highly variable and may extend laterally onto sides and towards chin – but not as extensive as in southern right whale)
- White scarring (from entanglement, vessel strikes and killer whale attacks) mainly on tailstock and flukes, but sometimes elsewhere on body

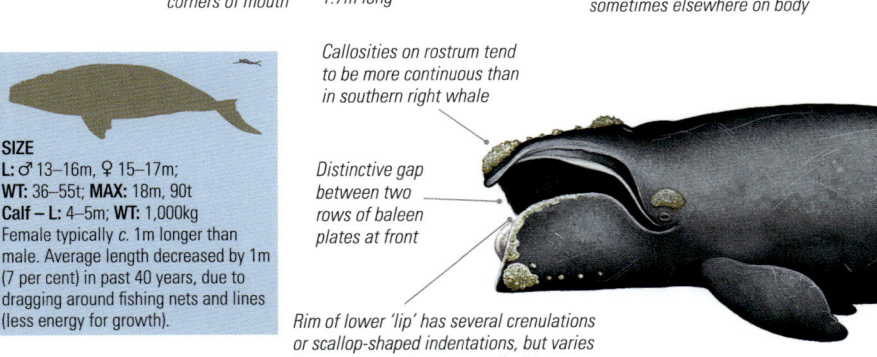

- Callosities on rostrum tend to be more continuous than in southern right whale
- Distinctive gap between two rows of baleen plates at front
- Rim of lower 'lip' has several crenulations or scallop-shaped indentations, but varies from many to a few or no callosities

**SIZE**
L: ♂ 13–16m, ♀ 15–17m;
WT: 36–55t; MAX: 18m, 90t
Calf – L: 4–5m; WT: 1,000kg
Female typically c. 1m longer than male. Average length decreased by 1m (7 per cent) in past 40 years, due to dragging around fishing nets and lines (less energy for growth).

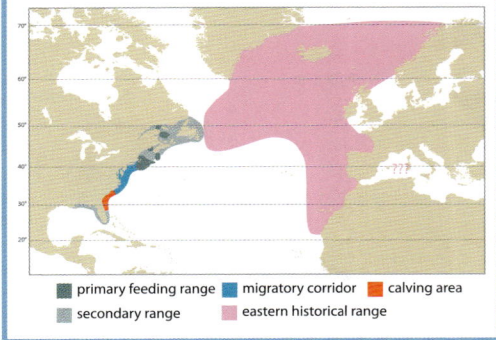

■ primary feeding range ■ migratory corridor ■ calving area
■ secondary range ■ eastern historical range

**AT A GLANCE** North Atlantic • Extra-large size • Extremely stocky body • Predominantly black • Smooth back with no dorsal fin or ridge • Low body profile at surface • Massive head covered in light-colored callosities • Very strongly arched jawline • No pleats or grooves on throat • V-shaped blow • Rectangular, broad, paddle-shaped flippers

**BEHAVIOR** Generally slow moving and may rest at the surface for long periods. Frequently engages in active surface behavior, however, and will breach, spyhop, lobtail and flipper-slap repeatedly. Often shows little or no avoidance behavior in the presence of boats, and can be inquisitive and approachable.

**FOOD AND FEEDING** Mostly calanoid copepod crustaceans, but also other small invertebrates, including smaller copepods, amphipods, krill, pteropods and larval barnacles. Normally skim-feeds; no feeding on winter breeding grounds.

**BALEEN** 205–270 plates (each side of the upper jaw). Long, thin plates averaging 2–2.8m long; gray-brown to black.

**GROUP SIZE AND STRUCTURE** Normally one to two, up to 12 in occasional loose aggregations; much larger aggregations may form on temporarily rich feeding grounds or in breeding groups.

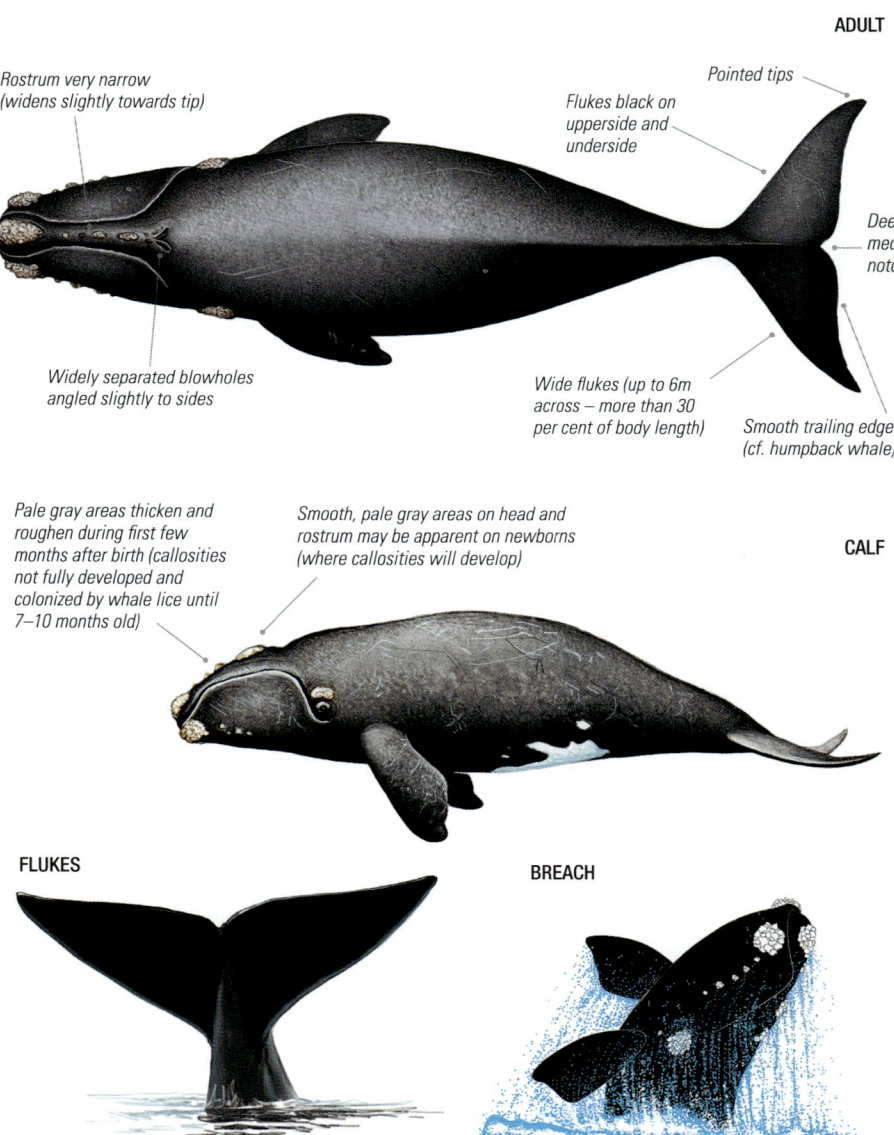

**CALLOSITIES** Scientists use the unique callosity patterns on the heads of right whales (as well as scars and other distinguishing features) to tell one individual whale from another.

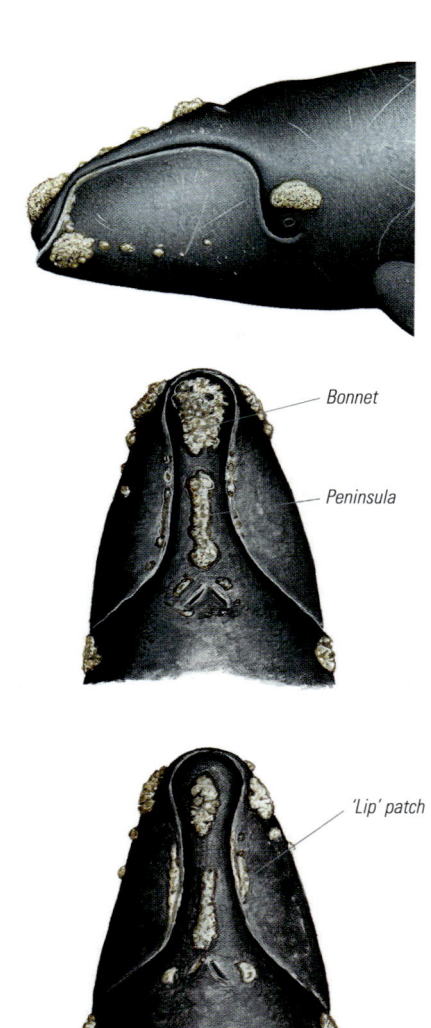

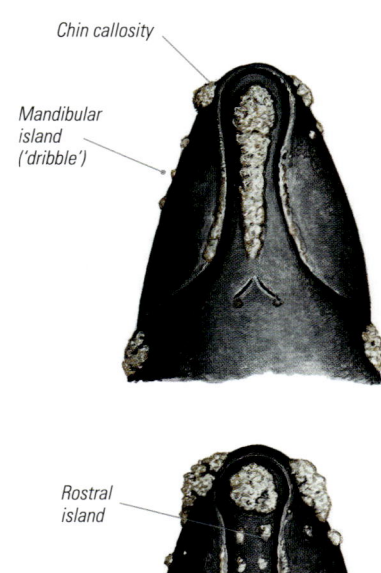

Chin callosity

Mandibular island ('dribble')

Bonnet

Peninsula

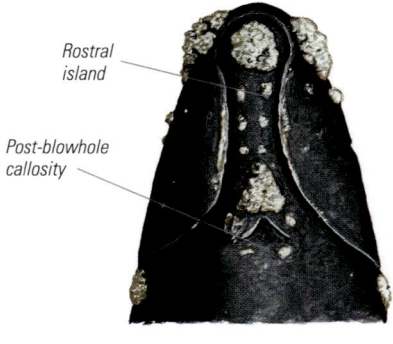

Rostral island

Post-blowhole callosity

'Lip' patch

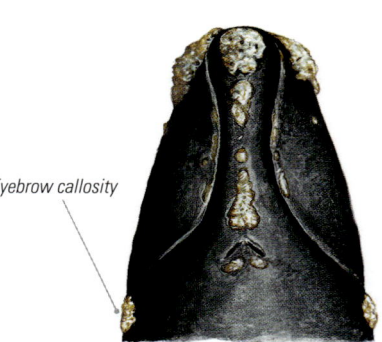

Eyebrow callosity

Coaming

**NORTH ATLANTIC RIGHT WHALE** 35

# NORTH PACIFIC RIGHT WHALE
## *Eubalaena japonica* (Lacépède, 1818)

In 1874, whaling captain and naturalist Charles Scammon remarked that North Pacific right whales were 'scattered over the surface of the water as far as the eye can discern'. But commercial whaling was so intense that they suffered one of the most dramatic and rapid depletions of all the great whales.

**IUCN status** Endangered (2017). Eastern North Pacific population Critically Endangered (2017).
**Population** Uncertain, but possibly low hundreds in the west and *c.* 30 in the east. An estimated 26,500–37,000 taken by commercial whalers 1839–1909. Pre-whaling population believed to have been 30,000+, though never documented in detail. Trend unknown.
**Classification** Mysticeti, family Balaenidae.
**Taxonomy** No recognized forms or subspecies, though two recognized populations occur, one on either side of the North Pacific. Split from the North Atlantic right whale in 2000 due to genetic differences between both the whales and their lice (the two were previously lumped as the 'northern right whale', *Eubalaena glacialis*).
**Other names** Pacific right whale, northern right whale.

**DISTRIBUTION** Formerly abundant in cold temperate waters across much of the North Pacific, but currently occupies only a fraction of this former range. There appear to be two distinct populations: one of several hundred individuals (western North Pacific, centered around the Sea of Okhotsk); and one of just tens of individuals (eastern North Pacific, primarily in the Bering Sea and Gulf of Alaska). There appear to be seasonal migrations from higher-latitude summer feeding grounds to potential lower-latitude winter breeding grounds. No calving grounds have been located. Tends to be more pelagic than the North Atlantic right whale. Most sightings in the past 20 years have occurred in the southeastern Bering Sea (where there has been greatest research effort). North Pacific, North Atlantic and southern right whales are separated by Arctic ice and warm equatorial waters, and it is estimated that there has been no interchange between the three populations for millions of years.

**Western population** Most of what we know about distribution is based on limited recent sightings and historical whale-catch data. Its known distribution in winter includes the waters of southern China, Taiwan and the Ogasawara Islands. There is a northerly migration in spring (March–May); however, it is uncertain whether some or all of the whales follow this seasonal movement and how these patterns might have changed in response to climate change (cf. North Atlantic right whales). The principal summer feeding grounds appear to be in the Sea of Okhotsk, around the Kuril and Commander Islands, off the Pacific coast of northern Honshu and along the south-east coast of the Kamchatka Peninsula. In fall, there is a southward shift in distribution. The lack of evidence of coastal winter-breeding grounds suggests that they may breed in open-ocean waters offshore.

**Eastern population** Historical whaling records indicate that the principal summer feeding grounds were in the eastern Bering Sea and the Gulf of Alaska – north of 40°N. In the fall, there was a southward shift in distribution to unknown wintering grounds (with relatively recent sightings as far south as Baja California and Hawai'i). Since the 1990s, most sightings of the eastern population have been concentrated in two areas during summer: particularly in the southeastern Bering Sea, west of Bristol Bay, Alaska (*c.* 57–59°N), where the whales appear to select relatively

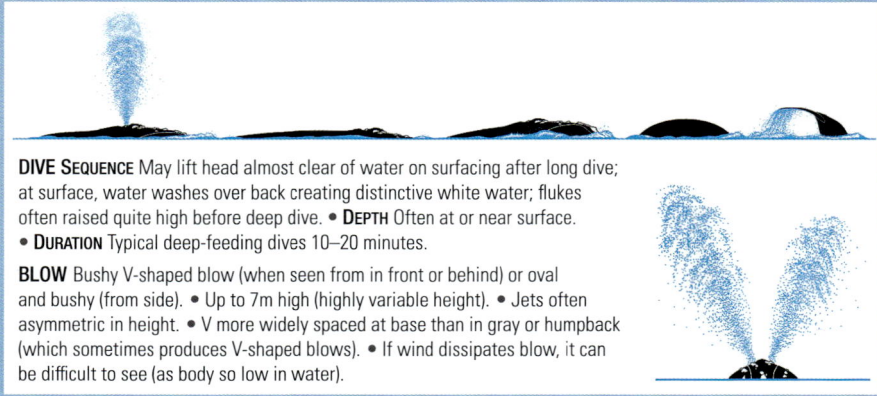

**DIVE SEQUENCE** May lift head almost clear of water on surfacing after long dive; at surface, water washes over back creating distinctive white water; flukes often raised quite high before deep dive. • **DEPTH** Often at or near surface. • **DURATION** Typical deep-feeding dives 10–20 minutes.

**BLOW** Bushy V-shaped blow (when seen from in front or behind) or oval and bushy (from side). • Up to 7m high (highly variable height). • Jets often asymmetric in height. • V more widely spaced at base than in gray or humpback (which sometimes produces V-shaped blows). • If wind dissipates blow, it can be difficult to see (as body so low in water).

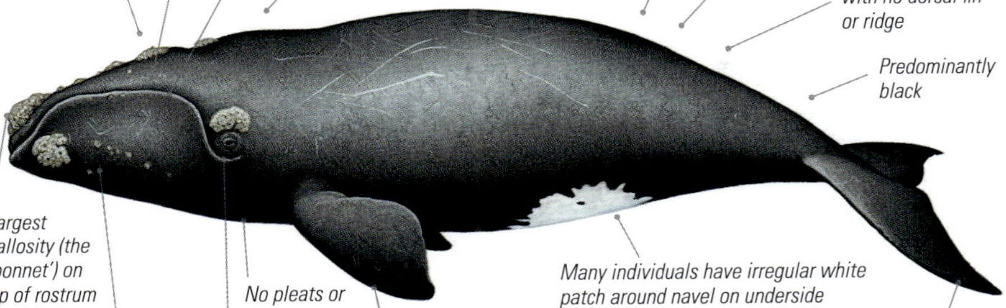

**ADULT**

- Top of head sprinkled with callosities (naturally light to dark gray, but appear white, cream or yellowish due to presence of whale lice)
- Very strongly arched jawline
- Upper margin of lower 'lip' serrated
- Massive head (up to one-third of body length)
- Extremely stocky body (bordering on rotund)
- Maximum girth can exceed c. 60 per cent of total length
- White blotches sometimes occur elsewhere on body
- Individuals sometimes appear mottled (caused by uneven sloughing of patches of skin)
- Smooth, broad back with no dorsal fin or ridge
- Predominantly black
- Largest callosity (the 'bonnet') on tip of rostrum
- Pattern of callosities varies between individuals but distributed in generally consistent locations
- No pleats or grooves on throat
- Large, broad flippers up to 1.7m long
- Eyes just above corners of mouth
- Many individuals have irregular white patch around navel on underside (highly variable and may extend laterally onto sides and towards chin)
- White scarring (from entanglement, vessel strikes and killer whale attacks) mainly on tailstock and flukes, but sometimes elsewhere on body

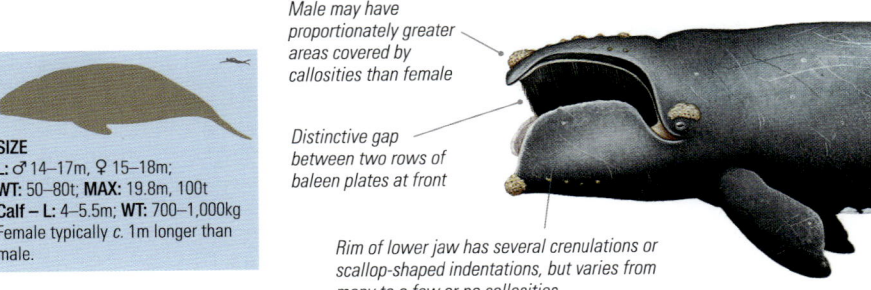

- Male may have proportionately greater areas covered by callosities than female
- Distinctive gap between two rows of baleen plates at front
- Rim of lower jaw has several crenulations or scallop-shaped indentations, but varies from many to a few or no callosities

**SIZE**
L: ♂ 14–17m, ♀ 15–18m;
WT: 50–80t; MAX: 19.8m, 100t
Calf – L: 4–5.5m; WT: 700–1,000kg
Female typically c. 1m longer than male.

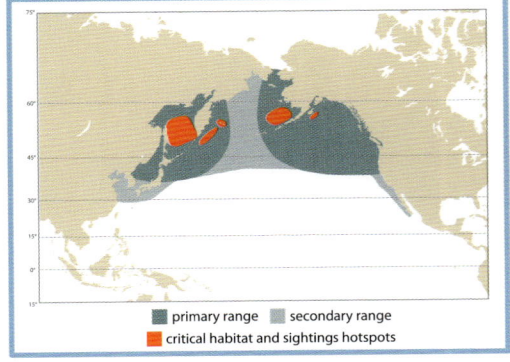

- primary range
- secondary range
- critical habitat and sightings hotspots

**AT A GLANCE** Northern North Pacific • Extra-large size • Extremely stocky body • Predominantly black • Smooth back with no dorsal fin or ridge • Low body profile at surface • Massive head covered in light-colored callosities • Very strongly arched jawline • No pleats or grooves on throat • V-shaped blow • Rectangular, broad, paddle-shaped flippers

shallow waters *c.* 70m deep over the mid-continental shelf; but also over the continental shelf and slope south of Kodiak Island in the Gulf of Alaska. Possible wintering grounds for the eastern population are obscure (calving likely takes place offshore).

**BEHAVIOR** There have been relatively few direct observations of living North Pacific right whales in recent decades. They are generally slow moving and may rest at the surface for long periods, but have been observed breaching, spyhopping, lobtailing and flipper-slapping. They show little or no avoidance behavior in the presence of boats and can be inquisitive and approachable, although a legacy of whaling means some individuals may still be very sensitive to vessels.

**FOOD AND FEEDING** Mostly calanoid copepods, but will take other small invertebrates, including smaller copepods, amphipods, krill, pteropods and larval barnacles. Normally skim-feeds (swimming slowly with mouth open through patches of concentrated prey at or near the surface); will also filter-feed at depth (up to 300m); at least one observation of lunge-feeding; no feeding on winter breeding grounds.

**BALEEN** 205–270 plates (each side of the upper jaw). Long, thin plates averaging 2–2.8m long; gray-brown to black.

**GROUP SIZE AND STRUCTURE** Normally one or two, though larger aggregations of 30+ may form on temporarily rich feeding grounds or breeding grounds.

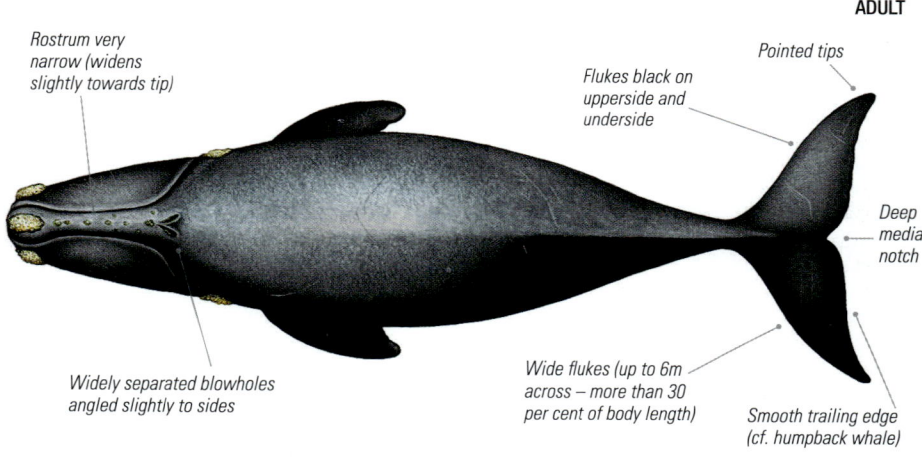

ADULT

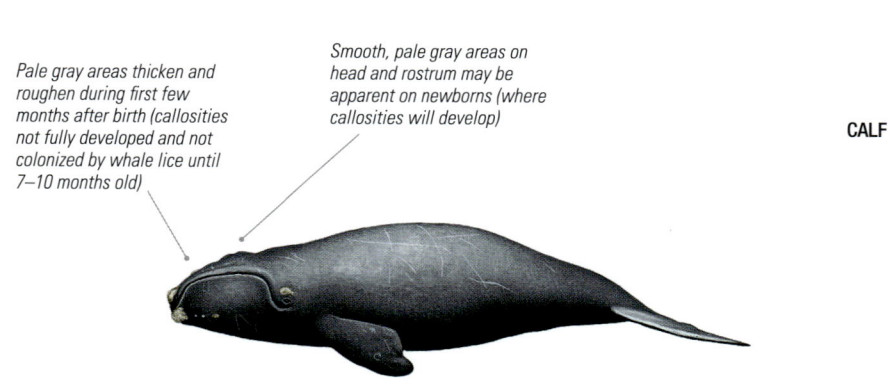

CALF

**FLUKES**

**BREACH**

## CALLOSITIES AND LICE

**CALLOSITIES** Callosities are areas of thick, irregular, calloused tissue – found only on right whales – that develop around the sparse hairs scattered about the whale's head. They are named after their resemblance to the 'callus' or thickened skin that occurs naturally in many animal species. Pockmarked with ridges and depressions, they feel like hard rubber to touch and, from a distance, look a little like barnacles. The callosity tissue is naturally light to dark gray, but it is home to colonies of thousands of creamy-white or yellowish cyamid crustaceans or 'whale lice', which obscure the underlying color. Callosities occur in approximately the same places that humans have facial hair: above the eyes (eyebrows), along the rostrum, between the blowholes and the tip of the snout (moustache), and along the margins of the lower 'lips' and jaw (beard). Their height can change throughout a whale's life (growing upwards and breaking off repeatedly), but their overall size and placement on the head remains the same; consequently, their shape and size serve as 'fingerprints' or 'distinguishable faces' that enable researchers to tell one individual right whale from another. The function of the callosities is unknown. One theory is that they are designed specifically to attract dense populations of whale lice, which stand on their hind legs to catch copepods – perhaps alerting the whale and helping to steer it towards denser concentrations of its tiny prey.

**LICE** Right whales carry large populations of three species of cyamid crustaceans or 'whale lice'. Two of these are host-specific (i.e. found on no other whale species) and one (*Cyamus ovalis*) is also found on sperm whales. Recent genetic evidence suggests that these three species should be split into nine species, because they are sufficiently different on their North Atlantic, North Pacific and southern hemisphere hosts. Whale lice occur among the callosities and are common in creases and folds elsewhere on the body.

*Cyamus gracilis*
6mm long
Predominantly yellow
Typically *c.* 500 per whale
Mainly in pits and grooves between elevated patches of callosity tissue

*Cyamus erraticus*
12–15mm long
Predominantly orange
Typically *c.* 2,000 per whale
Mainly on smooth skin in genital and mammary slits, and in large concentrations in wounds
Also found in large patches on heads (where there is no callosity tissue) of young calves (disappearing when calf is *c.* 2 months old)

*Cyamus ovalis*
12–15mm long
Predominantly white
Typically *c.* 5,000 per whale
Coat callosities at average density of one adult per cm² (main reason for pale color of callosities)

# BOWHEAD WHALE
## *Balaena mysticetus*  Linnaeus, 1758

The only large whale found exclusively in the Arctic, the bowhead is well adapted to life in its freezing home. With a layer of blubber up to 50cm thick and the ability to create its own breathing holes by breaking through thick ice, it can live at higher altitudes than any other baleen whale. Known to live to at least 200 years old.

**IUCN status** Least Concern (2018). Sea of Okhotsk sub-population Endangered (2018); East Greenland–Svalbard–Barents Sea sub-population Endangered (2018).
**Population** Minimum 10,000 mature individuals, likely 20,000–30,000 total (best estimates 16,800 Bering–Chukchi–Beaufort; low hundreds Sea of Okhotsk; 6,400 Eastern Canada–West Greenland; low hundreds East Greenland–Svalbard–Barents Sea). Pre-commercial whaling population 71,000–113,000. Increasing.
**Classification** Mysticeti, family Balaenidae.
**Taxonomy** No recognized forms or subspecies (though four separate stocks).
**Other names** Greenland/Arctic whale, Greenland/Arctic right whale.

**DISTRIBUTION** Circumpolar in the Arctic and sub-Arctic, mainly 54–85°N. Closely associated with the pack ice and its seasonal movements, migrating to the High Arctic in summer and retreating southward in winter with the advancing ice edge (the winter range is poorly known, but it is believed to live in areas near the ice edge, in polynyas and unconsolidated pack ice). May travel long distances (up to 200km per day) between high-productivity feeding areas. Mainly pelagic, but does occur in coastal waters. There is some geographic segregation by sex. Warming Arctic waters, due to climate change, are affecting seasonal distribution and timings.

Four 'stocks' (sub-populations) are currently recognized (mainly based on geographical separation): Bering–Chukchi–Beaufort (Alaska, Canada and Russia); Sea of Okhotsk (Russia); Eastern Canada–Western Greenland (formerly considered two stocks: Hudson Bay–Foxe Basin, Canada, and Baffin Bay–Davis Strait, Canada and Greenland); and East Greenland–Svalbard–Barents Sea (Greenland, Norway and Russia).

**BEHAVIOR** Generally a slow, deliberate swimmer, but capable of bursts of speed up to 21km/h. Frequently breaches, flipper-slaps, lobtails and spyhops, and may inspect or play with objects in the water. During breaches, up to 60 per cent of the body leaves the water and the whale usually falls back into the water on its back or side. Often quite approachable by boat and may closely investigate people standing on the floe edge. It can swim beneath ice, making breathing holes by breaking through ice up to 60cm thick with the raised part of its massive head. Often seen in association with belugas and narwhal.

**FOOD AND FEEDING** Catholic diet (more than 100 prey species known) but prefers small to medium-sized crustaceans (mostly 3–30mm-long), especially copepods and euphausiid krill; also feeds on mysids and gammarid amphipods. Feeds throughout water column, anywhere from surface to seabed, under ice as well as in open water (where may 'skim' through concentrated prey at surface, swimming slowly with mouth open).

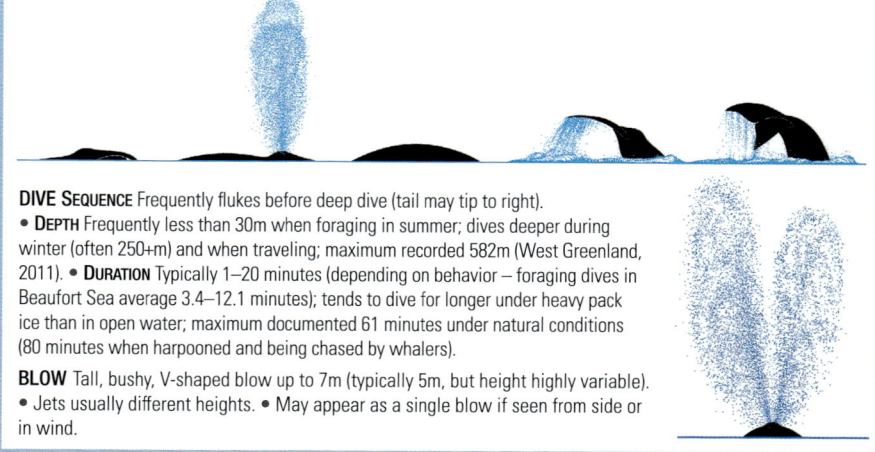

**DIVE SEQUENCE** Frequently flukes before deep dive (tail may tip to right).
• **DEPTH** Frequently less than 30m when foraging in summer; dives deeper during winter (often 250+m) and when traveling; maximum recorded 582m (West Greenland, 2011). • **DURATION** Typically 1–20 minutes (depending on behavior – foraging dives in Beaufort Sea average 3.4–12.1 minutes); tends to dive for longer under heavy pack ice than in open water; maximum documented 61 minutes under natural conditions (80 minutes when harpooned and being chased by whalers).

**BLOW** Tall, bushy, V-shaped blow up to 7m (typically 5m, but height highly variable).
• Jets usually different heights. • May appear as a single blow if seen from side or in wind.

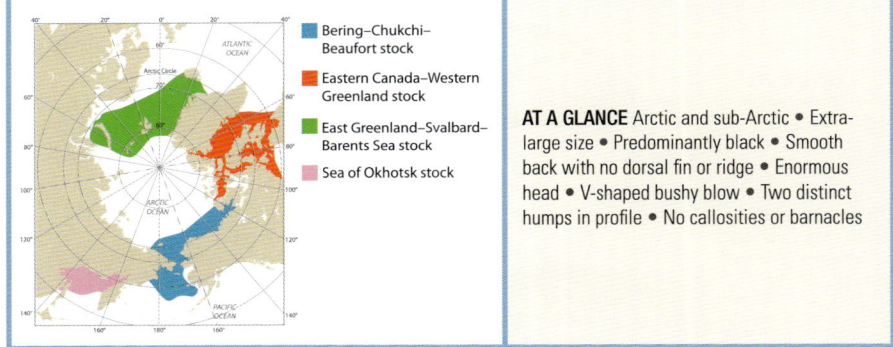

ADULT

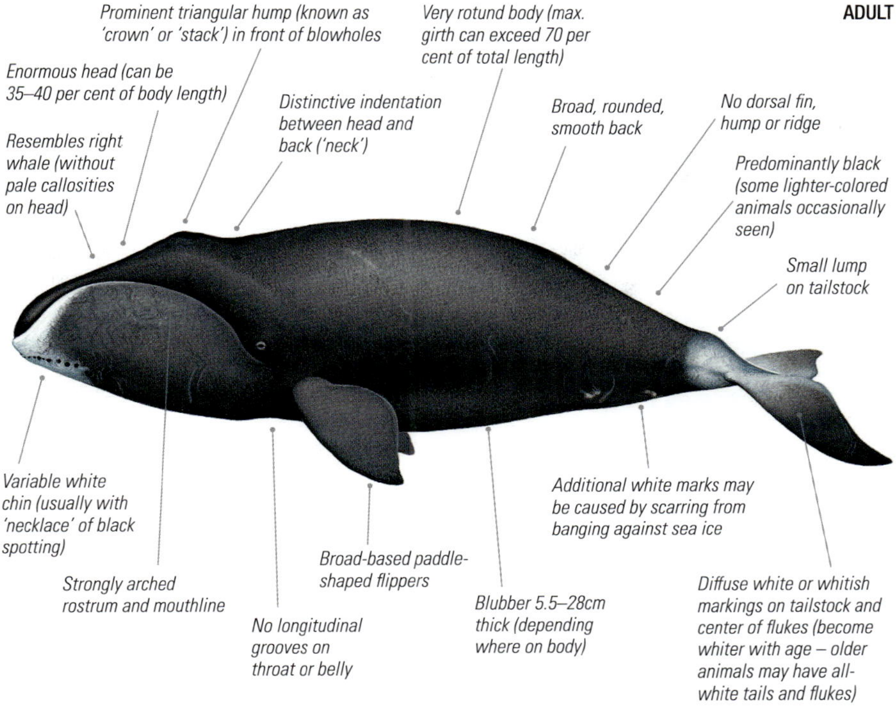

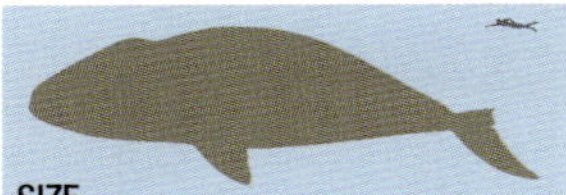

Prominent triangular hump (known as 'crown' or 'stack') in front of blowholes
Very rotund body (max. girth can exceed 70 per cent of total length)
Enormous head (can be 35–40 per cent of body length)
Distinctive indentation between head and back ('neck')
Broad, rounded, smooth back
No dorsal fin, hump or ridge
Resembles right whale (without pale callosities on head)
Predominantly black (some lighter-colored animals occasionally seen)
Small lump on tailstock
Variable white chin (usually with 'necklace' of black spotting)
Additional white marks may be caused by scarring from banging against sea ice
Broad-based paddle-shaped flippers
Strongly arched rostrum and mouthline
Diffuse white or whitish markings on tailstock and center of flukes (become whiter with age – older animals may have all-white tails and flukes)
Blubber 5.5–28cm thick (depending where on body)
No longitudinal grooves on throat or belly


**SIZE**
**L:** ♂ 14–17m, ♀ 16–18m;
**WT:** 60–90t; **MAX:** 19.8m, 107t
**Calf – L:** 4–4.5m (max. 5.2m);
**WT:** 900kg

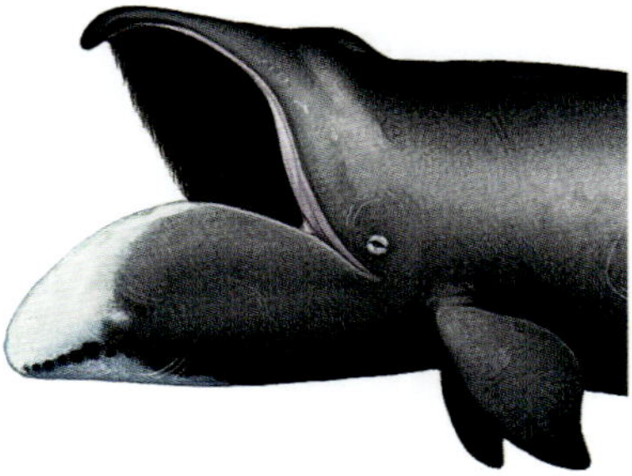

ATLANTIC OCEAN
Arctic Circle
ARCTIC OCEAN
PACIFIC OCEAN
Bering–Chukchi–Beaufort stock
Eastern Canada–Western Greenland stock
East Greenland–Svalbard–Barents Sea stock
Sea of Okhotsk stock


**AT A GLANCE** Arctic and sub-Arctic • Extra-large size • Predominantly black • Smooth back with no dorsal fin or ridge • Enormous head • V-shaped bushy blow • Two distinct humps in profile • No callosities or barnacles

**BALEEN** 230–360 plates (each side of the upper jaw). Baleen plates are up to 4m (maximum 5.2m) long – the longest of any whale; dark gray to brownish black, usually with lighter fringes.

**GROUP SIZE AND STRUCTURE** Bowheads are usually solitary, but are sometimes seen in small groups of two to three (up to 14). There are occasional loose aggregations of as many as 60 at productive feeding grounds and during migration. In summer, groups are often segregated by sex and age. Behavior synchronized acoustically within a range of up to 100km.

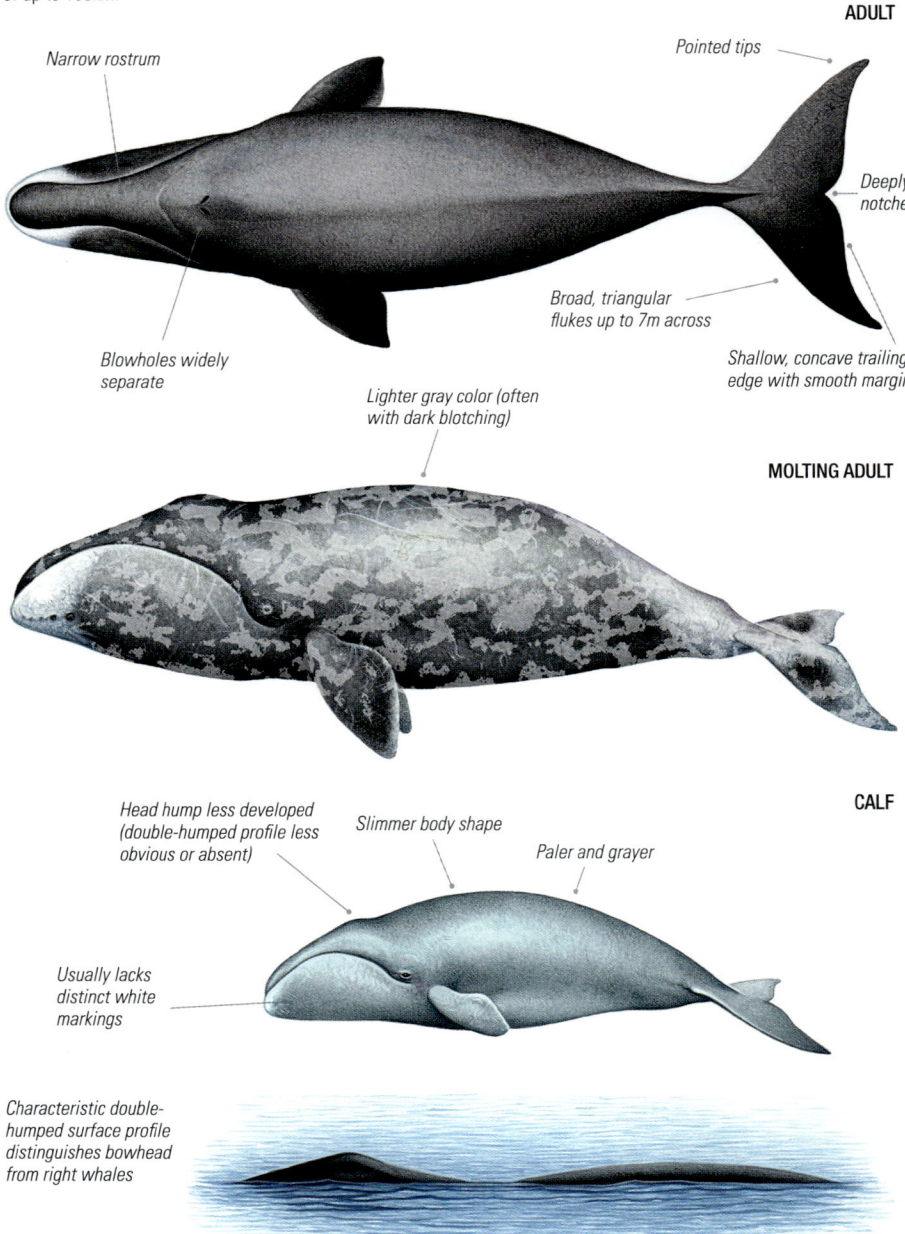

RIGHT AND BOWHEAD WHALES

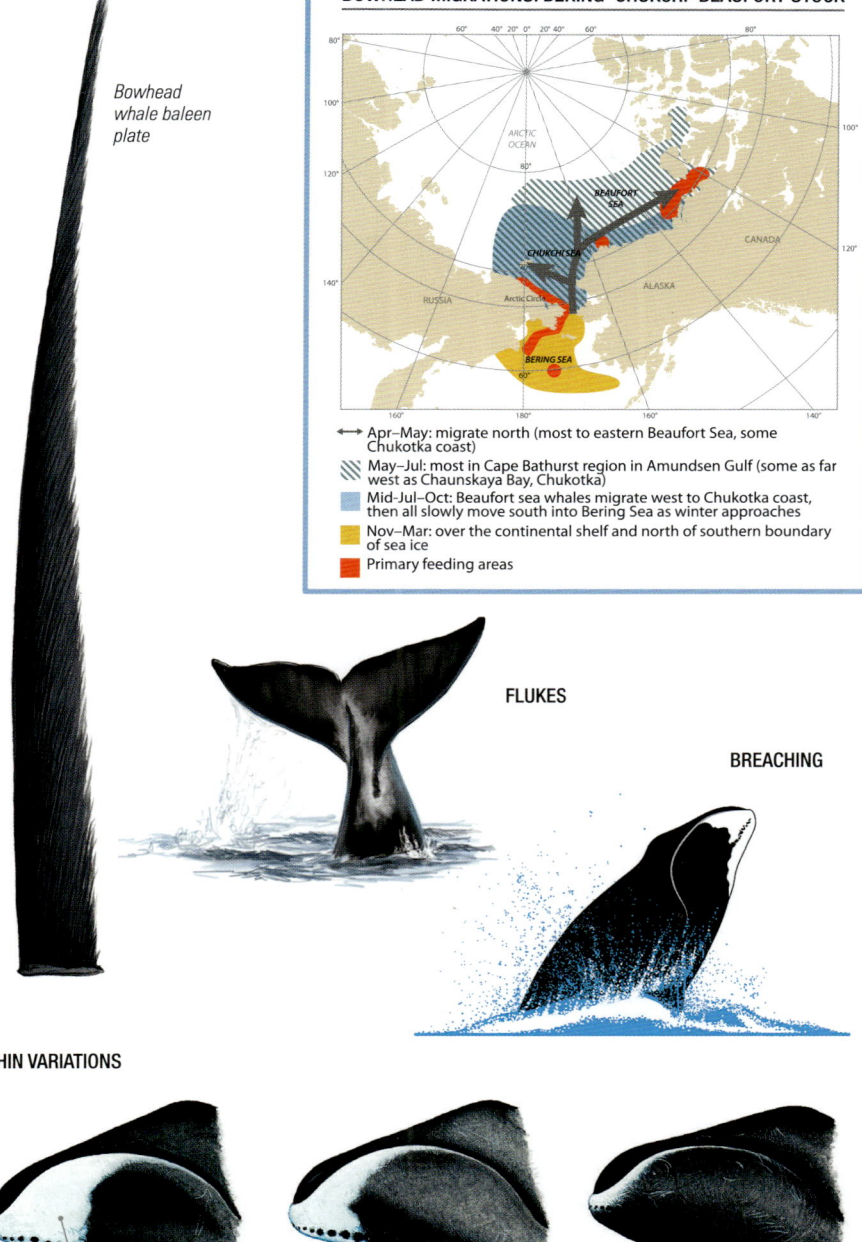

# GRAY WHALE
*Eschrichtius robustus* (Lilljeborg, 1861)

The gray whale is an inveterate traveler: the round-trip distance between its winter breeding grounds and summer feeding grounds can exceed 20,000km. Among the world's most-watched whales, it is instantly recognizable thanks to its mottled gray coloring and the small hump instead of a dorsal fin.

**IUCN status** Least Concern (2017). Western North Pacific sub-population Endangered (2018).
**Population** 19,260 (2024), 14,526 (2023) following an 'unusual' mortality event; 26,960 (2016). Western North Pacific population *c.* 175 (2020). Pre-whaling levels uncertain, but the most widely accepted estimate is 15,000–24,000 (although one DNA study estimated 76,000–118,000). Population stable (but fluctuates extensively).
**Classification** Mysticeti, family Eschrichtiidae.
**Taxonomy** No recognized forms or subspecies; two sub-populations (Eastern North Pacific or ENP and Western North Pacific or WNP), though there are no anatomical differences and there is evidence of mixing on the Mexican breeding grounds (more than 50 are known to have migrated from Russian summer feeding grounds to Baja winter breeding grounds).
**Other names** Grey whale (alternative spelling), grayback, California gray whale, Pacific gray whale; historically – mussel-digger, mud-digger, scrag whale, hardhead, devilfish.

**DISTRIBUTION** Mainly over shallow continental shelf waters of the North Pacific and adjacent seas. Primarily coastal, but does feed far from shore on the shallow flats of its feeding grounds and can navigate deep oceans on migration. The ENP population migrates between winter breeding grounds in Baja California, Mexico, and summer feeding grounds predominantly in the Bering, Chukchi and Beaufort seas; it appears to be expanding north-west as Arctic ice opens up. The WNP population migrates between winter breeding grounds (probably in the South China Sea), and summer feeding grounds in the Sea of Okhotsk and off southern and southeastern Kamchatka, Russia. There is evidence of some mixing between stocks during the winter breeding and summer feeding seasons. Historically, gray whales also occurred in the North Atlantic (see p. 47). There is one record from the South Atlantic – an individual spent nearly five weeks in Walvis Bay, Namibia, in 2013.

**Migration** Breeding and feeding grounds are widely separated, demanding long coastal migrations (spanning up to 50° of latitude). The ENP population makes an exceptionally long migration, hugging the length of the North America coast, usually within 10km of shore. The shortest return journey – between San Ignacio Lagoon, Mexico, and Unimak Pass, Alaska – is *c.* 12,000km, but many individuals swim considerably further; the longest documented migration of any mammal (excluding lost individuals) was a female gray whale that completed a 22,511km round trip between Sakhalin Island, Russia, and Baja California, Mexico. Since 2011, some individuals have remained in the Arctic for longer – exceptionally even year-round – to obtain sufficient food. This is presumed to result from reduced ice cover (due to global warming) affecting prey distribution and availability. With the retreat of sea ice, the summer distribution has also expanded in recent decades, especially to more northerly feeding areas (to at least 71°N); observations as far east

**DIVE SEQUENCE** On initial surfacing, head appears to slope downward from blowholes (giving appearance of shallow triangle); 'knuckles' clearly visible as back arches slightly to dive; flukes raised high into air before deep dive. • **DEPTH** Mainly seafloor feeder (prefers 30–60m; range of 3–120m, maximum known 170m) but feeds opportunistically in mid-water and at surface.
• **DURATION** On migration, 3–7 minutes; feeding dives typically 5–8 minutes; in breeding lagoons, 50 per cent of dives less than 1 minute; when resting, up to *c.* 26 minutes.

**BLOW** Bushy blow up to 5m (highly variable height). • May be V-shaped, tall and bushy, tree-shaped or heart-shaped (when spray falls inward) when seen from front or rear.

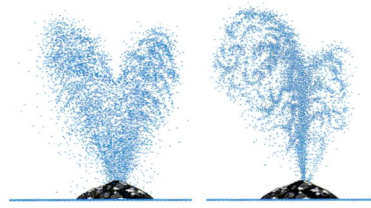

**ADULT**

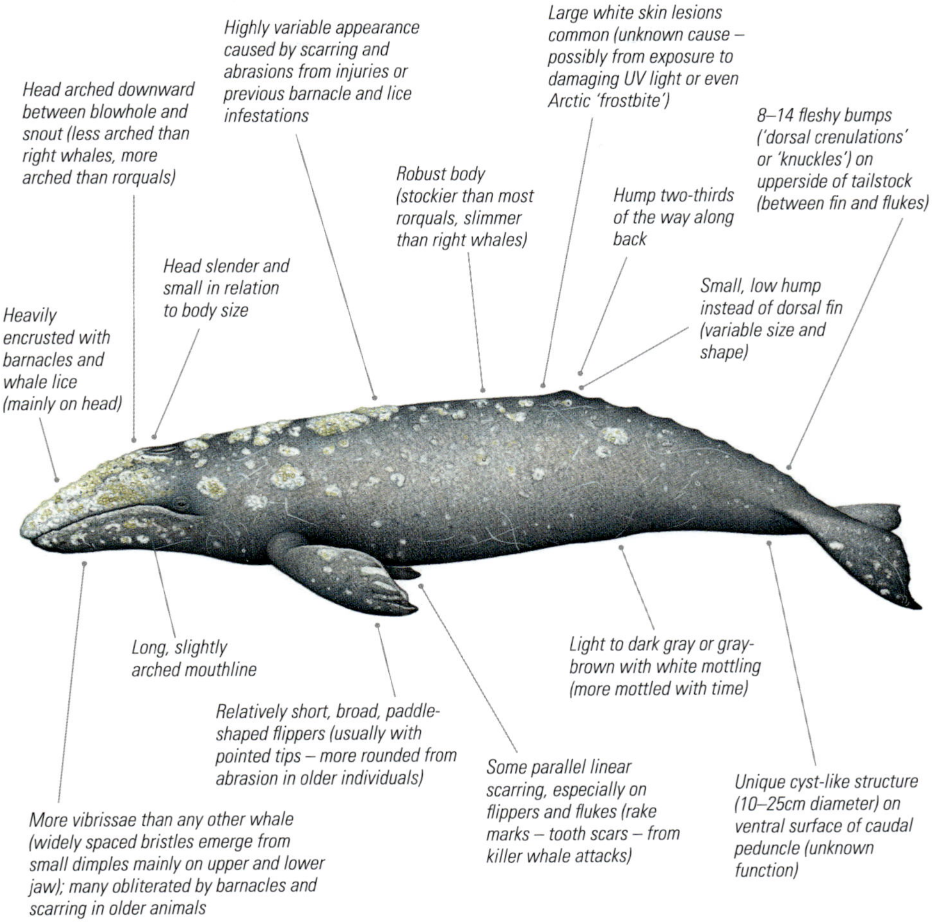

- Head arched downward between blowhole and snout (less arched than right whales, more arched than rorquals)
- Highly variable appearance caused by scarring and abrasions from injuries or previous barnacle and lice infestations
- Large white skin lesions common (unknown cause – possibly from exposure to damaging UV light or even Arctic 'frostbite')
- 8–14 fleshy bumps ('dorsal crenulations' or 'knuckles') on upperside of tailstock (between fin and flukes)
- Head slender and small in relation to body size
- Robust body (stockier than most rorquals, slimmer than right whales)
- Hump two-thirds of the way along back
- Small, low hump instead of dorsal fin (variable size and shape)
- Heavily encrusted with barnacles and whale lice (mainly on head)
- Long, slightly arched mouthline
- Light to dark gray or gray-brown with white mottling (more mottled with time)
- Relatively short, broad, paddle-shaped flippers (usually with pointed tips – more rounded from abrasion in older individuals)
- Some parallel linear scarring, especially on flippers and flukes (rake marks – tooth scars – from killer whale attacks)
- Unique cyst-like structure (10–25cm diameter) on ventral surface of caudal peduncle (unknown function)
- More vibrissae than any other whale (widely spaced bristles emerge from small dimples mainly on upper and lower jaw; many obliterated by barnacles and scarring in older animals)

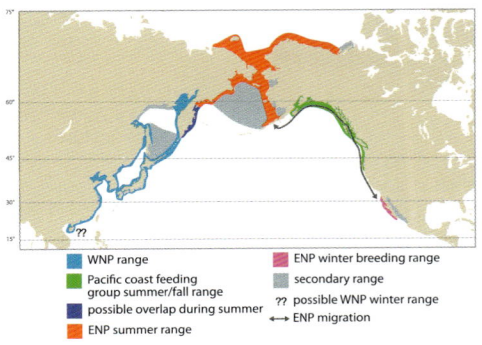

- WNP range
- Pacific coast feeding group summer/fall range
- possible overlap during summer
- ENP summer range
- ENP winter breeding range
- secondary range
- ?? possible WNP winter range
- ENP migration

**AT A GLANCE** Coastal or shallow waters of North Pacific and adjacent seas • Light to dark gray or gray-brown with white mottling • Large size • Low hump (instead of dorsal fin) • Head (and other parts of body) encrusted with barnacles and lice • 'Knuckles' on upper side of tailstock (between fin and flukes) • Low V-shaped or heart-shaped bushy blow • Frequently flukes upon deep dive

GRAY WHALE

as the Canadian Beaufort Sea and as far west as Wrangel Island in the Chukchi Sea (and even into the East Siberian Sea) are now common. There was also a sighting in Hawai'i in 2022. The southernmost record in the North Pacific is a stranded adult in El Salvador (c. 14°N) in 2010.

**PACIFIC COAST FEEDING GROUP** Some 200–240 gray whales – known as the Pacific Coast Feeding Group (Canada) or the Pacific Coast Feeding Aggregation (USA) – do not migrate all the way to the Arctic, and instead spend the summer and fall feeding in a well-defined coastal region between northern California and south-east Alaska.

**BEHAVIOR** One of the most active large whales at the surface, frequently breaching (often several times in a row, exceptionally 40–50 times), spyhopping, and waving tail or flippers in the air. May appear to 'play' in the surf and rub against pebble beaches, rocks, piers and boats, possibly to ease skin irritations caused by ectoparasites. Often inquisitive and may approach boats. 'Friendly' or 'curious' behavior is most common in Mexican breeding lagoons.

**FOOD AND FEEDING** Variety of benthic and planktonic prey; in northern seas, benthic amphipods usually account for 90 per cent of diet; south of Aleutian Islands, main prey often planktonic mysids but also benthic amphipods and other species. Will opportunistically take pelagic species such as red crabs and crab larvae, mysids, fish eggs and larvae, baitfish and squid. Most feeding during summer, fasting thereafter (except opportunistically on migration or by pregnant or lactating females in winter). Swims slowly along seabed, sucking up sediment, then filters out prey. May also skim-feed like right whales or gulp-feed like rorquals, to exploit free-swimming prey.

**BALEEN** 130–180 plates (each side of the upper jaw). Among the shortest and coarsest baleen plates of any whale, just 5–50cm long.

**GROUP SIZE AND STRUCTURE** Gray whales usually migrate alone or in twos or threes, but up to 16 can be seen in unstable groups. Mother–calf pairs tend to migrate alone. More than 1,000 can congregate in a single winter aggregation area and breeding lagoon. On summer feeding grounds, usually alone or in pairs, but several hundred may be scattered across food-rich areas. Towards the end of the feeding season, recently weaned young animals may form groups of up to 12 or more.

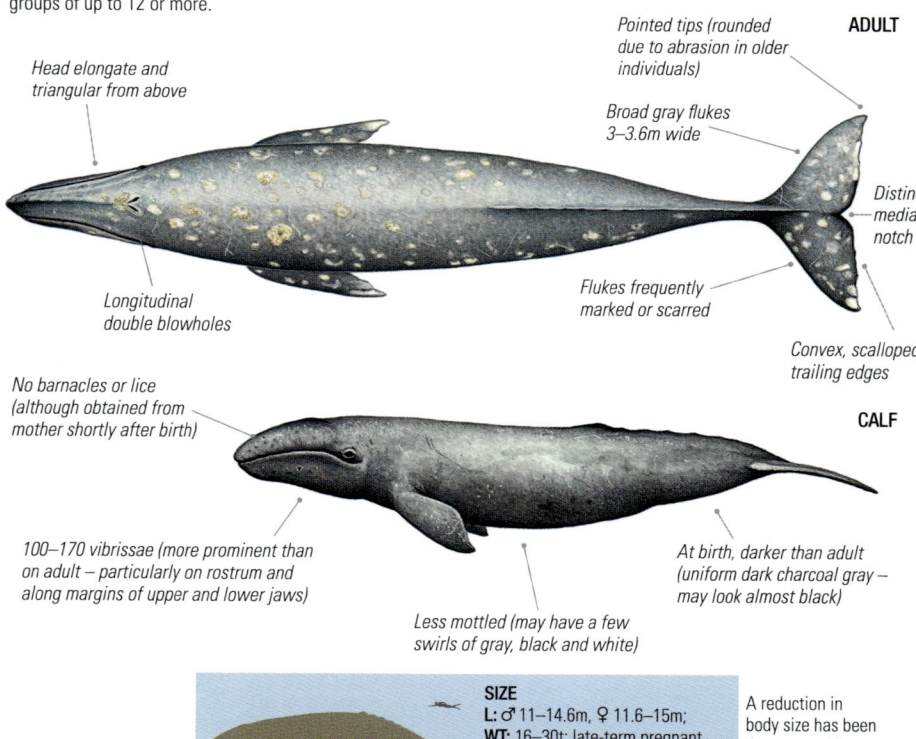

**ADULT**
- Pointed tips (rounded due to abrasion in older individuals)
- Head elongate and triangular from above
- Broad gray flukes 3–3.6m wide
- Distinct median notch
- Longitudinal double blowholes
- Flukes frequently marked or scarred
- Convex, scalloped trailing edges

**CALF**
- No barnacles or lice (although obtained from mother shortly after birth)
- 100–170 vibrissae (more prominent than on adult – particularly on rostrum and along margins of upper and lower jaws)
- Less mottled (may have a few swirls of gray, black and white)
- At birth, darker than adult (uniform dark charcoal gray – may look almost black)

**SIZE**
**L:** ♂ 11–14.6m, ♀ 11.6–15m;
**WT:** 16–30t; late-term pregnant females can weigh an additional 5t;
**MAX:** 15.6m, 40t
**Calf – L:** 4.2–4.9m; **WT:** c. 1.1t

A reduction in body size has been documented in recent years, in response to climate change.

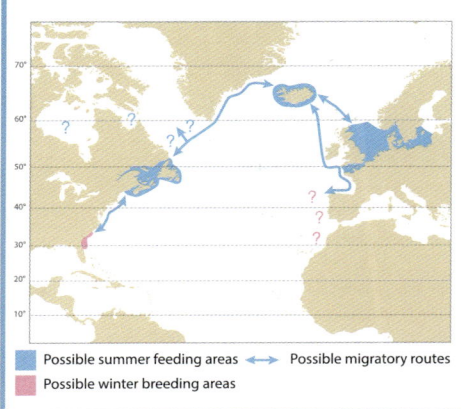

### GRAY WHALE HYPOTHETICAL HISTORICAL RANGE IN NORTH ATLANTIC

Gray whales once occurred on both sides of the North Atlantic (indeed, Lilljeborg's original description of the species in 1861 was from a subfossil skeleton found in Sweden). There is little information on their range or migratory routes, but there were possibly two discrete sub-populations (partly overlapping in Iceland). The population was predominantly extinct by the late 17th or early 18th centuries, probably due to (or at least hastened by) hunting by Basque, Icelandic and Yankee whalers. There have been five exceptional sightings of lone individuals in the Atlantic in recent years: Israel/Spain in 2010; Namibia in 2013; Morocco/Italy/France/Spain/Gibraltar in 2021; Florida in 2023; and Massachusetts in 2024.

Possible summer feeding areas ⬌ Possible migratory routes
Possible winter breeding areas

**ADULT HEAD** right side

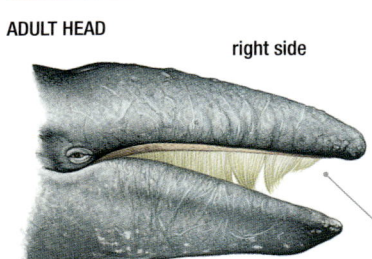

Baleen plates creamy to pale yellow (and shorter than in any other whale)

**ADULT HEAD UNDERSIDE**

Throat has 2–7 (usually 2–3) short, deep grooves (c. 1.5m long)

Most gray whales are right-dominant feeders and right side of head often differs from left side due to abrasion during seafloor feeding: more heavily scarred, fewer barnacles and whale lice, shorter and more worn baleen plates

**FLUKES**

**ADULT FLUKES** (showing shape change with age)

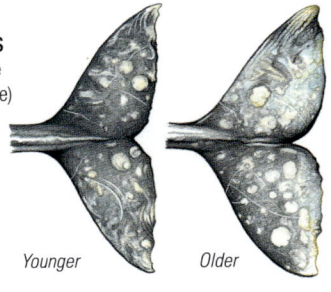

Younger    Older

**BARNACLES AND LICE**

*Cryptolepas rhachianecti* (whale barnacle) | *Cyamus scammoni* (found only on grays) | *Cyamus kessleri* (found only on grays) | *Cyamus ceti* (also found on bowheads) | *Cyamus eschrichtii* (found only on grays)

Gray whale calves are born free of external parasites, but rapidly acquire them as they grow. Adults have more than any other cetacean: consisting of one species of barnacle (up to 5.5cm in diameter and host-specific to gray whales) and four species of whale louse (1–2.7cm long).

# BLUE WHALE
*Balaenoptera musculus* (Linnaeus, 1758)

The largest animal known to have existed on Earth, the blue whale can be remarkably inconspicuous and difficult to see. But a close encounter with this true gargantuan is unforgettable. It was hunted relentlessly worldwide, until every population was severely depleted, and came dangerously close to extinction.

**IUCN status** Endangered (2018). Antarctic blue whale Critically Endangered (2018).
**Population** *c.* 10,000–25,000. Pre-whaling population *c.* 300,000, including 239,000 Antarctic blue whales. During the period 1868–1978, 382,595 were killed by commercial whalers. Increasing.
**Classification** Mysticeti, family Balaenopteridae.
**Taxonomy** Four subspecies are currently recognized: northern blue whale (*B. m. musculus*), Antarctic or 'true' blue whale (*B. m. intermedia*), northern Indian Ocean blue whale (*B. m. indica*), and pygmy blue whale (*B. m. brevicauda*); the Chilean blue whale (*B. m. chilensis*) has been proposed (but not formally accepted) as a fifth. The first two share a similar external form (though they differ genetically and acoustically).
**Other names** Sulphur-bottomed whale or sulphur-bottom (after the diatom film that can form on its body), Sibbald's rorqual.

**DISTRIBUTION** Occurs from the tropics to the edge of the pack ice in both hemispheres, though distribution is patchy and it is rare in most equatorial waters and in the center portions of major ocean basins. Most populations are migratory – moving between productive, higher-latitude summer and early fall feeding areas, and lower-latitude winter breeding and feeding areas – but at least one population (in the northern Indian Ocean) is largely resident year-round. Unlike most other baleen whales, blue whales feed year-round so food availability probably dictates distribution for the majority of the year; they will forage in productive areas anywhere. Seasonal movements can be extensive, but they are complex and poorly understood.

No specific breeding grounds have been discovered conclusively in any ocean (they do not appear to be as well defined as for humpback, gray and right whales) but they are believed to be in tropical and sub-tropical waters. One probable breeding ground is the Costa Rica Dome (or Papagayo upwelling) in the eastern tropical Pacific; there is another in the Galapagos Islands. The Gulf of California, Mexico, appears to be a nursing area and possible calving area (though far fewer have been seen in the Gulf in recent years).

The species is mainly oceanic and associated with waters deeper than the continental shelf, roaming widely across ocean basins, but also inhabits some shelf and coastal waters (such as in Mexico's Gulf of California, the southern California Bight in the US, Canada's Gulf of St Lawrence, and Iceland's Skjalfandi Bay). It prefers habitats marked by steep submarine topographic features that enhance upwelling. There are still many gaps in our knowledge of any overlap in distribution between different subspecies, especially in the southern hemisphere.

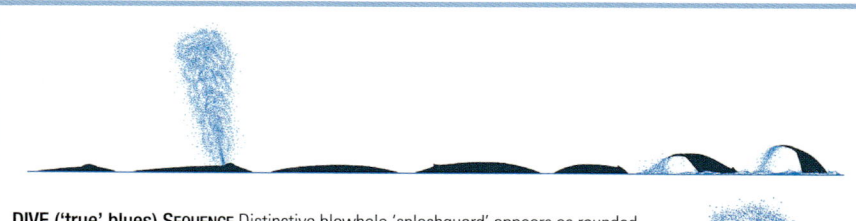

**DIVE ('true' blues) SEQUENCE** Distinctive blowhole 'splashguard' appears as rounded hump; massive elongated expanse of back rolls into view; many individuals raise flukes before a sounding dive. • **DEPTH** Foraging dives typically to maximum 250m but capable of 300+m (deepest recorded 370m off Isla Monserrat, Baja California); dives deeper during middle of the day, following diel vertical movements of prey to feed at shallower depths up to surface later in the day and at night; non-foraging dives typically shallower than *c.* 70m. • **DURATION** Feeding dives typically 8–15 minutes, but 20 minutes not uncommon (considerably shorter – average 3 minutes – in New Zealand); occasionally up to 30 minutes (maximum recorded 36 minutes).

**BLOW** Slender, columnar blow can be at least 12m high (highly variable height).
• Denser and broader than that of fin or sei whale.

**ADULT**

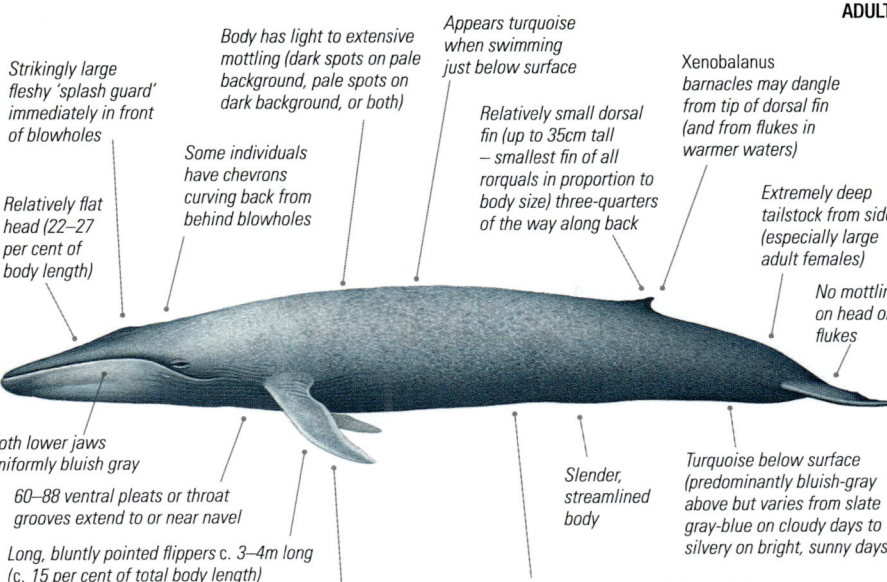

- Strikingly large fleshy 'splash guard' immediately in front of blowheads
- Body has light to extensive mottling (dark spots on pale background, pale spots on dark background, or both)
- Appears turquoise when swimming just below surface
- Xenobalanus barnacles may dangle from tip of dorsal fin (and from flukes in warmer waters)
- Relatively flat head (22–27 per cent of body length)
- Some individuals have chevrons curving back from behind blowholes
- Relatively small dorsal fin (up to 35cm tall – smallest fin of all rorquals in proportion to body size) three-quarters of the way along back
- Extremely deep tailstock from side (especially large adult females)
- No mottling on head or flukes
- Both lower jaws uniformly bluish gray
- 60–88 ventral pleats or throat grooves extend to or near navel
- Long, bluntly pointed flippers c. 3–4m long (c. 15 per cent of total body length)
- Slender, streamlined body
- Turquoise below surface (predominantly bluish-gray above but varies from slate gray-blue on cloudy days to silvery on bright, sunny days)
- Flippers with bluish-gray upperside (usually with thin white border or tip and occasionally mottled), whitish underside
- In cold waters, all or parts of body (especially underside) may be covered in yellow to greenish diatom films (when diatoms fresh) or opaque rust-colored diatom films (when diatoms mature and dying)

**ADULT ANTARCTIC/NORTHERN**

- Jaws capable of opening to nearly 90°

**AT A GLANCE** Worldwide (though patchy distribution) • Extra-large size • Streamlined body shape • Mottled bluish-gray color • Turquoise underwater (when viewed from surface) • Relatively small dorsal fin three-quarters of the way along back • Prominent blowhole 'splash guard' • Extremely deep tailstock • Often raises flukes on diving

**BEHAVIOR** Some blue whales raise their flukes when diving (c. 18 per cent in the north-west Atlantic and north-east Pacific, 25 per cent in Mexico's Gulf of California and 55 per cent off Sri Lanka). Capable of burst speeds of up to 35km/hr if being chased by boats or killer whales (when swimming very fast, it sometimes almost porpoises above the surface, throwing up a large rooster-tail and pushing a mass of water in front as it flees). There have been a few observations of breaching blue whales – usually youngsters – leaping out of the water at a c. 45° angle, but this is very rare. Behavior around boats varies from avoidance through indifference to inquisitiveness.

**FOOD AND FEEDING** Mainly euphausiid crustaceans (i.e. krill); some other crustaceans (including copepods, mysids and amphipods); occasionally small schooling fish and cephalopods. Unlike most baleen whales, probably does not fast during winter (continues to feed on breeding grounds); dives below dense layer of krill, turns upwards and lunges, rolling sideways or doing full barrel-roll and opening mouth, then drifts slowly forward as closes mouth; typically up to 6–7 lunge-feeds per dive (maximum 15); when feeding near surface, often surfaces slowly on one side or upside down (with one flipper and part of flukes above water).

**BALEEN** 260–400 plates (each side of the upper jaw). Baleen plates are black, broad-based, and each c. 1m long (slightly longer in 'true' blues, cf. pygmy blue).

**GROUP SIZE AND STRUCTURE** Usually alone or in pairs, though groups of 3–6 are known in some areas during summer. May be scattered in loose aggregations of 50+ on good feeding grounds.

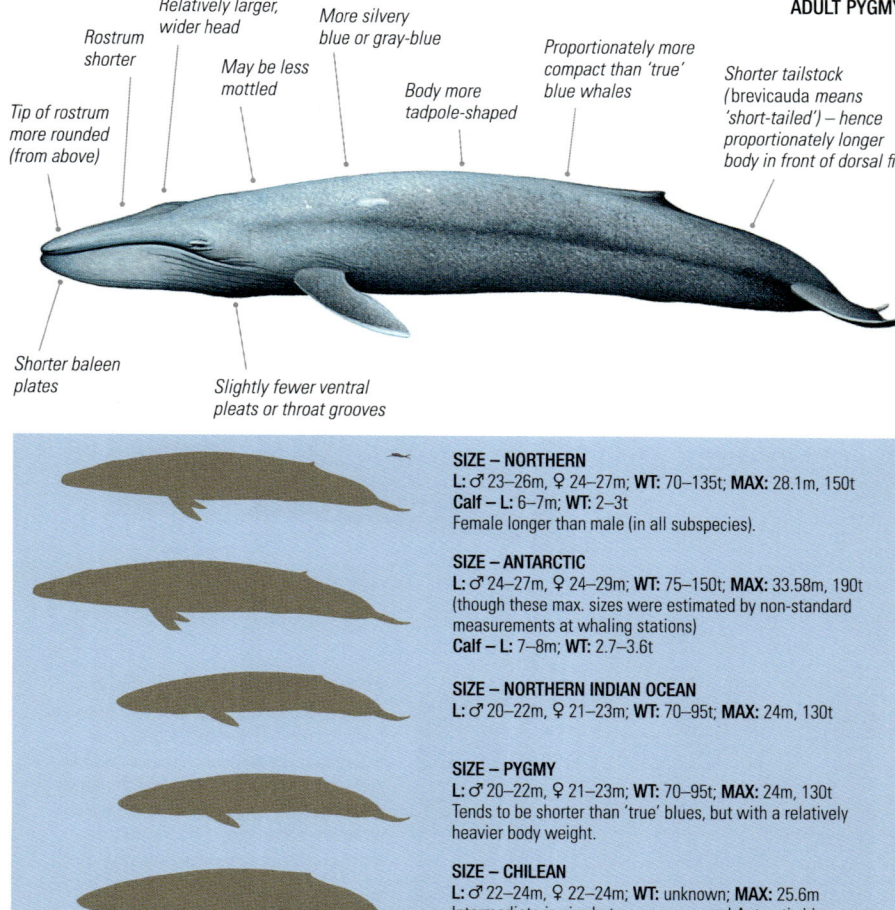

**ADULT PYGMY**

- Rostrum shorter
- Relatively larger, wider head
- More silvery blue or gray-blue
- Proportionately more compact than 'true' blue whales
- Shorter tailstock (brevicauda means 'short-tailed') – hence proportionately longer body in front of dorsal fin
- May be less mottled
- Body more tadpole-shaped
- Tip of rostrum more rounded (from above)
- Shorter baleen plates
- Slightly fewer ventral pleats or throat grooves

**SIZE – NORTHERN**
L: ♂ 23–26m, ♀ 24–27m; WT: 70–135t; MAX: 28.1m, 150t
Calf – L: 6–7m; WT: 2–3t
Female longer than male (in all subspecies).

**SIZE – ANTARCTIC**
L: ♂ 24–27m, ♀ 24–29m; WT: 75–150t; MAX: 33.58m, 190t (though these max. sizes were estimated by non-standard measurements at whaling stations)
Calf – L: 7–8m; WT: 2.7–3.6t

**SIZE – NORTHERN INDIAN OCEAN**
L: ♂ 20–22m, ♀ 21–23m; WT: 70–95t; MAX: 24m, 130t

**SIZE – PYGMY**
L: ♂ 20–22m, ♀ 21–23m; WT: 70–95t; MAX: 24m, 130t
Tends to be shorter than 'true' blues, but with a relatively heavier body weight.

**SIZE – CHILEAN**
L: ♂ 22–24m, ♀ 22–24m; WT: unknown; MAX: 25.6m
Intermediate in size between pygmy and Antarctic blue whales.

**ADULT ANTARCTIC/NORTHERN**

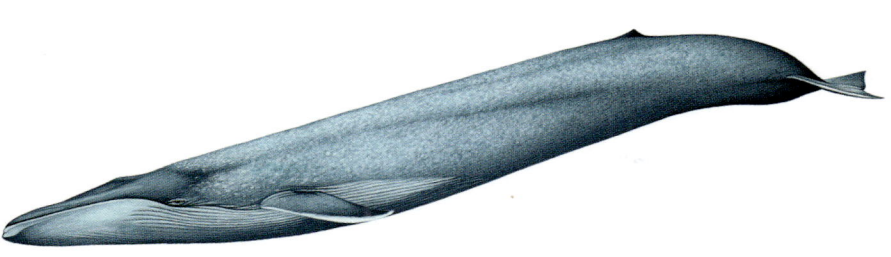

**ADULT ANTARCTIC/NORTHERN**

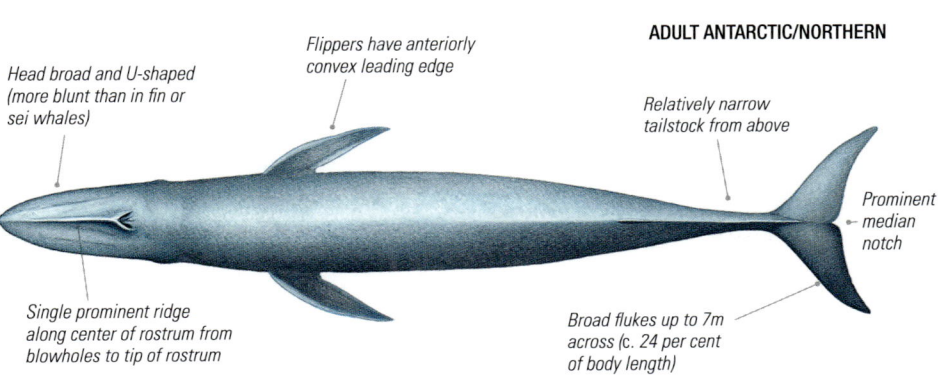

Head broad and U-shaped (more blunt than in fin or sei whales)

Flippers have anteriorly convex leading edge

Relatively narrow tailstock from above

Prominent median notch

Single prominent ridge along center of rostrum from blowholes to tip of rostrum

Broad flukes up to 7m across (c. 24 per cent of body length)

**ADULT ANTARCTIC/NORTHERN**

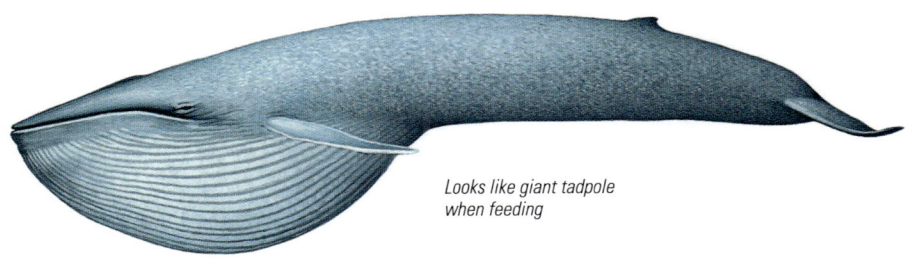

Looks like giant tadpole when feeding

**CALF**

Similar in appearance and shape to adult

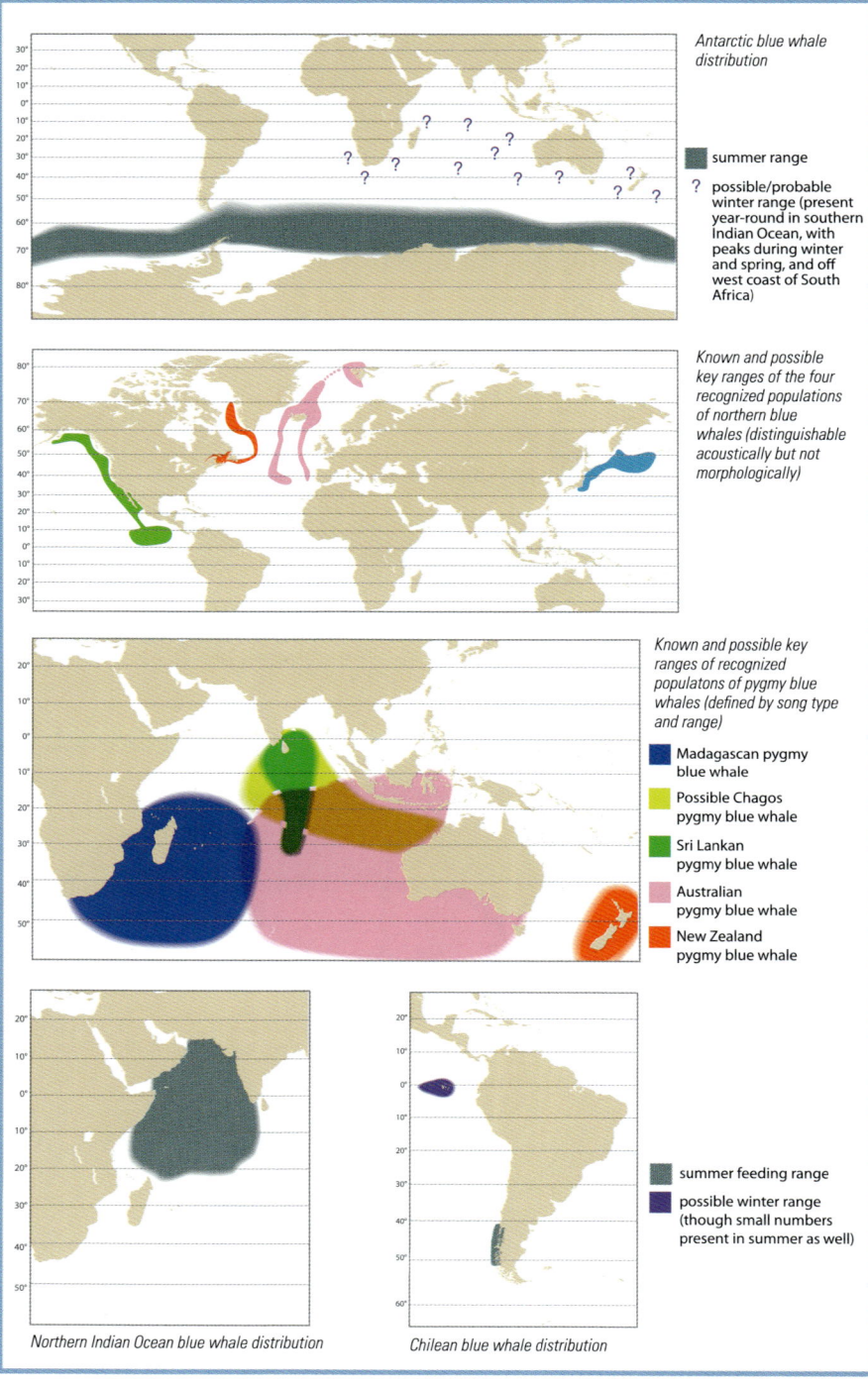

## WHY ARE BLUE WHALES SO BIG?

Blue whales feed on krill – small animals that swarm in mind-boggling numbers (up to 1,500 individuals per square meter) but live in patches that might be hundreds or thousands of kilometers apart. Anything that feeds on krill needs to be able to swim great distances quickly and efficiently (the relative energy cost of traveling declines as body size increases); able to store energy for days, weeks or even months at a time, to fill the gaps between krill patches (about a quarter of a blue whale's body mass consists of blubbery fat reserves); and able to ingest huge numbers of the tiny prey when it is available (feeding on a few krill at a time would be far too inefficient). Therefore, the ultimate krill-eating predator is a large-bodied, big-gulping, filter-feeding whale. And there are no size constraints thanks to the buoyancy of water.

**FLUKES**

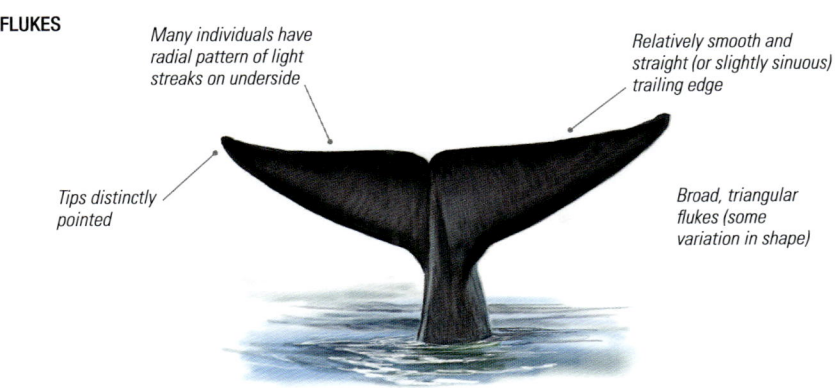

Many individuals have radial pattern of light streaks on underside

Relatively smooth and straight (or slightly sinuous) trailing edge

Tips distinctly pointed

Broad, triangular flukes (some variation in shape)

**DORSAL FIN VARIATIONS**

Highly variable dorsal fin (from small nubbin to triangular, hooked or falcate)

## PSEUDO-STALKED BARNACLE *Xenobalanus globicipitis*

This curious dark, worm-like animal (up to 5cm long) hangs from the trailing edges of the tails, dorsal fins, and flippers of at least 34 cetacean species – especially baleen whales such as blues – in tropical, sub-tropical and temperate waters worldwide. It is sometimes on the rostrum, and even on baleen plates and teeth, as well. There can be just one or as many as 100 in a cluster. They burrow into the skin (and blubber) to various depths and, once attached with the shell base embedded in the host, do not move.

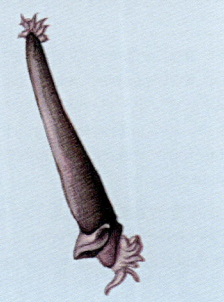

# FIN WHALE
## *Balaenoptera physalus* (Linnaeus, 1758)

The second-longest whale, after the blue, the fin whale is also one of the fastest (it's been dubbed the 'greyhound of the sea'). The distinctive asymmetrical pigmentation on its lower jaw – largely dark on the left and white on the right – has never been satisfactorily explained (it is also a feature of Omura's whale, and some sei and dwarf minke whales, but is more marked on the fin whale).

**IUCN status** Vulnerable (2018); Mediterranean sub-population Endangered (2021).
**Population** Minimum 150,000. At least 915,000 killed by commercial whalers. Increasing.
**Classification** Mysticeti, family Balaenopteridae.
**Taxonomy** Three subspecies are recognized: North Atlantic fin whale (*B. p. physalus*) in the North Atlantic, North Pacific fin whale (*B. p. velifera*) in the North Pacific, and southern fin whale (*B. p. quoyi*) in most of the southern hemisphere; pygmy fin whale (*B. p. patachonica*) off the west coast of South America has been proposed (but not formally accepted) as a fourth.
**Other names** Finback, finner, razorback, common rorqual, herring whale, finfish.

**DISTRIBUTION** In summer, found in cool temperate to polar waters worldwide, in all major oceans in both hemispheres. It is rarely found in the tropics (except in certain cool-water areas such as off Peru) or in high latitudes near the ice edge. Movements are complex: some populations appear to be migratory (especially in the southern hemisphere), with a general shift to higher latitudes for feeding in summer and lower latitudes for breeding (and less feeding) in winter, but they do not follow a simple pattern and breeding grounds remain uncertain (assuming such areas exist). Resident or semi-resident populations occur in the Gulf of California (Mexico), the Gulf of Alaska (USA), the East China Sea (off Japan), southern California and the central and western Mediterranean Sea (where there appears to be some mixing with seasonal visitors from the North Atlantic).

Density tends to be higher near or seaward of the continental shelf edge, but it is frequently seen over the shelf and close to shore where the water is deep enough. Typically, it frequents water deeper than 200m (100m in some regions) wherever topographic and oceanographic conditions concentrate prey.

**BEHAVIOR** Capable of swimming exceptionally fast, reaching 37km/h for short bursts. Rarely breaches. Often forms mixed schools with blue whales and sometimes associates with pilot whales and dolphins; often seen in large feeding aggregations with humpback whales, minke whales, and other species. Typically, neither avoids boats nor approaches them, but it can be quite approachable and is sometimes curious.

**FOOD AND FEEDING** Opportunistic, depending on locality, season and availability. Northern hemisphere: mainly krill, also copepods, schooling fish (including herring, mackerel, cod, pollock, capelin, sardines, sand lance and blue whiting), some small squid. Southern hemisphere: almost exclusively krill, but also other planktonic crustaceans. Feeds

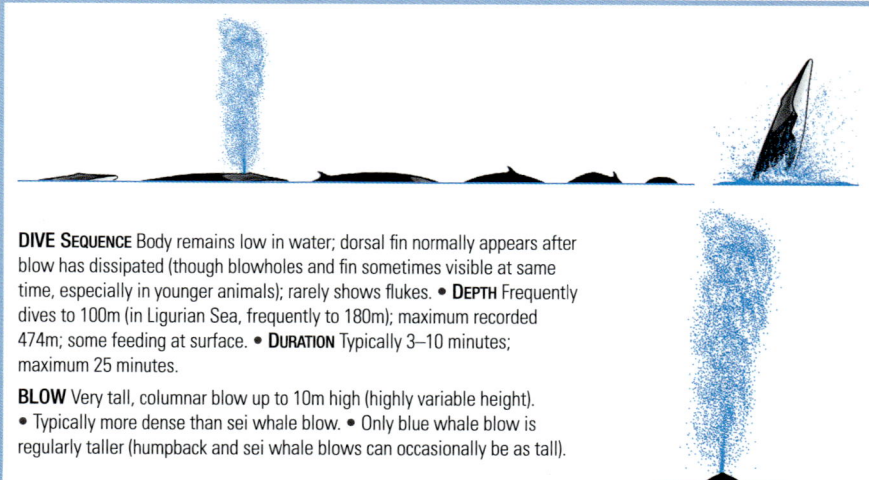

**DIVE SEQUENCE** Body remains low in water; dorsal fin normally appears after blow has dissipated (though blowholes and fin sometimes visible at same time, especially in younger animals); rarely shows flukes. • **DEPTH** Frequently dives to 100m (in Ligurian Sea, frequently to 180m); maximum recorded 474m; some feeding at surface. • **DURATION** Typically 3–10 minutes; maximum 25 minutes.

**BLOW** Very tall, columnar blow up to 10m high (highly variable height).
• Typically more dense than sei whale blow. • Only blue whale blow is regularly taller (humpback and sei whale blows can occasionally be as tall).

**ADULT – LEFT SIDE**

- Chevrons more visible in good light (sometimes faint or obscured by diatoms)
- No blaze on left side
- Dark gray or brownish-gray upperside and sides (often appears dark with silvery sides)
- Rear half especially may be heavily pockmarked with light oval cookiecutter shark and lamprey bites (varies with locality)
- Raised 'splashguard' around blowholes
- Usually one or more light gray V-shaped 'chevrons' on back behind head (cf. U-shaped on sei whale) with apex pointing forward
- Sleek, streamlined body (slimmer than blue whale)
- Relatively flattened head
- No dark eye or ear stripes on left side
- No mottling (cf. blue whale)
- Tall, falcate dorsal fin (up to 60cm) seven-tenths of the way along back
- Leading edge of dorsal fin highly variable but averages 33° angle from horizontal (cf. sei whale)

- Left lower 'lip' largely dark gray or dark brownish-gray (highly variable extent towards throat)
- May be distinct flipper 'shadows' on lower sides
- Long, tapered flippers with pointed tips (shorter and broader in northern fin whales)
- Upperside of flippers can be dark gray, brownish-gray or creamy white
- Underside of flippers light gray to white
- Fades to creamy white underside (sometimes yellowish or brownish in colder waters due to film of diatoms – can become blotchy as diatom layer sloughed off)

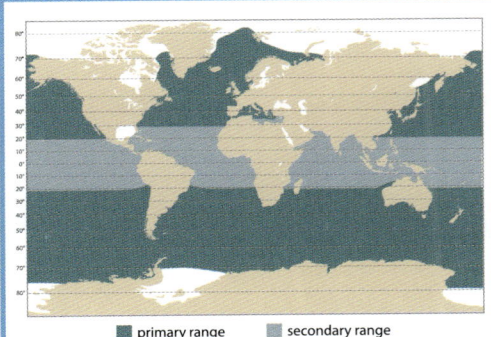

**SIZE (NORTHERN)**
L: ♂ 18–22m, ♀ 20–23m;
WT: 40–50t; MAX: 24m, 90t
Calf – L: 6–6.5m; WT: 1–1.7t

**SIZE (SOUTHERN)**
L: ♂ 23–25m, ♀ 24–26m;
WT: 60–80t; MAX: 27m, 120t
Calf – L: 6–7m; WT: 1–1.9t
Females are 5–10 per cent longer than males; northern hemisphere animals are smaller.

■ primary range   ■ secondary range

**AT A GLANCE** Worldwide • Dark gray or brownish-gray upperside • Extra large size • Light gray V-shaped chevrons on back • Asymmetrical lower 'lip' coloration • Single prominent ridge on rostrum • Backward-sloping dorsal fin • Rarely raises flukes on diving • Alone or in pairs or small groups

intensively in summer, consumes much less in winter; lunge-feeder (often rolling on side – typically to the right); mouth opens to almost 90° angle; no evidence of cooperative feeding.

**BALEEN** 260–480 (average *c.* 350–390) plates (each side of the upper jaw). Northern fin whales have slightly more plates on average. Longest plates *c.* 80cm.

**GROUP SIZE AND STRUCTURE** Frequently seen alone, but often in small groups of 2–7; large, loose aggregations of several dozen (up to 100 in exceptional cases) may occur in highly productive areas. Group composition tends to be dynamic (with individuals frequently moving between groups).

## ADULT – RIGHT SIDE

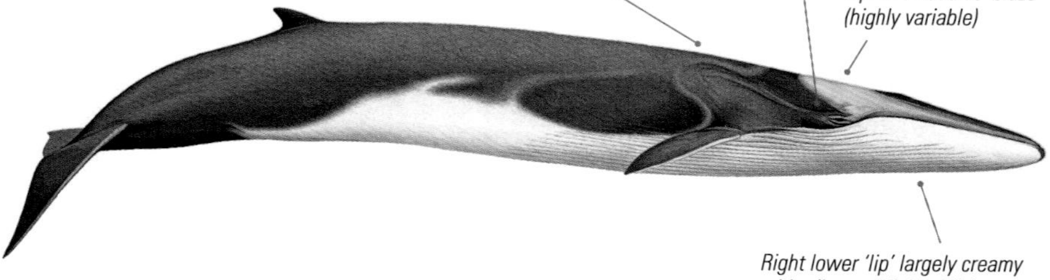

May be other streaks or swirls of light gray extending from underside upward (especially over right flipper) and from eye

Typically has dark eye and ear stripes on right side

White coloration can extend onto right upper 'lip' and head as 'blaze' (highly variable)

Right lower 'lip' largely creamy white (in good light visible even several meters below surface)

### ADULT PYGMY

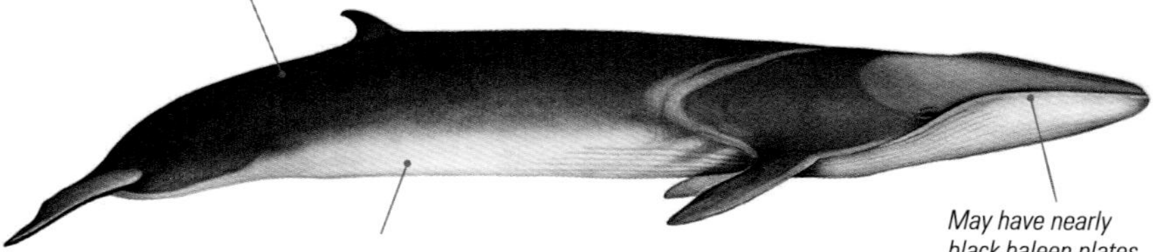

Darker coloration

Noticeably smaller than southern fin whale (c. 18–24m)

May have nearly black baleen plates

## ADULT

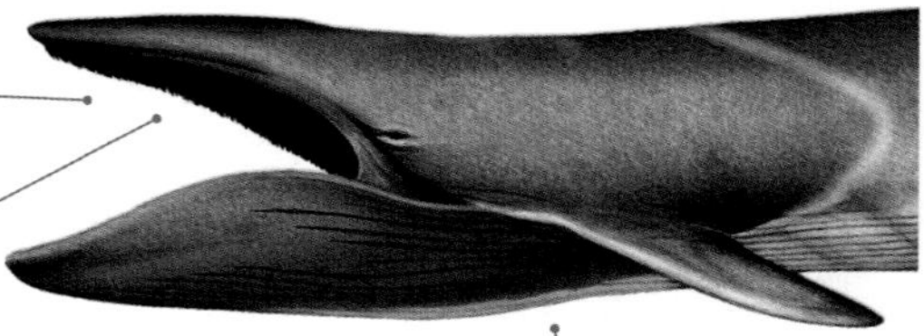

Baleen plates predominantly dark blue-gray to nearly black (except front 20–30 per cent on right side, which are all whitish or yellowish)

Baleen plates often striated with gray bands and fringed with horizontal lines of yellowish-white, brownish-gray or olive-green

50–100 longitudinal throat pleats (extend slightly beyond umbilicus)

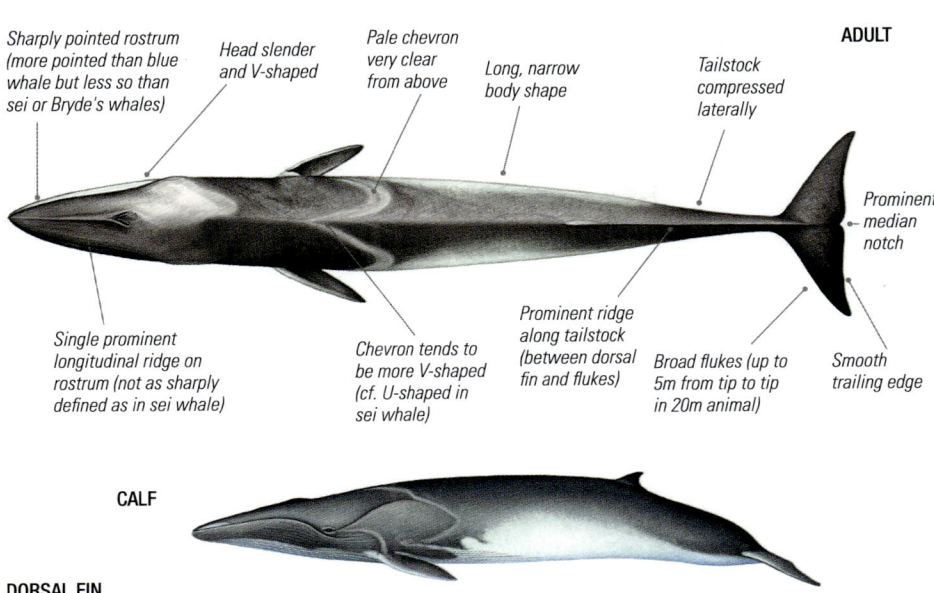

**ADULT**

- Sharply pointed rostrum (more pointed than blue whale but less so than sei or Bryde's whales)
- Head slender and V-shaped
- Pale chevron very clear from above
- Long, narrow body shape
- Tailstock compressed laterally
- Prominent median notch
- Single prominent longitudinal ridge on rostrum (not as sharply defined as in sei whale)
- Chevron tends to be more V-shaped (cf. U-shaped in sei whale)
- Prominent ridge along tailstock (between dorsal fin and flukes)
- Broad flukes (up to 5m from tip to tip in 20m animal)
- Smooth trailing edge

**CALF**

## DORSAL FIN

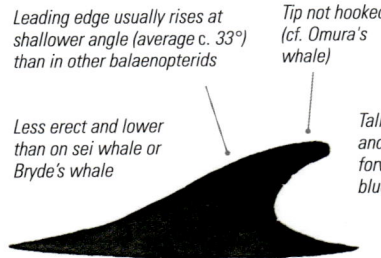

- Leading edge usually rises at shallower angle (average c. 33°) than in other balaenopterids
- Tip not hooked (cf. Omura's whale)
- Less erect and lower than on sei whale or Bryde's whale
- Taller, more falcate and set further forward than on blue whale
- Shape varies from falcate and rounded to triangular and pointed (tip always strongly directed backward)

## FLUKES

Underside of flukes light gray to white with dark gray border (rarely raised above surface)

### FIN/BLUE HYBRIDIZATION

Hybrids between fin whales and blue whales have been reported since the 19th century, most originating from the successful mating of male fin whales with female blue whales. The offspring share characteristics from both species. Fin/blue hybridization is believed to be more common than the limited genetic evidence suggests, likely due to reduced population sizes causing a significant disruption to the whales' reproductive dynamics. There is limited evidence that, in certain circumstances, some first-generation hybrids might be fertile and able to breed with one of the parental species.

### COOKIECUTTER SHARK BITES

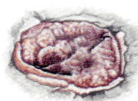 

Raw bite wound      Healed bite wound

**FIN WHALE** 57

# SEI WHALE
*Balaenoptera borealis*                                           Lesson, 1828

The enigmatic sei whale is the third-longest whale, yet it is surprisingly poorly known. This is partly because in the past it was often confused in whaling records and scientific accounts with Bryde's (and possibly Omura's) whales.

**IUCN status** Endangered (2018).
**Population** *c.* 80,000. Pre-whaling population *c.* 230,000. At least 325,000 killed by whalers. Increasing.
**Classification** Mysticeti, family Balaenopteridae.
**Taxonomy** Two subspecies are recognized: northern sei whale (*B. b. borealis*) and southern sei whale (*B. b. schlegelii*).
**Other names** Coalfish whale, sardine whale, lesser fin whale, pollack whale, northern rorqual.

**DISTRIBUTION** Ranges from the tropics to the poles in both hemispheres, but most abundant in mid-latitude temperate zones (20–55° latitude). Distribution is poorly documented and most information comes from whaling catches. Migrates between higher-latitude (cold temperate to sub-polar) summer and fall feeding grounds and lower-latitude (warm temperate to sub-tropical) winter breeding grounds. Compared with some other rorquals, migrations are less extensive, feeding and breeding grounds are less distinct, and it generally does not range as far north or south. It tends to be less predictable than other rorquals. It may abruptly disappear from areas where it had occurred regularly for years, and suddenly appear in other areas where it had been absent for years (or even decades); irruptions of sei whales are known as 'sei whale years' or 'invasion years'.

Generally considered a pelagic species with an offshore distribution along and beyond the continental shelf edge, especially in areas characterized by complex submarine topography such as seamounts and ridges. However, in some areas (e.g. Chile and the Falkland Islands) it regularly enters shelf waters and can be found at relatively shallow depths (less than 40m) and close to shore, including inside inlets and channels. Prefers sea surface temperatures of 8–18°C (occasionally up to 25°C).

**BEHAVIOR** One of the swiftest of the rorquals, capable of swimming at 25km/h (even 55km/h in short bursts, according to some whaling records). Breaching is rare but usually at a low angle and ends in a belly-flop. The sei whale has been seen in association with Peale's dolphins in the Falkland Islands. Most individuals avoid boats, or are indifferent, but some can be curious, repeatedly approaching and swimming alongside.

**FOOD AND FEEDING** Diverse diet varies regionally; mainly dense concentrations of minuscule copepods and krill, but also amphipods, squid, schooling fish (including sand lance, lumpfish, capelin, anchovy, herring, saury, lanternfish); in the North Atlantic it prefers pelagic copepods; around the Falkland Islands, mainly lobster krill. In some areas, especially when feeding near the surface, often associated with large flocks of feeding seabirds. Unusually among baleen whales, has two modes of feeding: normally 'skims' like right whales, but sometimes 'lunges and gulps' like

**DIVE SEQUENCE** Surfaces at shallow angle; tip of rostrum usually just breaks the surface; blowhole and dorsal fin often (but not always) visible at same time; tends to sink below surface (back relatively flat, though sometimes arches before deep dive); fin disappears last; very rarely, if ever, raises flukes; sometimes dives and surfaces in predictable line (remaining visible just below surface), frequently leaving long series of flukeprints on surface – but often impossible to predict and can be erratic in surfacing behavior. • **DEPTH** Varies according to vertical migrations of prey: limited evidence from Japan suggests averages *c.* 10–12m at night, *c.* 16–19m in day. • **DURATION** On feeding grounds typically 1–3 breaths at the surface over 20–30 seconds, then longer dives of up to 13 minutes.

**BLOW** Blow up to 9–10m high and columnar to bushy (highly variable height – typically 3–5m). • Generally more diffuse than fin whale blow.

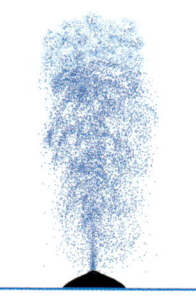

**ADULT (NORTHERN)**

- Often has fin whale-like lighter-colored chevron (less extensive than on fin whale and U-shaped – highly variable in brightness and extent – more visible in good light)
- Relatively tall, erect dorsal fin (average 55cm) slightly less than two-thirds of the way along back (further forward than on other rorquals)
- Some individuals have white streak behind eye
- Lighter 'brush stroke' on sides (highly variable between individuals and more visible in good light)
- Dark brown-gray upperside (can appear bluish-gray to steely black in poor light)
- Leading edge of fin rises at steep angle from back (steeper than in fin whale – typically c. 46° from horizontal)
- Head 21–25 per cent of body length
- Mid brown-gray sides
- Fin often has distinct backward bend halfway to two-thirds up (cf. Omura's whale – though some overlap with Bryde's whale)
- Somewhat arched head with slightly downturned tip
- Deep tailstock
- Lower 'lips' dark to light gray on both sides (varies according to light conditions and some individuals may show mild form of fin whale asymmetry)
- Lighter brown-gray (sometimes creamy white) underside
- Sleek, streamlined body
- May be heavily pockmarked (especially rear half) with light oval cookiecutter shark and lamprey bites
- 32–65 (average 50) relatively short longitudinal throat pleats on underside (short for rorqual – terminate midway between flippers and umbilicus)
- Relatively small, slender, pointed flippers (c. 9 per cent of body length)

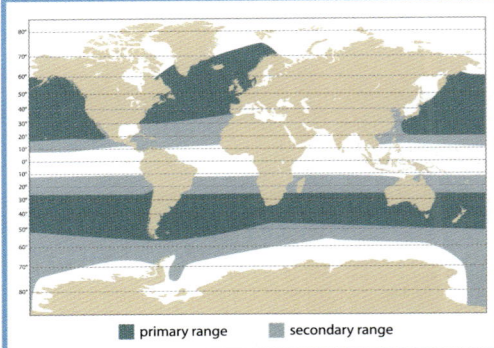

primary range | secondary range

**AT A GLANCE** Sub-tropical to sub-polar offshore waters worldwide • Large size • Sleek body • Dark upperside, lighter underside • Pale 'brush strokes' on sides • May have fin-whale-like chevron (U-shaped) • Single prominent ridge on rostrum • Rostrum has downturned tip • Tall and erect dorsal fin (highly variable) • Symmetrical head coloring • Dorsal fin and blowholes may be visible simultaneously

other rorquals; most feeding occurs during summer (winter consumption is low); individuals may remain in a specific feeding area for several weeks if prey densities are sufficient; there are no reports of cooperative feeding.

**BALEEN** 219–402 (average *c.* 350) plates (each side of the upper jaw). Longest plates *c.* 80cm; tend to be narrower than in other rorquals.

**GROUP SIZE AND STRUCTURE** Varies according to location and season – often seen alone or in small, fluid groups of 2–5. Larger groups may travel together, and loose aggregations numbering tens of individuals can form in productive feeding areas. Sightings of apparent social groups have been recorded, perhaps engaged in courtship behavior, involving high-speed chases and swimming on their sides with tail flukes emerging from the water.

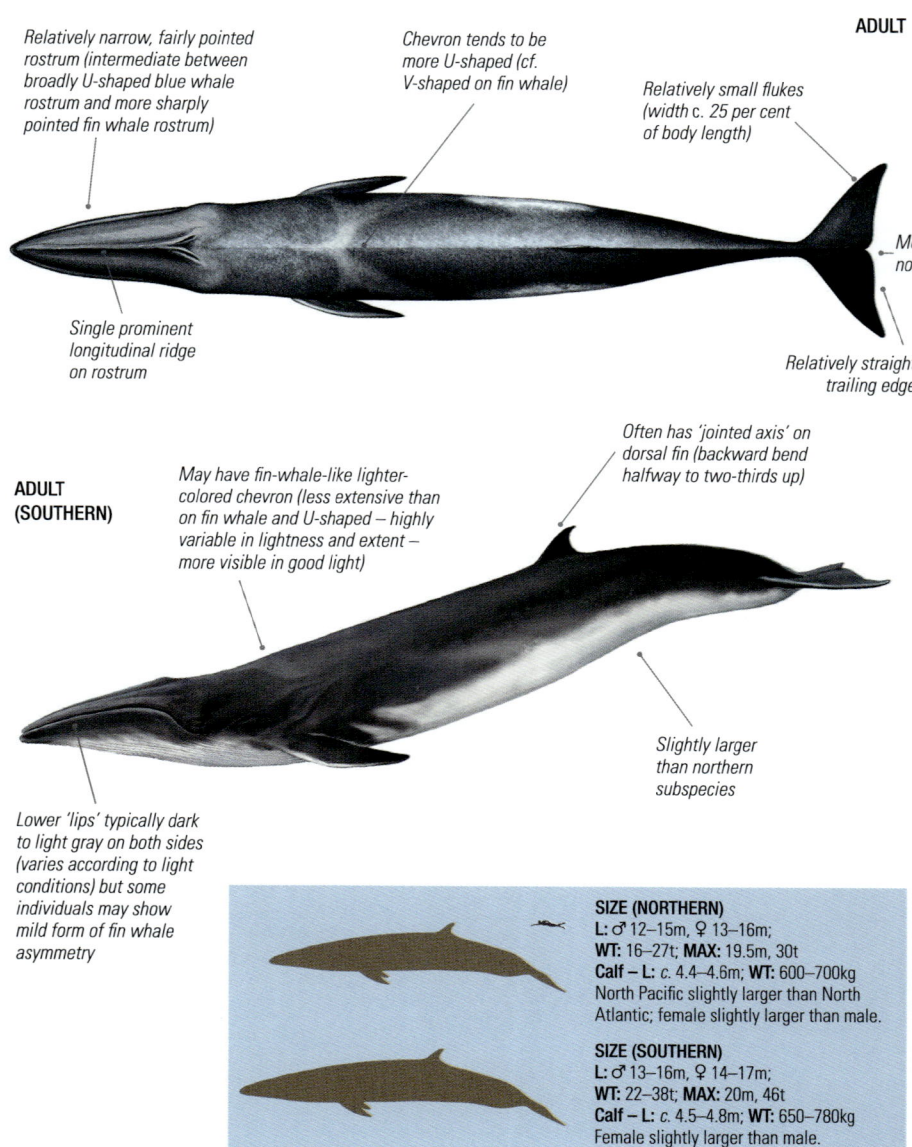

**ADULT**

*Relatively narrow, fairly pointed rostrum (intermediate between broadly U-shaped blue whale rostrum and more sharply pointed fin whale rostrum)*

*Chevron tends to be more U-shaped (cf. V-shaped on fin whale)*

*Relatively small flukes (width c. 25 per cent of body length)*

*Median notch*

*Single prominent longitudinal ridge on rostrum*

*Relatively straight trailing edge*

**ADULT (SOUTHERN)**

*May have fin-whale-like lighter-colored chevron (less extensive than on fin whale and U-shaped – highly variable in lightness and extent – more visible in good light)*

*Often has 'jointed axis' on dorsal fin (backward bend halfway to two-thirds up)*

*Slightly larger than northern subspecies*

*Lower 'lips' typically dark to light gray on both sides (varies according to light conditions) but some individuals may show mild form of fin whale asymmetry*

**SIZE (NORTHERN)**
**L:** ♂ 12–15m, ♀ 13–16m;
**WT:** 16–27t; **MAX:** 19.5m, 30t
Calf – **L:** *c.* 4.4–4.6m; **WT:** 600–700kg
North Pacific slightly larger than North Atlantic; female slightly larger than male.

**SIZE (SOUTHERN)**
**L:** ♂ 13–16m, ♀ 14–17m;
**WT:** 22–38t; **MAX:** 20m, 46t
Calf – **L:** *c.* 4.5–4.8m; **WT:** 650–780kg
Female slightly larger than male.

**CALF**

**ADULT**

Some anterior plates may be nearly white (may also be paler markings on upper 'lip')

Baleen generally dark gray or black (may have diffuse, longitudinal yellowish-brown streaks) with very fine bristles

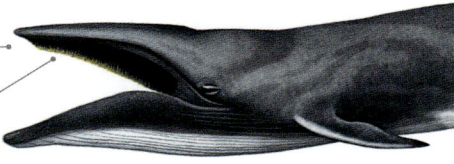

**FIN VARIATIONS**

Tip can be pointed or rounded

Shape and curve can vary from triangular to falcate to backswept

**FLUKES**

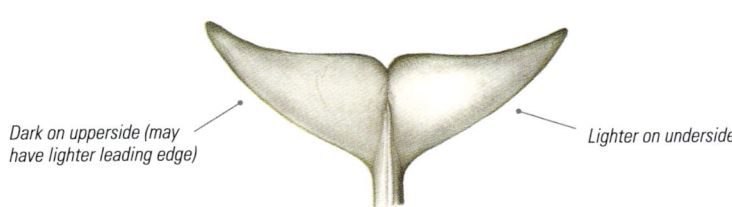

Dark on upperside (may have lighter leading edge)

Lighter on underside

### WHAT'S IN A NAME?

'Sei whale' comes from the Norwegian 'seihval' – 'seje' for a fish known in English as pollock, saithe, coley or coalfish (a close relative of codfish) and 'hval' for 'whale'. The two species often appeared off northern Norway at the same time (presumably feeding on the same prey). It is normally pronounced 'say' or 'sigh' (the Norwegian pronunciation is halfway between the two).

# BRYDE'S WHALE
*Balaenoptera edeni*  Anderson, 1879

One of the least known and more elusive of the large baleen whales, 'Bryde's whale' is actually a complex of subspecies and possible species with taxonomic issues that are yet to be resolved. They all have one particular characteristic in common: three parallel longitudinal ridges on the rostrum (all other rorquals, except Rice's whale and some Omura's whales, have a single ridge). The name is pronounced 'bree-duss' (the correct Norwegian pronunciation) or, frequently, 'broo-duss'.

**IUCN status**  Least Concern (2017).
**Population**  No overall global estimate, though broad guesstimate might be 90,000–100,000. Whalers killed at least 30,000 during 1911–87. Trend unknown.
**Classification**  Mysticeti, family Balaenopteridae.
**Taxonomy**  Two subspecies are recognized: the larger, more pelagic and globally distributed Bryde's whale (*B. e. brydei*), otherwise known as the 'large-form Bryde's whale', 'offshore Bryde's whale' or 'ordinary Bryde's whale'; and the smaller, predominantly coastal Eden's whale (*B. e. edeni*), of the western Pacific and Indian Oceans. Given strong genetic and morphological differences, and habitat partitioning, it is highly likely that these should be given full species status (as Bryde's whale and Eden's whale, respectively). The term 'pygmy Bryde's whale' was erroneously used for whales now known to be Omura's whale, described as a new species in 2003 (but originally considered part of the Bryde's whale complex).
**Other names**  Eden's whale, tropical whale, sittang.
**DISTRIBUTION**  Circumglobal distribution in tropical, sub-tropical and some warm temperate waters in the Atlantic, Pacific and Indian oceans, primarily between 40°N and 40°S. Tends to concentrate in water warmer than 16°C, in areas with exceptionally high productivity. It occurs in some semi-enclosed seas, such as the Red Sea and Persian Gulf, but is not found in the Mediterranean. Primarily pelagic or coastal, depending on the subspecies (albeit with exceptions – e.g. the New Zealand coastal population is confirmed as *brydei* subspecies). No extensive north–south migrations are known, although at least some offshore animals make shorter, general movements towards lower latitudes in winter and mid-latitudes in summer. Other populations – especially inshore in mid-latitudes – remain year-round in highly productive waters (e.g. Mexico's Gulf of California, New Zealand's Hauraki Gulf and the Gulf of Thailand).
**BEHAVIOR**  Occasionally breaches (typically coming out of the water vertically), sometimes multiple times in a row (70 times on one exceptional occasion off Ogata, Japan). When feeding, it typically makes sudden changes in direction, both underwater and at the surface. Behavior around vessels ranges from taking flight to unconcerned, or even sometimes curious.
**FOOD AND FEEDING**  Mainly small schooling fish (including pilchard, anchovy, mackerel, herring, sardine, lanternfish);

**DIVE SEQUENCE**  Surfaces at shallow angle; tip of rostrum usually breaks surface first; dorsal fin usually seen after blowholes submerged (sometimes simultaneously, especially in younger individuals); tends to arch back before deep dive (cf. sei whale); does not raise flukes; often impossible to predict and can be erratic in surfacing behavior; frequently does not leave telltale flukeprints behind on surface (cf. sei whale). • **DEPTH**  Often feeds at or close to surface; maximum 300m. • **DURATION**  5–15 minutes; maximum 20 minutes.

**BLOW**  Blow up to 9–10m high and columnar to bushy (highly variable height – often only 3–4m and even shorter in Eden's); tends to be lower in coastal *edeni* than pelagic *brydei*. • Often exhales under water, then surfaces with little or no visible blow (especially if frightened by killer whales or vessel approaching too close).

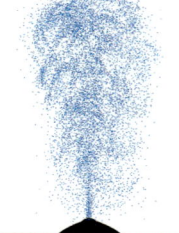

**ADULT**

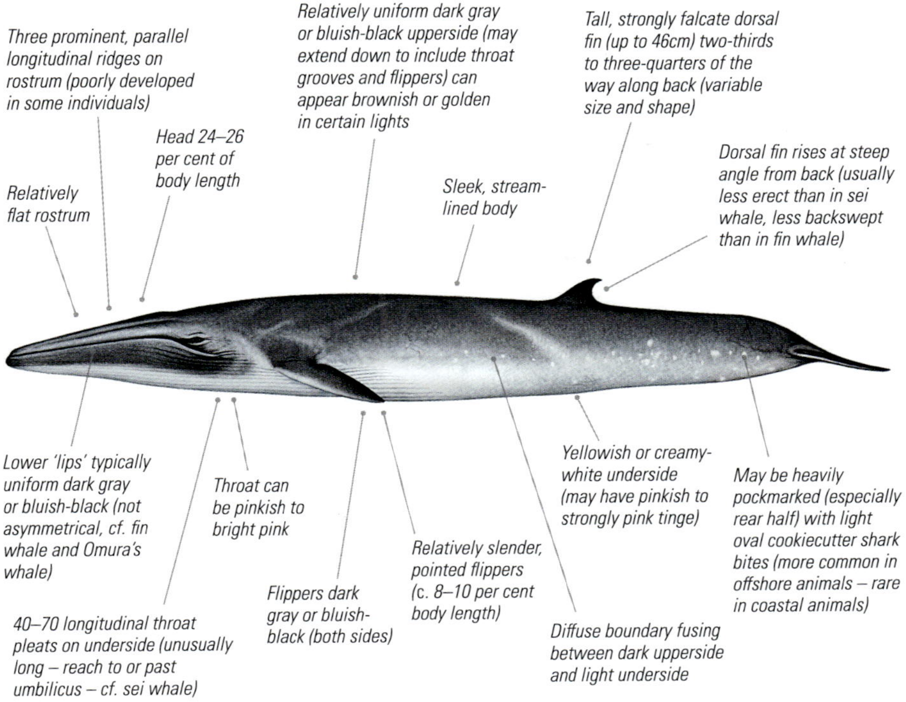

- Three prominent, parallel longitudinal ridges on rostrum (poorly developed in some individuals)
- Head 24–26 per cent of body length
- Relatively uniform dark gray or bluish-black upperside (may extend down to include throat grooves and flippers) can appear brownish or golden in certain lights
- Tall, strongly falcate dorsal fin (up to 46cm) two-thirds to three-quarters of the way along back (variable size and shape)
- Relatively flat rostrum
- Sleek, streamlined body
- Dorsal fin rises at steep angle from back (usually less erect than in sei whale, less backswept than in fin whale)
- Lower 'lips' typically uniform dark gray or bluish-black (not asymmetrical, cf. fin whale and Omura's whale)
- Throat can be pinkish to bright pink
- Flippers dark gray or bluish-black (both sides)
- Relatively slender, pointed flippers (c. 8–10 per cent body length)
- Yellowish or creamy-white underside (may have pinkish to strongly pink tinge)
- Diffuse boundary fusing between dark upperside and light underside
- May be heavily pockmarked (especially rear half) with light oval cookiecutter shark bites (more common in offshore animals – rare in coastal animals)
- 40–70 longitudinal throat pleats on underside (unusually long – reach to or past umbilicus – cf. sei whale)

**SIZE (LARGER, PELAGIC FORM – BRYDE'S)**
L: ♂ 12–14.5m, ♀ 12.5–15m;
WT: c. 15–25t; MAX: 15.6m, 25t
Calf – L: 3.8–4m; WT: 600–750kg

**SIZE (SMALLER, COASTAL FORM – EDEN'S)**
L: ♂ 10–11.5m, ♀ 11–13m;
WT: 12–17t; MAX: 11.7m, 17t
Calf – L: 3.4–4m; WT: 600–700kg

Female slightly larger than male in both forms.

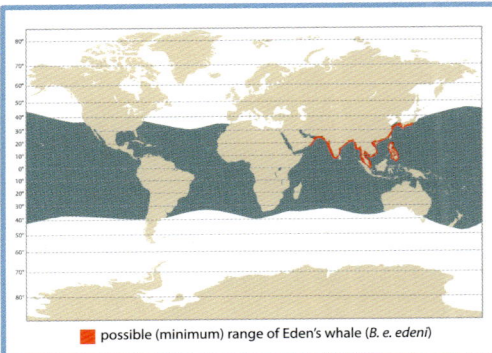

■ possible (minimum) range of Eden's whale (*B. e. edeni*)

**AT A GLANCE** Tropical to warm temperate waters worldwide • Large size • Sleek, streamlined body • Uniform dark gray upperside, lighter underside • Throat sometimes pinkish • Three parallel longitudinal ridges on rostrum • Tall, strongly falcate dorsal fin two-thirds to three-quarters of the way along back • Dorsal fin usually visible after blowholes submerged • Symmetrical lower 'lip' coloration • Typically arches back and tailstock on diving

**BRYDE'S WHALE**

also squid, krill, pelagic red crabs, other zooplankton; mostly an opportunistic feeder, switching prey preference according to availability, geographical location, season and year. Wide variety of foraging techniques; active lunge-feeder (often attracting seabirds and other pelagic predators); may skim-feed at surface like right whale; sometimes uses bubble nets to corral prey; in Gulf of Thailand, uses passive feeding technique among schooling fish called 'trap-feeding' or 'tread-water feeding' (there are several subtle variations in this technique but, basically, the whale hangs nearly vertically for several seconds, with mouth wide open at surface, allowing fish to swim or wash inside, then lifts head up and closes mouth); in New Zealand, uses 'chin-slaps' to aggregate zooplankton prey, then side-lunges through concentrated patch.

**BALEEN** 250–280 plates (each side of upper jaw) but up to 365 (including many rudimentary plates). Longest plates c. 50cm; may be more slender in Eden's whale.

**GROUP SIZE AND STRUCTURE** Generally seen alone, but sometimes in small groups of 2–3, and occasionally loose aggregations of 10–20 on prime feeding grounds.

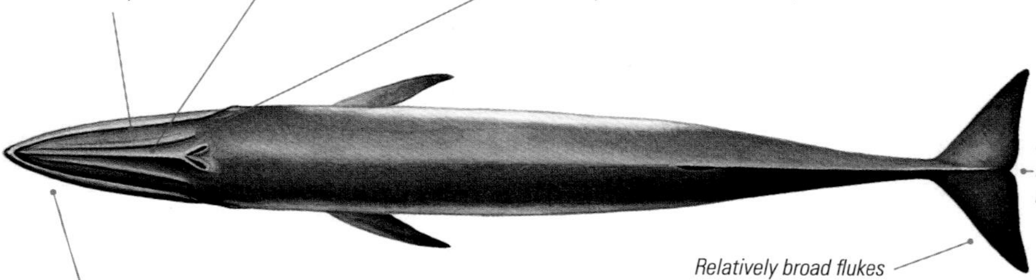

*Three ridges not always easy to see (especially with water washing over rostrum)*

*Prominent median ridge (running from blowholes to tip of rostrum)*

*Shorter auxiliary ridge on each side of median ridge (poorly developed in some individuals)*

*Slight median notch*

*Relatively broad flukes (width c. 23–24 per cent of body length)*

*Relatively narrow, fairly pointed V-shaped rostrum (intermediate between broadly U-shaped blue whale rostrum and more sharply pointed fin whale rostrum)*

**ADULTS**

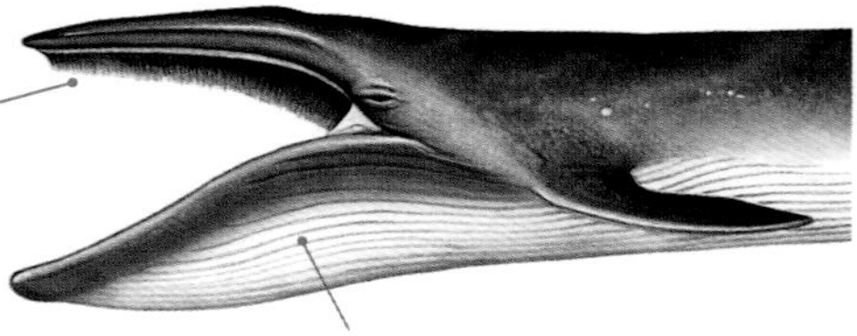

*Plates tend to be yellowish or creamy white in the anterior quarter to one-third of mouth, often darkening to slate-gray or dark gray (particularly on outside) in posterior three-quarters to two-thirds of mouth*

*Some individuals have asymmetrical coloration of throat and baleen plates*

**CALF**

**FIN VARIATIONS**

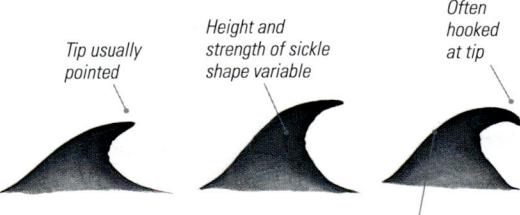

*Tip usually pointed*

*Height and strength of sickle shape variable*

*Often hooked at tip*

Fin of some individuals may have slight backward bend halfway to two-thirds up (cf. Omura's whale – not as obvious as in sei whale)

**FLUKES**

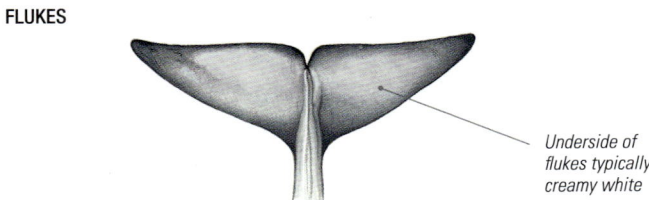

*Underside of flukes typically creamy white*

## THE BRYDE'S WHALE COMPLEX

The so-called 'Bryde's whale complex' originally comprised a single species, Bryde's whale. But research has revealed that it consists of several different species: the original Bryde's whale, Omura's whale (formally separated in 2003) and Rice's whale (formally separated in 2021). Bryde's whale itself currently consists of two subspecies – or species – and the precise taxonomic status of these is still in dispute.

### Eden's whale
Appears to be restricted to the northern Indian Ocean and western Pacific Ocean, roughly between the equator and 40°N. Countries where genetic studies confirm its presence include India, Bangladesh, Myanmar, Singapore, Thailand, Indonesia, the Philippines, China, Hong Kong and as far as south and southwestern Japan. It is uncertain if its range extends to Australia. There are no verified records from the Atlantic Ocean. Primarily in coastal waters (with some records very close inshore) and over the continental shelf; it has not been recorded offshore. It seems to be resident year-round and there is no evidence of long-distance migrations.

### Bryde's whale
Circumglobal distribution in tropical and sub-tropical waters of the Pacific, Atlantic, and Indian Oceans, and the Caribbean Sea. Genetic studies confirm its presence in many countries outside the range of Eden's whale. The subspecies in Australia is still uncertain, but in New Zealand it has been confirmed as *brydei*. All Bryde's whales in the Atlantic Ocean are believed to be this form. Primarily offshore, but its distribution is more cosmopolitan than previously thought and appears to include some coastal habitat (the resident population in the Hauraki Gulf, New Zealand, for example, is primarily coastal). Some offshore populations are known to migrate, but these migrations appear to be relatively short for baleen whales (typically over 20°–30° latitude). Inshore populations tend to be resident year-round. Recent genetic studies demonstrate that broadly sympatric populations of migratory offshore Bryde's whales and resident inshore Bryde's whales off South Africa – which also differ in size and prey preference – are this large-form Bryde's.

# RICE'S WHALE
## *Balaenoptera ricei*

Rosel, Wilcox, Yamada and Mullin, 2021

A small population of whales belonging to the 'Bryde's whale complex' has been known in the Gulf of Mexico since an individual stranded there in 1965. But recent research reveals that these whales are genetically and morphologically distinct (and geographically separated from other Bryde's whale populations). They were identified as belonging to a new species in 2021.

**IUCN status** Critically Endangered (2021). With such a small population and limited range, arguably the most endangered whale species in the world.
**Population** Current best estimate 51, based on surveys conducted in 2017–18 (which indicates that the number of mature Rice's whales could be as few as 26). There are no historical estimates, but the population is believed to be decreasing).
**Classification** Mysticeti, family Balaenopteridae.
**Taxonomy** The genetic differences between Rice's whale and Bryde's whale are two to three times greater than the differences between the three recognized species of right whale. No recognized forms or subspecies.
**Other names** Gulf of Mexico Bryde's whale, Gulf of Mexico whale.

**DISTRIBUTION** Appears to be focused predominantly around the continental shelf break near the De Soto Canyon, especially off the northwestern coast of Florida, in the northeastern Gulf of Mexico. The majority of sightings occur where the seafloor varies between 100m and 400m in depth. There have also been two possible sightings (in the early 1990s) and one confirmed sighting (in 2017) in the western Gulf of Mexico off Texas, and unique vocalizations have been heard south of Louisiana and Texas. These records suggest three possibilities: some individuals venture out of their core range into the western side; there is another, as yet unidentified, population; or, most likely, these are the remnants of a formerly more broadly distributed population. Whaling data does suggest a wider past distribution (records of 'finback' whales from the north-central Gulf, south of the Mississippi, and in the southern Gulf on the Campeche Banks are almost certainly misidentified Rice's whales). There have been no confirmed sightings outside the Gulf of Mexico, and it is believed to be non-migratory. Single strandings in South Carolina and North Carolina are believed to have been extralimital strays. It is the only year-round resident baleen whale in the Gulf of Mexico and there is no genetic evidence for any other Bryde's-like whale species or subspecies in the Gulf. The nearest confirmed populations of other members of the Bryde's whale complex are in the southern Caribbean south to Venezuela and Brazil, the eastern North Atlantic and the eastern South Atlantic off South Africa. Sightings and strandings of all other baleen whale species in the Gulf are rare and considered extralimital. During the 181 sightings, the whales were observed almost entirely in water 151–352m deep (with two exceptions, in 117m and 408m respectively). A satellite-tagged individual remained in waters 100–400m deep for a month.

**BEHAVIOR** Unknown, but likely similar to other members of the Bryde's whale complex.

**FOOD AND FEEDING** Feeds primarily on small schooling fish, especially silver-rag driftfish, which occur in large shoals in water depths of 50–500m but particularly near the muddy seabed.

**BALEEN** *c.* 264 plates (each side of the upper jaw).

**GROUP SIZE AND STRUCTURE** Generally seen alone, or in pairs. May occasionally form larger groups in productive feeding areas.

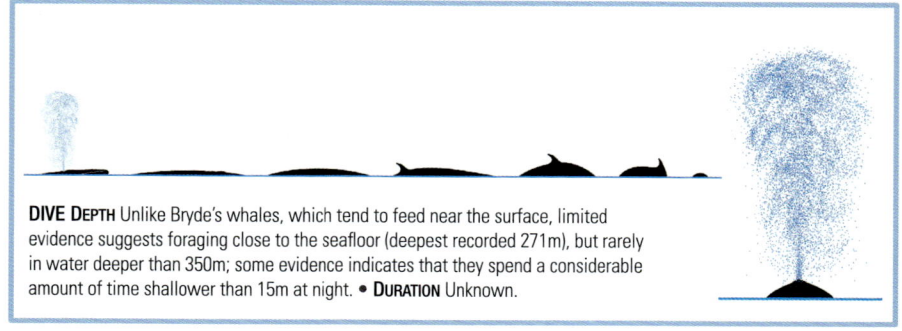

**DIVE DEPTH** Unlike Bryde's whales, which tend to feed near the surface, limited evidence suggests foraging close to the seafloor (deepest recorded 271m), but rarely in water deeper than 350m; some evidence indicates that they spend a considerable amount of time shallower than 15m at night. • **DURATION** Unknown.

## Rice's Whale

**Relatively flat rostrum**

**Relatively uniform dark charcoal gray upperside** (may appear brownish in some lighting conditions)

**Sleek, streamlined body**

**Tall, strongly falcate dorsal fin two-thirds of the way along back**

**Lower 'lips' typically uniform dark charcoal gray on both sides** (not asymmetrical cf. fin whale and Omura's whale)

**c. 54 longitudinal throat pleats on underside** (unusually long – some reach to or past umbilicus)

**Relatively slender, pointed flippers uniformly dark gray (both sides)**

**Lighter (sometimes pinkish) underside**

**ADULTS**

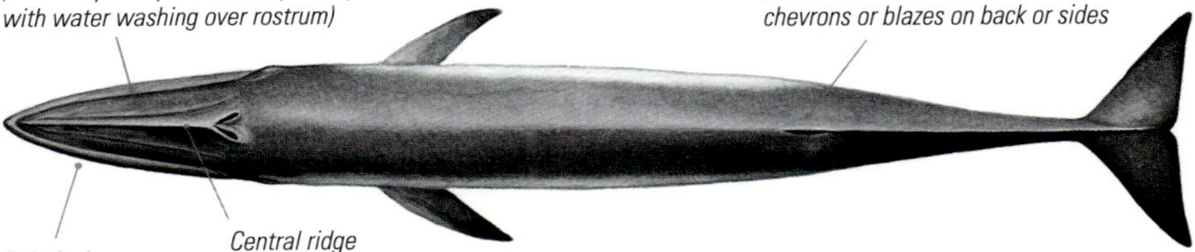

**Three prominent, parallel longitudinal ridges from blowhole to tip of rostrum** (not always easy to see – especially with water washing over rostrum)

**May be diffuse white washes around base of dorsal fin and/or along sides** (some individuals only) but no white chevrons or blazes on back or sides

**Relatively narrow, fairly pointed V-shaped rostrum**

**Central ridge larger, lateral ridges smaller**

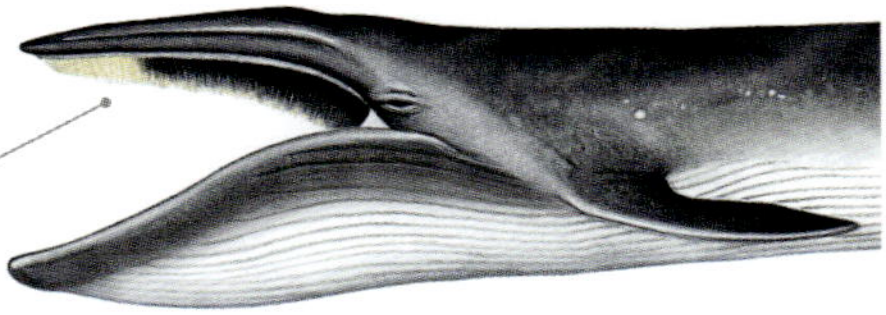

**Cream-colored anterior baleen plates on both sides, posterior plates black with cream-colored fringe**

**FLUKES**

**Underside of flukes lighter**

### SIZE
**L:** ♂ 11.26m (largest adult measured; an immature was 11.05m); ♀ 12.65m (largest adult measured);
**WT:** possibly c. 12–17t
**Calf – L:** 4.7m (one calf of unknown age that stranded alive); **WT:** unknown
Females presumed to be slightly larger than males (as in other rorquals).

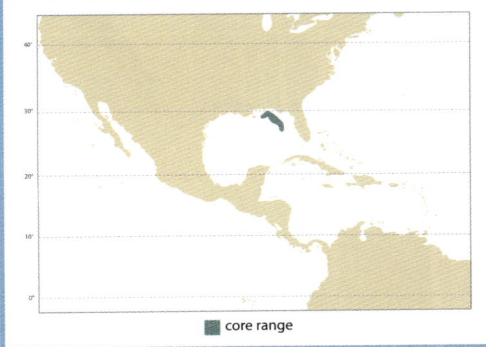

core range

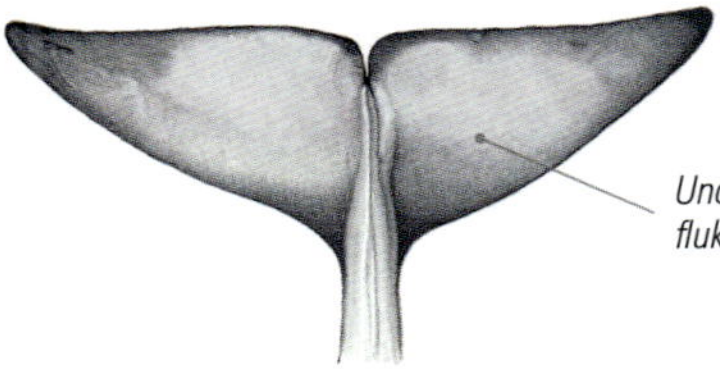

**AT A GLANCE** Core range in the northeastern Gulf of Mexico • Larger than Omura's whales, smaller than Bryde's whales (roughly the same size as Eden's whales) • Sleek, streamlined body • Uniform dark charcoal gray upperside, lighter to pinkish underside • Three parallel longitudinal ridges on rostrum • Tall, strongly falcate dorsal fin two-thirds of the way along back • Symmetrical lower 'lip' coloration

RICE'S WHALE  **67**

# COMMON MINKE WHALE
*Balaenoptera acutorostrata*     Lacépède, 1804

The common minke whale is the smallest rorqual and the second smallest of all the baleen whales (after the pygmy right whale). It has three disjunct populations: in the North Atlantic, the North Pacific and the southern hemisphere.

**IUCN status** Least Concern (2018).
**Population** Minimum *c.* 200,000 mature individuals. At least 170,000 killed by whalers. Trend unknown.
**Classification** Mysticeti, family Balaenopteridae.
**Taxonomy** Two subspecies are currently recognized: North Atlantic minke whale (*B. a. acutorostrata*); and North Pacific minke whale (*B. a. scammoni*). The 'dwarf minke whale' may be a third valid subspecies.
**Other names** Northern minke whale, dwarf minke whale; formerly lesser/least rorqual, little piked whale, pikehead, lesser finback, sharp-headed finner, little finner.

**DISTRIBUTION** In the northern hemisphere, the North Atlantic minke whale ranges to at least *c.* 80°N during summer. The wintering grounds – probably in the southern North Atlantic – are poorly known but extend at least to the Caribbean and possibly West Africa. The North Pacific minke whale ranges to at least 70°N during summer. The wintering grounds – probably in the southern North Pacific – are also poorly known but extend to at least 15°N. Northern minke migrations are not as well defined as in some other baleen whales. There is a tendency for the distribution to shift from high-latitude summer feeding areas to lower-latitude winter breeding areas (although some individuals are resident in cold temperate regions year-round). During summer, they appear to be most abundant in cold temperate to polar waters (where they are known to penetrate areas with extensive ice floes and polynyas). At this time of year, they occur in inshore coastal waters more frequently than any other rorqual, and will enter bays, inlets, fjords and even some large rivers (such as the St Lawrence River, Canada). Winter sightings are uncommon, suggesting that when in lower latitudes they are mainly offshore. Some populations – such as around the Isle of Mull (Scotland, UK) and the San Juan Islands (Washington, USA) – have high site fidelity, with certain individuals returning each year to feed in particular locations.

The dwarf minke occurs only in the southern hemisphere and may or may not be circumglobal (relatively little is known about its distribution). Occurs in both coastal and offshore waters off South Africa, southern Mozambique, Australia, New Zealand (North and South Islands), New Caledonia, eastern South America (from northern Brazil to northern Argentina), and Chilean Patagonia. Records cover most of the year (March–December), but there are strong indications that at least some populations are migratory. The only known predictable aggregation of dwarf minke whales is in winter off the northern Great Barrier Reef in Australia, predominantly in June–July. The most northerly confirmed records are from 2°S off the northern coast of Brazil and 11°S in the western Pacific off Australia. It partly overlaps with the Antarctic minke whale during summer in the sub-Antarctic but is not as polar. Most sightings in the sub-Antarctic have been in December–March south of Australia and New Zealand – between 55°S and 60°S, with one record as far as 65°S – probably because this is where there has been most research effort. However, it is also likely to occur in sub-Antarctic waters south of South America and South Africa. It is not known from the northern Indian Ocean.

**BEHAVIOR** Breaches fairly frequently, sometimes completely clearing the water, and performs other aerial behaviors such as head rises and spyhops (particularly in icy areas). Rarely lobtails or fipper-slaps. Can be quite curious towards

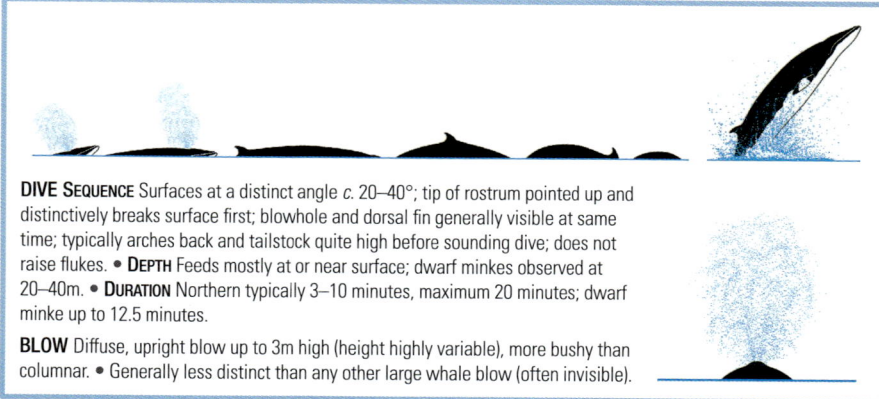

**DIVE SEQUENCE** Surfaces at a distinct angle *c.* 20–40°; tip of rostrum pointed up and distinctively breaks surface first; blowhole and dorsal fin generally visible at same time; typically arches back and tailstock quite high before sounding dive; does not raise flukes. • **DEPTH** Feeds mostly at or near surface; dwarf minkes observed at 20–40m. • **DURATION** Northern typically 3–10 minutes, maximum 20 minutes; dwarf minke up to 12.5 minutes.

**BLOW** Diffuse, upright blow up to 3m high (height highly variable), more bushy than columnar. • Generally less distinct than any other large whale blow (often invisible).

**ADULT NORTHERN HEMISPHERE**

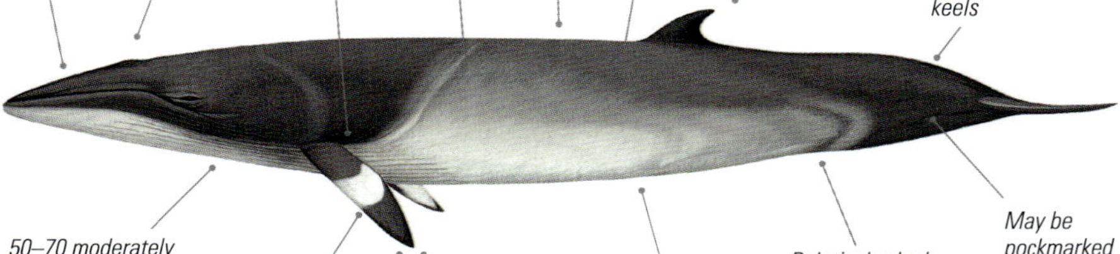

- Variable pale shoulder streak behind flippers (typically well-defined leading edge) may extend to form chevron pattern over back (roughly symmetrical with other side)
- Brush strokes and intermediate shades of gray on sides and back (highly variable)
- Relatively tall, falcate dorsal fin slightly less than two-thirds of the way along back (variable shape)
- Head 22–23 per cent of body length
- Dark gray, brownish-gray or blackish upperside
- Dorsal fin may be hooked at tip
- Sharply pointed, flattened rostrum
- No white shoulder patch (cf. dwarf minke)
- May have distinct dorsal and ventral keels
- 50–70 moderately short longitudinal throat pleats on underside (terminate between flippers and umbilicus)
- Pointed tip
- Relatively sleek, streamlined body
- May be pockmarked with light round or oval cookiecutter shark bites
- Sharply demarcated brilliant white band in middle of flipper (often clearly visible when animal just below surface)
- Snowy-white to creamy-white underside (may flush pink when active)
- Slender, pointed flippers (c. 12 per cent of body length)

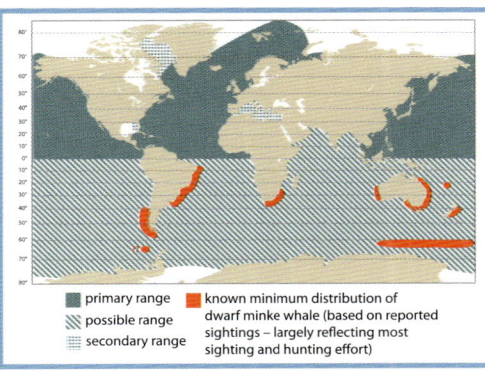

**SIZE (NORTH PACIFIC AND NORTH ATLANTIC)**
**L:** ♂ 7.5–8.5m, ♀ 8.5–9.5m;
**WT:** 6–8t; **MAX:** 9.8m, 9.2t
**Calf – L:** 2.2–2.8m; **WT:** 350–450kg
The female is longer than the male in all subspecies.

**SIZE (DWARF)**
**L:** ♂ 6–7m, ♀ 6.5–7.2m;
**WT:** 4–5t; **MAX:** 7.8m, 6.4t
**Calf – L:** 2–2.3m; **WT:** 250–350kg

- primary range
- possible range
- secondary range
- known minimum distribution of dwarf minke whale (based on reported sightings – largely reflecting most sighting and hunting effort)

**AT A GLANCE** Tropics to poles worldwide • Medium size • Dark gray, brownish-gray or blackish upperside, white underside • Variable swathes of lighter gray on sides and back • Sharply pointed rostrum breaks surface first • Single longitudinal ridge on rostrum • Relatively tall, falcate dorsal fin two-thirds of the way along back • Unique, bright white flipper bands • Indistinct or invisible blow

COMMON MINKE WHALE

boats in some parts of the world – and will swim around stationary vessels or alongside moving vessels for minutes or even hours at a time. Elsewhere, it can be difficult to approach.

**FOOD AND FEEDING** Northern hemisphere minkes feed on wide variety of small schooling fish (including sand lance, salmon, capelin, cod, mackerel, sprat, pollack, whiting, herring, haddock, anchovy and lanternfish) and small invertebrates (including euphausiids and copepods). Dwarf minkes prefer lanternfish, but opportunistically feed on other fish and possibly krill. Feeding technique varies significantly according to prey and location; entrapment maneuvers include circles, gyres, ellipses, figures-of-eight, hyperbolas, head-slaps and underwater blows; engulfing maneuvers include oblique, lateral, vertical and ventral lunges.

**BALEEN** 231–290 plates in North Pacific, 270–325 in North Atlantic, 200–300 in dwarf (each side of the upper jaw). Longest plates *c.* 21cm; in northern animals, usually white, creamy or yellowish; in dwarf minke, about half of the plates posteriorly appear dark gray or brown (due to a narrow, dark fringe); all have symmetrical plate coloration.

**GROUP SIZE AND STRUCTURE** Typically solitary, sometimes in twos or threes, but there can be larger, temporary aggregations in good feeding areas. Social structure appears to be complex, with evidence for some segregation by age, sex and/or reproductive class.

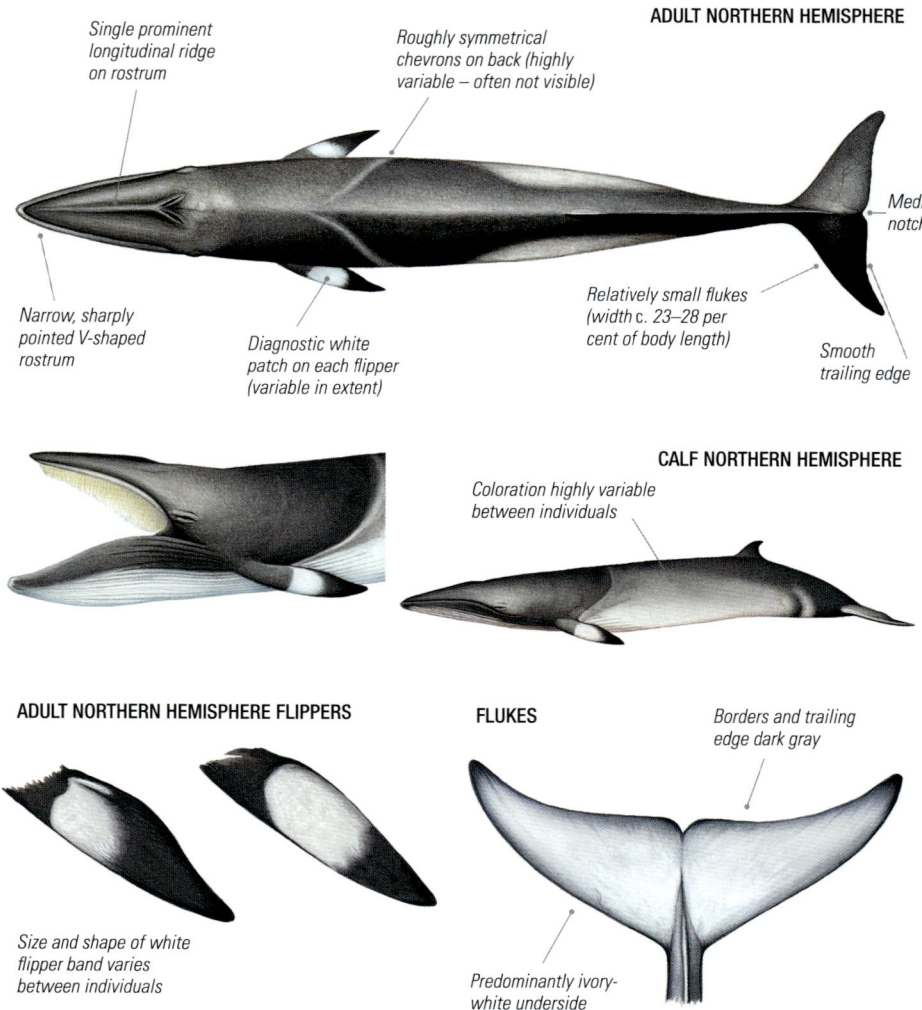

**ADULT DWARF**

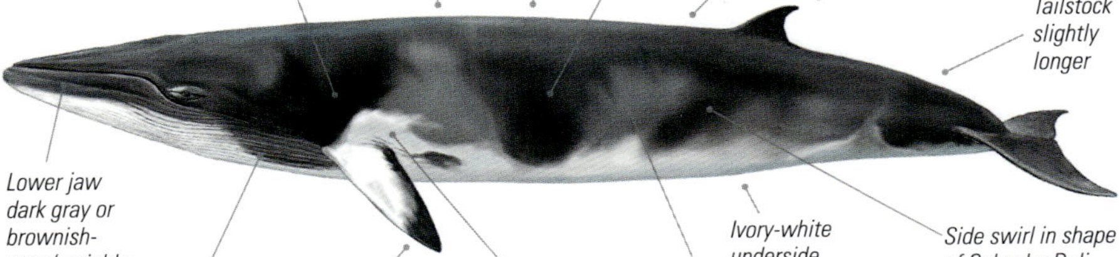

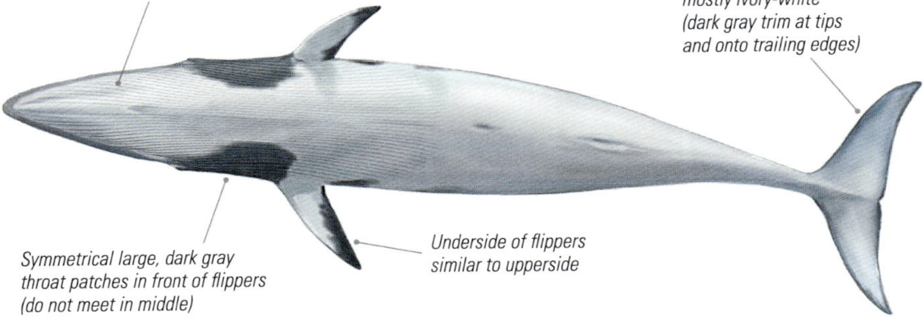

- Dark gray upperside
- Dark gray band on neck continues as dark throat patch
- Most complex coloration of any baleen whale
- Complex lateral coloration with three dark gray fields descending from back
- Coloration highly variable between individuals
- Relatively larger dorsal fin than common minke
- Tailstock slightly longer
- Lower jaw dark gray or brownish-gray (variable extent)
- Roughly triangular, pale gray throat patch
- Tip of flipper usually remains dark
- Flipper tends to have much longer band of ivory-white extending to base and onto body as highly distinctive broad white shoulder patch
- Ivory-white underside
- Ivory-white side streaks and blazes typically extend from underside onto back
- Side swirl in shape of Salvador Dali moustache

- 55–75 throat pleats
- Symmetrical large, dark gray throat patches in front of flippers (do not meet in middle)
- Underside of flippers similar to upperside
- Underside of flukes mostly ivory-white (dark gray trim at tips and onto trailing edges)

**ADULT DWARF FLIPPERS**

**ADULT DWARF**

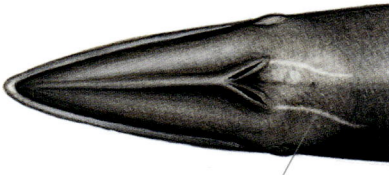

May have light gray 'blowhole streaks' visible from above (left streak consistently curves to left)

**COMMON MINKE WHALE**

# HUMPBACK WHALE
## *Megaptera novaeangliae* (Borowski, 1781)

Renowned for its spectacular breaching, lobtailing and flipper-slapping, its complex and melodious song, and its remarkably long flippers, the humpback whale is one of the most familiar and best known of all the large whales. Black-and-white markings on the undertail are distinctive and readily identifiable, enabling researchers to tell one individual from another.

**IUCN status** Least Concern (2018). Arabian Sea sub-population Endangered (2008). Oceania sub-population Endangered (2008).
**Population** Roughly 190,000 worldwide (compared to an all-time low of 5,000–10,000 due to whaling). Pre-whaling population *c.* 240,000. At least 300,000 killed by whalers. Increasing.
**Classification** Mysticeti, family Balaenopteridae.
**Taxonomy** Three subspecies are recognized: North Atlantic (*M. n. novaeangliae*); North Pacific (*M. n. kuzira*); and southern (*M. n. australis*). A fourth subspecies, the Arabian Sea humpback whale (*M. n. indica*), has been proposed.
**Other names** Hump-backed whale.

**DISTRIBUTION** Worldwide. Migrates between mid- to high-latitude summer feeding grounds and low-latitude winter breeding grounds (most populations migrate through deep oceanic waters). During summer, frequents coastal, continental shelf and offshore waters. During winter, breeds around oceanic islands, offshore seamounts and reef systems. Significant numbers also occur at mid- to high latitudes during winter – in British Columbia, Norway, Iceland and other locations – but it is unclear whether these are overwintering or simply very late-leaving migrants. Some populations are responding to climate change by expanding their range into higher polar latitudes (significant declines in sea ice are also resulting in longer feeding seasons). The breeding area in Central America used by southern hemisphere humpbacks during the austral winter is also used by North Pacific humpbacks during the boreal winter (it's unclear whether they ever overlap at the same time).
**Winter breeding distribution** There are 15 known and two suspected winter breeding grounds: two (plus one suspected) in the North Atlantic, five (plus one suspected) in the North Pacific, seven in the southern hemisphere and one in the Arabian Sea; the isolated population of *c.* 90 individuals in the Arabian Sea is unique in being resident. All are between 30°N and 40°S, mostly centered around *c.* 20° latitude. They are typically in warm, relatively shallow water (less than 200m) surrounded by much deeper water. The preferred sea temperature on the breeding grounds is *c.* 25°C (in the range of 21.1–28.3°C). There is high fidelity to natal breeding grounds and little interchange between them.
**Summer feeding distribution** Individuals usually return to the same feeding grounds used by their mothers, with relatively little interchange between feeding areas. Habitat preferences include the continental shelf break, submarine channels, oceanic fronts, eastern boundary currents and ice-edge zones (areas of upwelling); in the southern hemisphere, feeding habitat is often closely linked to regions of marginal sea ice. The preferred sea temperature on feeding grounds is typically below 14°C.
**North Atlantic** Most North Atlantic humpbacks breed in the West Indies, primarily on the oceanic side of many Caribbean islands, and in northern Venezuela (the most populous areas are off the northern Dominican Republic and

**DIVE SEQUENCE** Blowholes appear first and remain in view as dorsal fin appears; distinctive sloping back forms shallow triangle with surface of sea; body arches, forming high triangle, making hump on back especially evident; flukes lifted high on many dives. • **DEPTH** On summer feeding grounds often follows diel movement cycle of prey (deeper during day); most foraging in upper 120m (upper 25m during bubble-netting), but capable of 400+m; on breeding grounds usually shallow. • **DURATION** Depends on season, location and behavior; singing dives up to 20 minutes; resting dives on breeding grounds 15–30 minutes; foraging dives typically 3–10 minutes (up to 15); maximum *c.* 40 minutes.

**BLOW** More variable than in any other large whale (bushy, columnar or, very occasionally, V-shaped). • Usually tall and columnar (may be bushier at top), up to 10m high (height highly variable – often only 4–5m).

## ADULT MALE NORTHERN HEMISPHERE

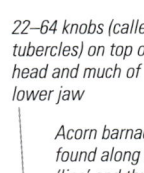
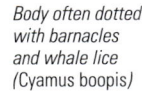

- 22–64 knobs (called tubercles) on top of head and much of lower jaw
- Acorn barnacles usually found along rostrum, 'lips' and throat
- Body often dotted with barnacles and whale lice (Cyamus boopis)
- Stocky body
- Dark gray to black upperside and sides
- Dorsal fin sits on raised fleshy hump
- Low, broad-based dorsal fin (up to 30cm) slightly more than two-thirds of the way along back
- Dorsal fin highly variable (from small and blunt to tall and falcate)
- Distinctive cluster of acorn barnacles on tip of chin (known as the 'cutwater')
- 14–35 ventral pleats or throat grooves (extend to umbilicus or beyond)
- Less extensive white on underside compared to southern hemisphere animals
- Relatively narrow tailstock
- Black-and-white pigmentation on underside of flukes
- White circular scarring on lower jaw (scars from when acorn barnacles have fallen off)
- Flippers up to 5m long (23–33 per cent of total body length) and weigh up to 1t
- Scalloped leading edge with knobs or tubercles (including two more prominent ones dividing margin into thirds) often encrusted with acorn barnacles
- Underside varies from black to white or mottled black and white (variable between individuals and by population – more white in North Atlantic, less in North Pacific)

**SIZE**
**L:** ♂ 11–15m, ♀ 12–16m;
**WT:** 25–35t; **MAX:** 18.6m (historically – rarely more than 16m today), 40t
**Calf – L:** 4–4.6m; **WT:** 0.6–1t
Adult females typically 1–1.5m longer than males.

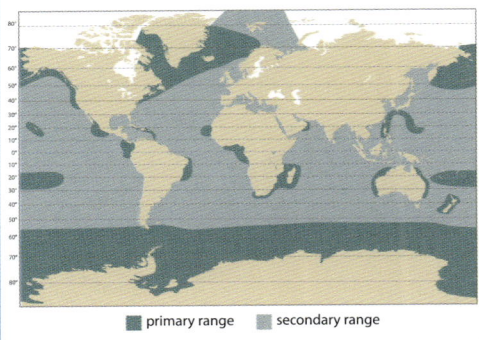

primary range   secondary range

**AT A GLANCE** Worldwide distribution
- Large size • Predominantly dark gray to black upperside • Variable amount of white on underside • Stocky body • Small dorsal fin sits on (variable) hump on back • Exceptionally long white (or black-and-white) flippers • Distinctive knobs on head • Strongly arches back when diving • Usually flukes on sounding dive • Variable (and individually distinctive) black-and-white pigmentation on underside of flukes

HUMPBACK WHALE

over the offshore reef systems of Silver, Navidad and Mouchoir banks). Historical whaling records reveal Cape Verde (possibly extending to the continental shelf of Senegal and Western Sahara) as another breeding ground, with up to 4,000 whales before whaling commenced. However, there are relatively few whales today – far fewer than are known to exist in the eastern North Atlantic (and unaccounted for in the West Indies), suggesting the existence of a third, as yet undetermined, breeding ground.

**North Pacific** There are five known and one suspected breeding populations in the North Pacific: the main Hawai'ian islands, accounting for about half of all North Pacific humpbacks (about half feeding off northern British Columbia and south-east Alaska, and half in the northern Gulf of Alaska and the Bering Sea); mainland Mexico (feeding mainly off California); Revillagigedo Islands (feeding mostly from northern California to Alaska); Central America, along the Pacific coast from southern Mexico to Costa Rica (feeding almost exclusively off California and Oregon); southern Japan (mainly the Okinawa Islands) to Taiwan and the northern Philippines, and east to the Mariana and Marshall Islands (feeding mainly off eastern Kamchatka, but also across a broad band in the western Pacific). There is also an unknown breeding ground somewhere else in the western North Pacific – possibly in the Northwestern Hawai'ian Islands and/or the Mariana Archipelago (between the Philippines and Japan); this is inferred from sightings in the Russian Far East and around the Aleutian Islands that cannot be linked to any known breeding population. Baja California and the Ogasawara Islands are not believed to be primary migratory destinations but are more likely to be transiting areas.

**Southern hemisphere** Seven separate breeding populations migrate to feed in the Southern Ocean (in six specific areas designated by the International Whaling Commission as I–VI): Pacific coasts of Central and South America, from northern Peru to Costa Rica (with records as far north as Nicaragua) but particularly Colombia (feeding in Antarctic Area I, particularly off the western Antarctic Peninsula, as well as in southern Chile); coastal waters off Brazil from 3–23°S, particularly around Abrolhos Bank (feeding in the Scotia Sea, i.e. Antarctic Area II); western Africa, centered around the Gulf of Guinea (feeding south of 18°S, in waters off Namibia and western South Africa, and in Antarctic Areas II and III); southeastern Africa, Madagascar and archipelagos of the western Indian Ocean (feeding in Antarctic Area III); northwestern Australia (feeding in Antarctic Area IV); northeastern Australia (feeding in Antarctic Area V); and Oceania or the South Pacific islands (feeding in Antarctic Areas V, VI and I, including the western Antarctic Peninsula).

**BEHAVIOR** More demonstrative at the surface than any other large whale. Breaching, flipper-slapping and lobtailing are common. May lie on side or back, holding one or both flippers in the air. All these activities occur year-round and in a

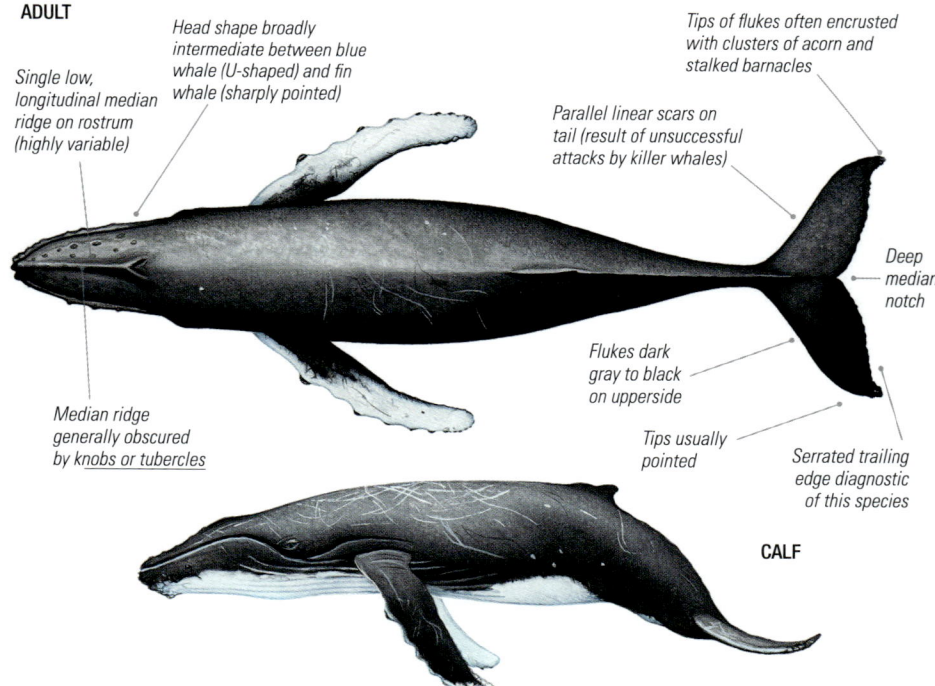

variety of contexts, so clearly perform various functions (probably including communication, mate attraction, parasite removal, prey corralling, expression of excitement or annoyance, and even play). Both sexes and all ages frequently breach (sometimes fully) on both the feeding grounds and the breeding grounds, alone or in small groups, once or many times consecutively. Shows little fear of boats: frequently very inquisitive (especially juveniles).

**FOOD AND FEEDING** Krill; wide variety of schooling fish; occasionally squid; some mysids, copepods and benthic amphipods; generalist in northern hemisphere (with regional preferences); in southern hemisphere, mainly Antarctic krill. Adaptable gulp- or lunge-feeder using diverse techniques; uniquely among large whales, uses bubble nets (circular nets of large bubbles) or bubble clouds (bursts of tiny bubbles) to corral schooling prey (individually or in groups); one of few baleen whales to group-feed. Other feeding techniques include flick-feeding (upside down, repeatedly sweeping tail forward at surface before surfacing and lunging with mouth open), lobtail-feeding (slapping surface above school of fish, then diving to blow bubble screen) and trap-feeding (at surface, opening mouth wide, using flippers to push fish into the 'trap'); little or no feeding on winter breeding grounds; opportunistic feeding during migration.

**BALEEN** 270–400 plates (each side of the upper jaw). Plates dark gray to black, often with white or brownish-white longitudinal streaks (anterior-most plates may be lighter); max. length 85–107cm.

**GROUP SIZE AND STRUCTURE** On winter breeding grounds, usually occurs singly or in small groups, with seven main groupings: singer (usually a lone male); adult female with a male escort; competitive or surface-active group (multiple males following a single female and competing with one another for the position of female's primary escort); lone (non-singing) whales traveling to join others; groups of juveniles; mother–calf pair (sometimes accompanied by the previous year's juvenile); and a mother–calf pair plus male escort. Five main groupings on summer feeding grounds: single individual or pair of either sex; mother–calf pair; small, usually temporary feeding association; larger (and, for most individuals, ephemeral) feeding association of up to 24 whales, working together to corral and capture prey by bubble-net feeding; and (much rarer) super-groups of up to 200 whales. Migrates typically in small, fluid groups with individuals joining and leaving on a regular basis.

**ADULT SOUTHERN HEMISPHERE**

*A few anomalous all-white (possibly albino or leucistic) individuals observed*

*White on underside often more extensive than most northern hemisphere animals (can extend up sides and even onto back)*

*Typically more white on underside of tail (cf. northern hemisphere)*

**ADULT FEMALE NORTHERN HEMISPHERE**

*Grapefruit-sized hemispherical lobe underneath tailstock behind genital slit (absent in male)*

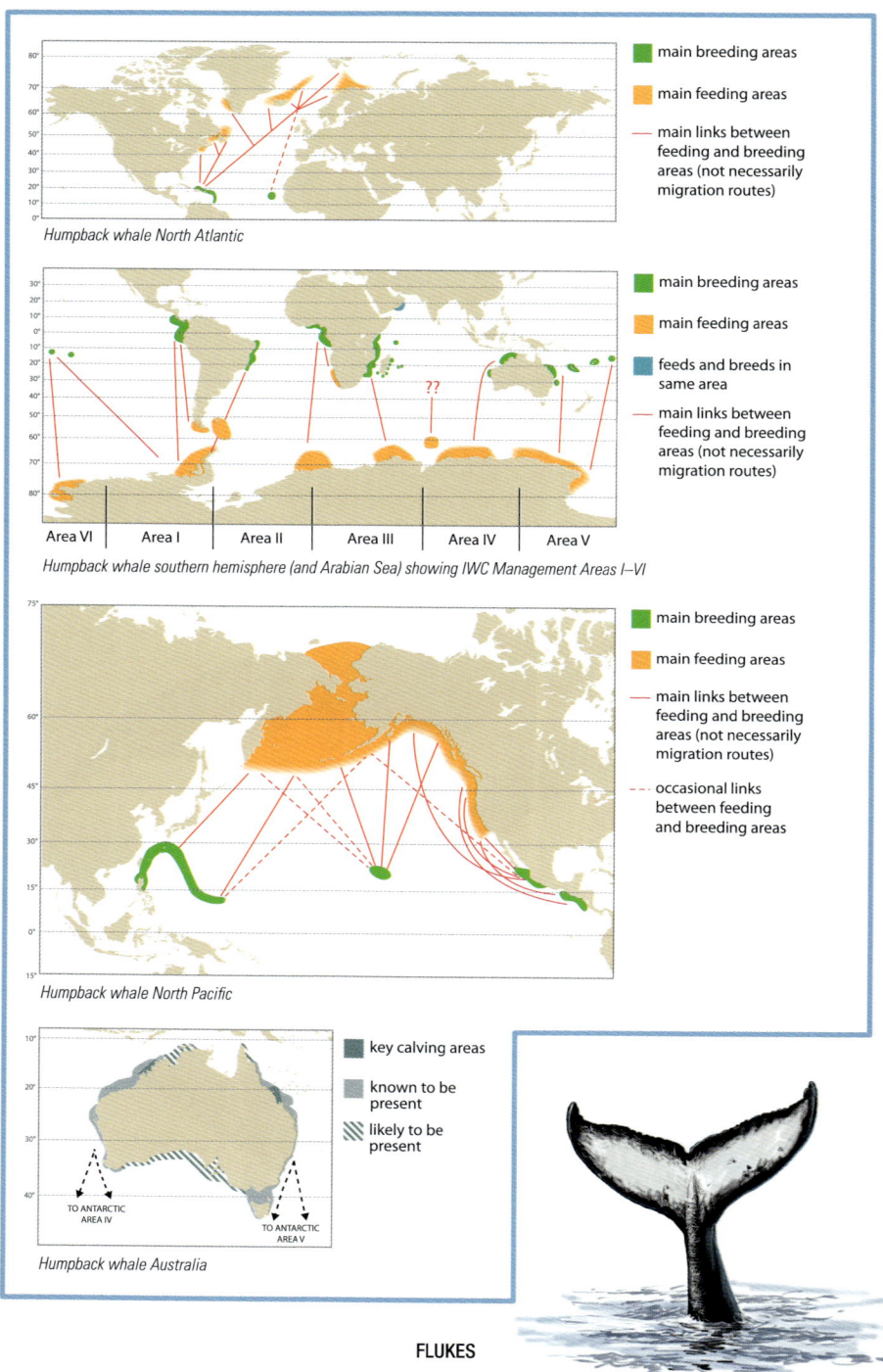

FLUKES

## FLUKE COMPARISONS

Underside of flukes varies from virtually all black to virtually all white with countless combinations of black and white in between

General proportion of black and white varies geographically

Coloration of underside of flukes ranked from one (nearly all white) to five (nearly all black) — more than 80 per cent of humpbacks in Australia are in category one (cf. less than 10 per cent in North Pacific)

Black-and-white patterns individually distinctive

## FLIPPER COMPARISONS

North Atlantic, North Pacific type one (one in three individuals) and Antarctic Peninsula — white underside, mostly white upperside (with varying amounts of black)

North Pacific type two — white underside, mostly black upperside (two in three individuals)

Western Australia — white underside, mostly black upperside (with varying amounts of white)

## TUBERCLES

Barnacles tend to attach to tubercles

Each tubercle is a hair follicle, with a single coarse sensory hair about 1–3cm long growing out of the center

A tubercle is about the size of golf ball

# SPERM WHALE
## *Physeter macrocephalus*

Linnaeus, 1758

The largest odontocete, or toothed whale, the iconic sperm whale – well known from *Moby Dick* – is designed for life in the ocean depths. This animal of extremes shows the greatest sexual size difference among cetaceans; moreover, its brain is the world's largest, and it dives deeper, and for longer, than almost any other whale.

**IUCN status**  Vulnerable (2008). Mediterranean sub-population Endangered (2020).
**Population**  *c.* 850,000. Pre-whaling population *c.* 2 million. 1.03 million killed by whalers. Trend unknown.
**Classification**  Odontoceti, family Physeteridae.
**Taxonomy**  No recognized forms or subspecies; the scientific name was highly controversial in the past (*Physeter macrocephalus* vs *P. catodon*), but *macrocephalus* is now used almost universally.
**Other names**  Cachalot.

**DISTRIBUTION** One of the most widely distributed marine mammals (after killer whales). Found in the deeper parts of all oceans, from the tropics to the edge of the polar pack ice. Occurs in many deep, semi-enclosed seas, including the Mediterranean Sea, Gulf of Mexico, Caribbean Sea, Sea of Japan and Gulf of California; mostly absent from enclosed seas and semi-enclosed seas with shallow entrances (e.g. the Black Sea, Red Sea and Persian Gulf). Generally more abundant where there is high productivity, usually from upwelling (in areas dubbed the 'grounds' by Yankee whalers). Prefers bathymetric and oceanographic features that concentrate prey, including frontal boundaries, eddies, submarine canyons and steep continental shelf edges. Daily movements depend on prey abundance: 10–20km when plentiful to 90–100km in poor feeding conditions.

For most of their adult lives, male and female sperm whales are widely separated. Females and young males tend to remain in the tropics, sub-tropics and warm temperate waters year-round, usually between 40°S and 50°N (sometimes higher in the North Pacific), corresponding roughly to sea surface temperatures above 15°C. They are usually in waters deeper than *c.* 1,000m (less around oceanic islands and in Mexico's Gulf of California). After leaving natal groups, young males gradually move to higher latitudes: the larger and older the male, the higher the average latitude.

Large adult males usually frequent waters deeper than 300m, with sea surface temperatures down to *c.* 0°C; they will approach close to shore where there is sufficiently deep water (such as in submarine canyons). They spend much time in more productive latitudes above 40°, often near the edge of the polar pack ice. However, they do return sporadically to warm-water breeding grounds to mate (on an unknown schedule). Over time, they range widely – often moving across entire ocean basins and sometimes between them.

**BEHAVIOR** The two main behavioral states comprise foraging (*c.* 75 per cent of their time) and resting/socializing. When foraging, they make repeated deep dives. Groups of family units spread out over 1+km, diving for extended periods between breathing intervals at the surface. Adult males typically forage alone. While resting or socializing, often in the afternoon, females and young gather at or near the surface, close together, when they may lie still and quiet (sometimes for hours), or they may be active, vocalizing, rolling, touching one another, breaching and lobtailing. Large males also lie quietly at the surface, or socialize if with females.

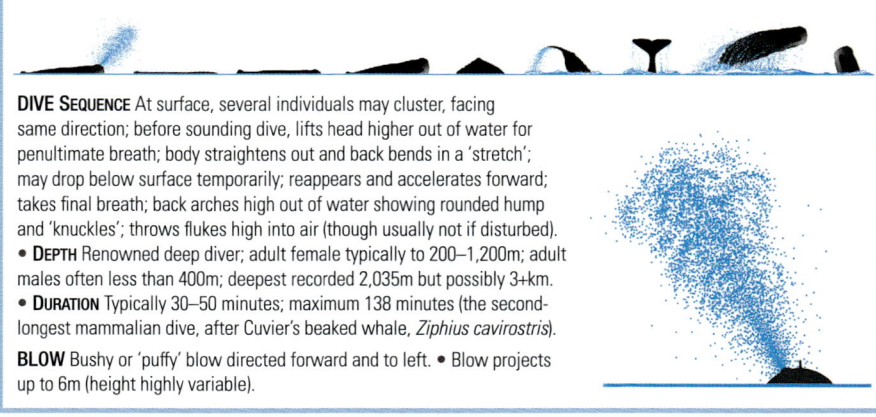

**DIVE SEQUENCE** At surface, several individuals may cluster, facing same direction; before sounding dive, lifts head higher out of water for penultimate breath; body straightens out and back bends in a 'stretch'; may drop below surface temporarily; reappears and accelerates forward; takes final breath; back arches high out of water showing rounded hump and 'knuckles'; throws flukes high into air (though usually not if disturbed).
• **DEPTH** Renowned deep diver; adult female typically to 200–1,200m; adult males often less than 400m; deepest recorded 2,035m but possibly 3+km.
• **DURATION** Typically 30–50 minutes; maximum 138 minutes (the second-longest mammalian dive, after Cuvier's beaked whale, *Ziphius cavirostris*).

**BLOW** Bushy or 'puffy' blow directed forward and to left. • Blow projects up to 6m (height highly variable).

# ADULT MALE

- Single blowhole at front of head and to left
- Surface of head smooth (no wrinkles)
- Some mature males have white patches around head
- Disproportionately large, squarish head (25–36 per cent of total length)
- Margin between head and trunk may appear as distinct crease (making spermaceti organ appear swollen, especially in larger males)
- Primarily dark gray (sometimes dark bluish-gray, may appear dark brownish-gray in bright sunlight)
- Dorsal fin a low, thick and usually rounded hump
- Dorsal fin two-thirds of the way along back (slightly further back than in female)
- c. 30 per cent of young males have rough white or yellowish calluses on dorsal fin (usually absent in adult male)
- Series of large bumps (often called knuckles or crenulations) along dorsal ridge (behind dorsal fin)
- Upper jaw tends to overhang tip of lower jaw
- Narrow, underslung lower jaw (barely visible from side)
- Upper 'lips', lower jaw and interior of mouth often white or creamy-white
- 2–10 short, deep throat grooves (less obvious in older animals)
- Large wrinkles or 'corrugations' cover most of body behind eyes
- May have prominent post-anal keel
- Flukes all dark on both sides
- May have white blotches on underside (variable extent and pattern)
- Short, wide, spatulate flippers held close to body (possibly as protection from predators)
- Skin on flippers and flukes smooth (no wrinkles)
- White scratches and scars common (especially on head) of large adults (made by other male sperm whales and cephalopod prey)

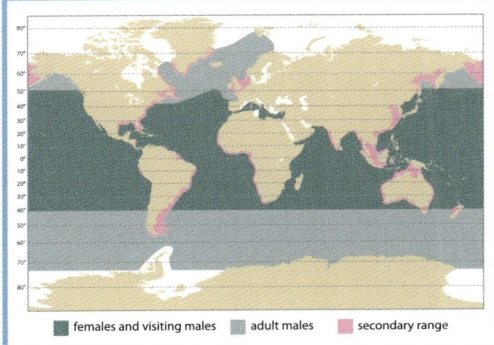

females and visiting males ■ adult males ■ secondary range

**AT A GLANCE** Deep, ice-free oceanic waters worldwide • Large to extra-large size • Primarily dark gray • Huge squarish head • Thick, low, rounded dorsal fin • 'Knuckles' from dorsal fin to flukes • Wrinkly, prune-like skin • Bushy blow directed forwards and to left • Often motionless (or swims leisurely) at surface • Flukes usually raised on diving

SPERM WHALE

Breaching is most common when groups come together or split, and between periods of foraging, and tends to occur in bouts (with individuals breaching repeatedly). Females breach more than lone adult males. Frequently spyhops – particularly females when there are mature males around during the breeding season, or in response to killer whale vocalizations. A behavior apparently unique to sperm whales is 'driftdiving'. They hang passively and upright in the water, with their heads up or down, just below the surface; they are probably sleeping (in one study for up to 31.5 minutes). Generally seems oblivious to boats, but will dive prematurely if a boat approaches too rapidly or too closely. Juveniles are often curious and may come close to investigate.

**FOOD AND FEEDING** Mainly deepwater squid (25+ species, including giant squid and jumbo squid); 60+ species of medium- to large-sized deep-sea fish; male typically takes larger individuals than females; occasionally takes octopuses, crustaceans, jellyfish etc. Feeds mainly in water column but some evidence of also feeding along seabed; details of how it catches prey unknown (probably draws it into mouth by suction; unlikely to use powerful clicks for acoustic stunning); males take fish from some longline fisheries.

**TEETH** Upper jaw 0; lower jaw 36–52.

**GROUP SIZE AND STRUCTURE** Six main types of social group: family unit or nursery group (about 10 females and their young); temporary cohesive group, consisting of 2+ family units sharing the same dialect (and therefore belonging to the same vocal clan); clan, containing many family units (hundreds or thousands of females and their calves); bachelor school (loose aggregations of young males of similar size and age, 4–21 years old); lone adult male (usually from late 20s onwards); and associations of adult males (long-term preferred companionships).

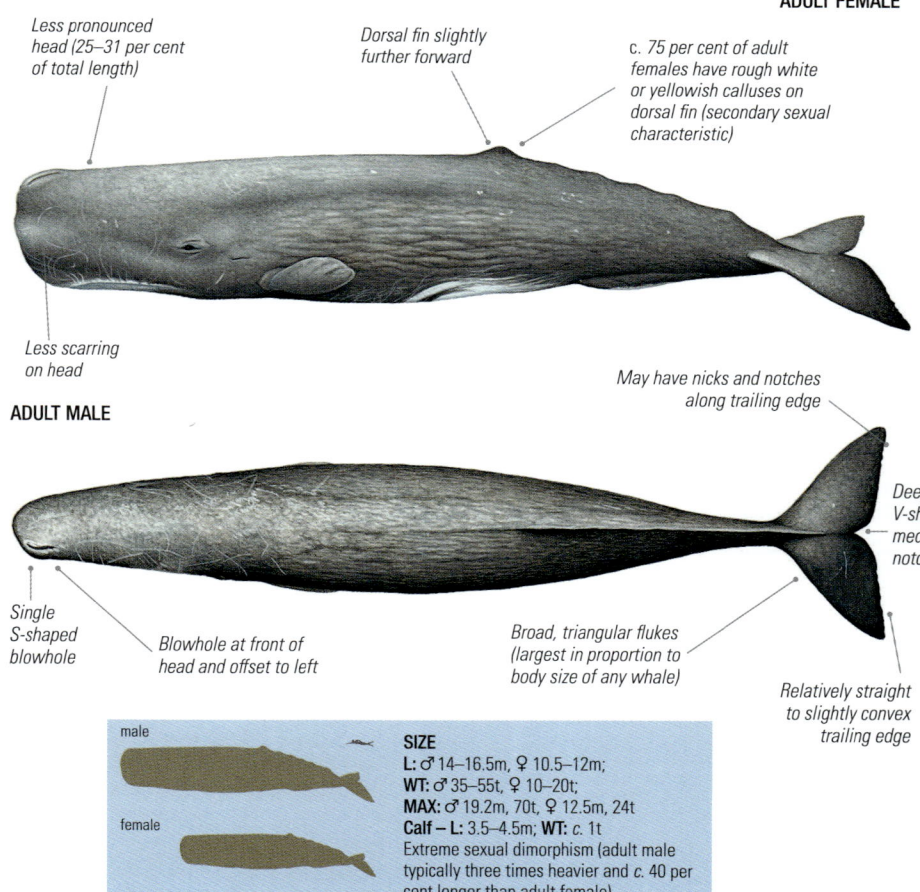

**ADULT FEMALE**
- Less pronounced head (25–31 per cent of total length)
- Dorsal fin slightly further forward
- c. 75 per cent of adult females have rough white or yellowish calluses on dorsal fin (secondary sexual characteristic)
- May have nicks and notches along trailing edge

**ADULT MALE**
- Less scarring on head
- Single S-shaped blowhole
- Blowhole at front of head and offset to left
- Broad, triangular flukes (largest in proportion to body size of any whale)
- Deep, V-shaped medial notch
- Relatively straight to slightly convex trailing edge

**SIZE**
L: ♂ 14–16.5m, ♀ 10.5–12m;
WT: ♂ 35–55t, ♀ 10–20t;
MAX: ♂ 19.2m, 70t, ♀ 12.5m, 24t
Calf – L: 3.5–4.5m; WT: c. 1t
Extreme sexual dimorphism (adult male typically three times heavier and c. 40 per cent longer than adult female).

**ADULT MALE**

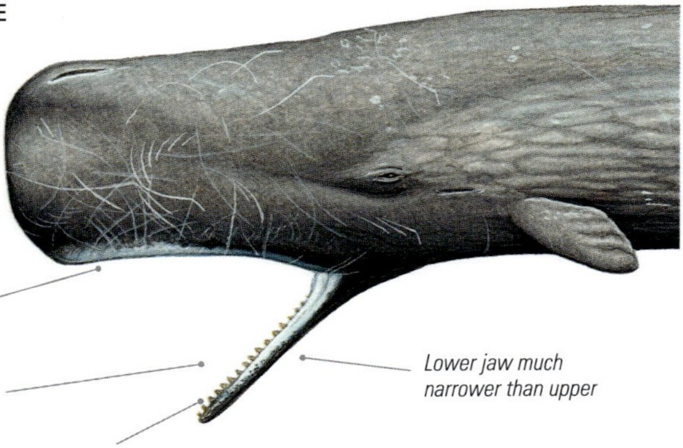

- Upper jaw has small vestigial teeth (rarely break through gums)
- Lower jaw can be opened nearly perpendicular to body
- Large, conical teeth on lower jaw (fit into sockets in upper jaw when mouth closed)
- Lower jaw much narrower than upper

**CALF**

- Same dark gray as adult
- Mouthline relatively short
- Skin wrinkled all over at birth (skin of head and flippers becomes smooth with age)
- Little or no scarring
- Flukes very long

**FLUKES**

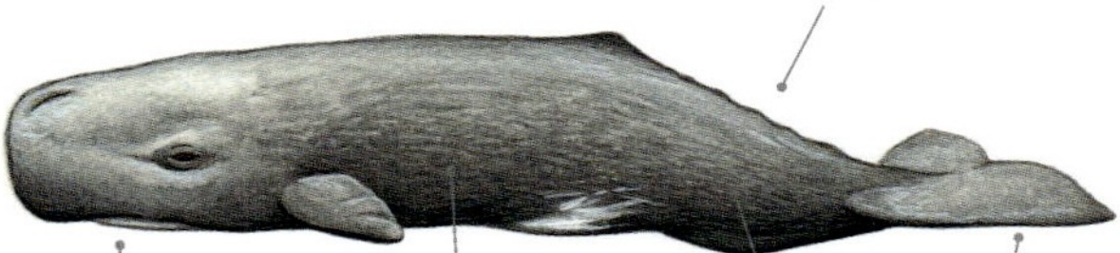

## THE SPERMACETI ORGAN

The spermaceti organ complex – which dominates the sperm whale's head – is the world's most powerful natural sonar or echolocation system. It consists of a complex array of soft structures, cradled above the lower jaw and in front of the skull, in the whale's highly modified and asymmetric head. Known by whalers as the 'case', the spermaceti organ itself is up to c. 5m long, and enclosed by a tough muscular sheath. It consists of a white spongy tissue soaked in a liquid wax called spermaceti oil. The organ is used mainly for forming, focusing and broadcasting extremely powerful, highly directional clicks (recent evidence suggests that it is not used to debilitate prey with sound, as was previously thought).

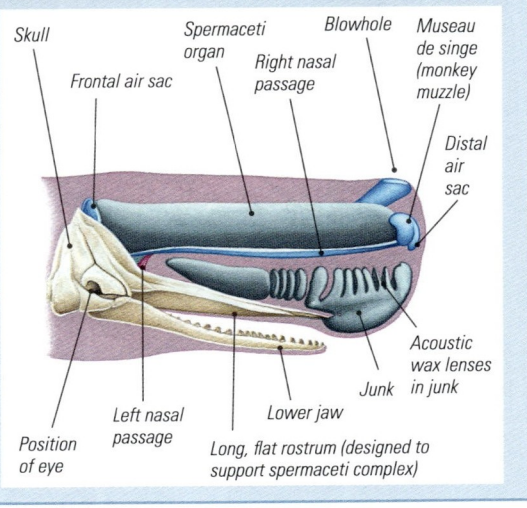

Skull · Spermaceti organ · Blowhole · Museau de singe (monkey muzzle) · Frontal air sac · Right nasal passage · Distal air sac · Acoustic wax lenses · Junk in junk · Position of eye · Left nasal passage · Lower jaw · Long, flat rostrum (designed to support spermaceti complex)

**SPERM WHALE**

# PYGMY SPERM WHALE
## *Kogia breviceps* (Blainville, 1838)

Pygmy and dwarf sperm whales are typically seen floating on the surface, with just the top of the head and back (as far as the dorsal fin) exposed. They are very difficult to spot in anything but calm conditions, and sightings of both species tend to be brief.

**IUCN status** Least Concern (2019).
**Population** Unknown, although frequent strandings in some areas implies that it is more common than the lack of sightings suggests. Trend unknown.
**Classification** Odontoceti, family Kogiidae.
**Taxonomy** No recognized forms or subspecies.
**Other names** Lesser cachalot, short-headed cachalot, lesser sperm whale, short-headed sperm whale.
**DISTRIBUTION** Tropical to warm temperate waters worldwide. Generally inhabits waters along the outer continental shelf and beyond, particularly over and near the continental slope. In the North Atlantic, it is closely associated with the Gulf Stream. Prefers more temperate seas and relatively deeper, more pelagic waters than the dwarf sperm whale. There is no evidence of long-distance migrations. Most information comes from strandings.
**BEHAVIOR** Difficult to spot, except in extremely calm seas. Surfacing patterns are hard to predict, and it tends to be shy, undemonstrative and tricky to approach closely. Aerial behavior is extremely rare, though it does sometimes breach. Between dives, it tends to raft motionless at the surface – from a distance resembling a piece of driftwood – with the top of the head, back and dorsal fin exposed, and the tail hanging down underwater.
**FOOD AND FEEDING** Mostly deepwater squid; will take some fish and shrimps; more diverse diet and averages larger prey than dwarf sperm whale. Feeding mostly on or near seabed; anatomy suggests powerful suction feeding.
**TEETH** Upper jaw 0; lower jaw 20–32.
**GROUP SIZE AND STRUCTURE** Usually solitary, but up to six of varying age and sex composition. Strandings usually involve single animals (maximum known was three: one male and two females).

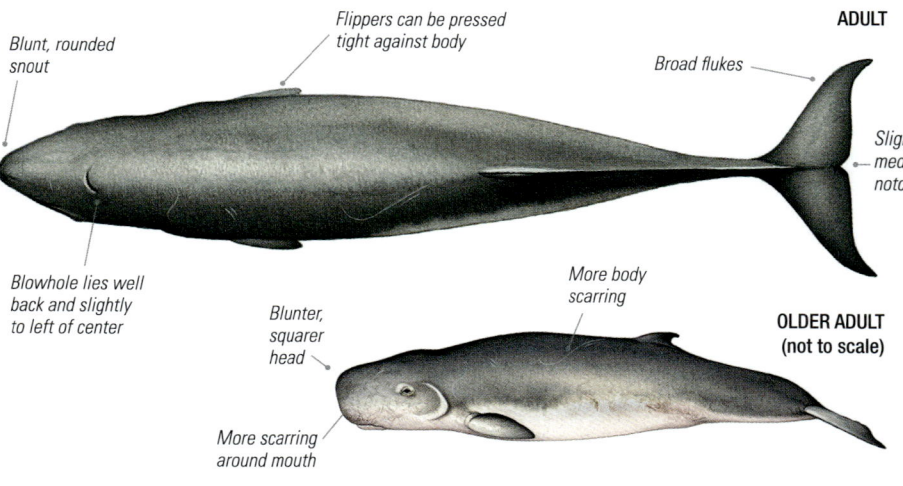

**DIVE SEQUENCE** Inconspicuous blow; floats motionless in same position on surface (with front of melon to dorsal fin visible); usually sinks vertically out of sight but (particularly if startled) may roll forward with little arching of back; does not show flukes. • **DEPTH** Unknown, but believed to forage at greater depths than dwarf sperm whale.
• **DURATION** 12–15 minutes (based on limited evidence); maximum recorded 18 minutes.

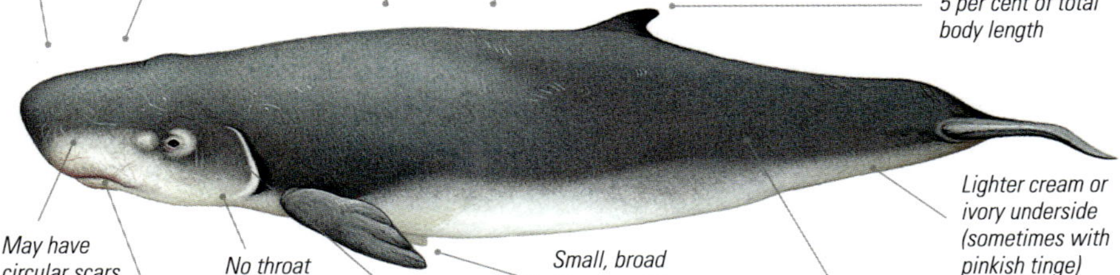

**ADULT**

- Blunt, squarish head (proportionately larger, less square and with longer snout than on dwarf sperm whale)
- Dark bluish-gray to brownish-black upperside
- Tip of dorsal fin rounded and usually below highest point (variable) – appears hooked
- Robust body (not unlike small sperm whale)
- Small, falcate dorsal fin well behind midpoint of back (slightly further back than on dwarf sperm whale)
- Head becomes blunter and squarer with age
- Dorsal fin less than 5 per cent of total body length
- Lighter cream or ivory underside (sometimes with pinkish tinge)
- May have circular scars around mouth (caused by squid bites)
- No throat grooves
- Small, broad flippers set far forward near head
- Scarring on both sexes (attributed to fighting during mating season or shark attacks)
- Narrow underslung lower jaw
- Crescent-shaped, light-colored marks usually present on sides of head (dubbed 'false gills')
- Tiny, underslung lower jaw with long, sharp teeth (fit into sockets in upper jaw)

**CALF**

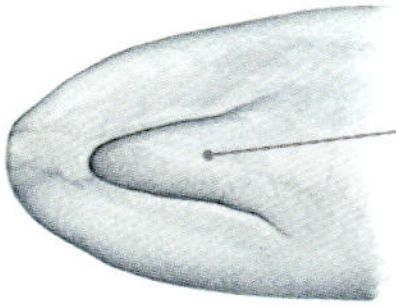

**SIZE**
L: ♂ 2.7–3.5m, ♀ 2.7–3.5m;
**WT:** 315–450kg; **MAX:** 3.8m, 515kg
**Calf – L:** *c.* 1–1.2m; **WT:** *c.* 50–55kg

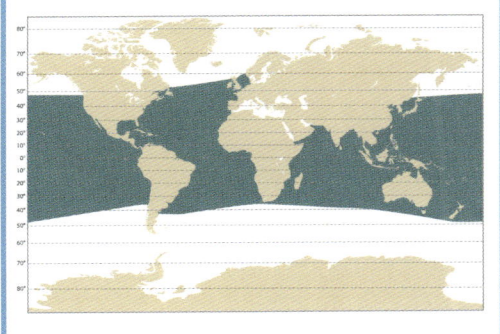

**AT A GLANCE** Deep tropical to warm temperate waters worldwide • Small size • Generally appears dark gray at sea • Blunt, squarish head • Small, falcate dorsal fin slightly further back than on dwarf sperm whale • Floats motionless on surface between dives • Back has distinctive bulge when logging

# DWARF SPERM WHALE
*Kogia sima* (Owen, 1866)

With their long, sharp teeth, underslung lower jaws and gill-like markings on either side of the head, dwarf and pygmy sperm whales are often mistaken for sharks when they strand. Their appearance may be a form of mimicry, to help avoid predation.

**IUCN status** Least Concern (2020).
**Population** Unknown, although frequent strandings in some areas implies that it is more common than the lack of sightings suggests. Trend unknown.
**Classification** Odontoceti, family Kogiidae.
**Taxonomy** No recognized forms or subspecies; however, recent genetic studies suggest there may be two distinct subspecies or even species, one in the Atlantic, the other in the Indo-Pacific.
**Other names** Owen's pygmy whale, snub-nosed cachalot.
**DISTRIBUTION** Tropical to warm temperate waters worldwide. Generally inhabits deep waters over or near the edge of the continental shelf. In the North Atlantic, it is closely associated with the Gulf Stream. Prefers more tropical seas and relatively shallower, less pelagic waters than the pygmy sperm whale (sometimes frequenting more coastal areas) and probably does not range as far into high latitudes. Most information comes from strandings.
**BEHAVIOR** Difficult to spot, except in extremely calm seas. Surfacing patterns are hard to predict. Rarely allows close approach. Aerial behavior is extremely rare, though it does sometimes breach. Between dives, it tends to raft (or log) motionless at the surface – where, from a distance, it looks like a piece of driftwood – with the top of the head, back and dorsal fin exposed, and the tail hanging down underwater.
**FOOD AND FEEDING** Mostly mid- and deepwater squid; will take some fish and shrimps; less diverse diet and averages smaller prey than pygmy sperm whale. Feeding mostly on or near seabed; anatomy suggests powerful suction feeding.
**TEETH** Upper jaw 0–6; lower jaw 14–26.
**GROUP SIZE AND STRUCTURE** Usually solitary, but up to 12 (maximum 16 recorded) of varying age and sex composition. Group size varies according to location and season: e.g. in the Bahamas 1–8 in summer, 1–12 in winter; average in Hawai'i is 2.7. Individuals in a group appear to be loosely associated – when one appears, there are often others several hundred meters away. Strandings usually involve single animals (maximum known was four immatures: one male and three females).

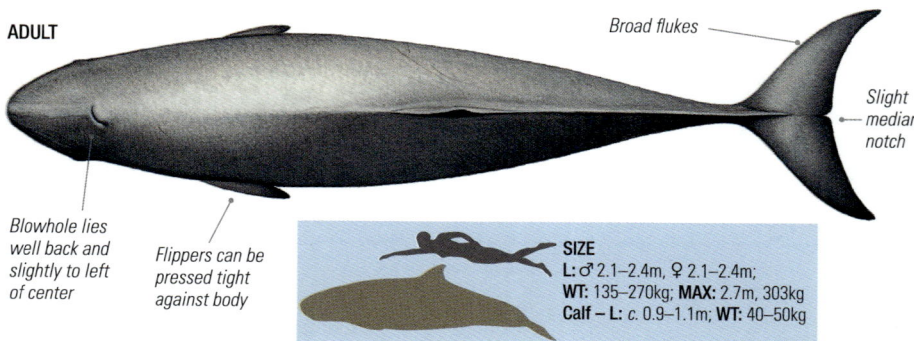

ADULT
Broad flukes
Slight median notch
Blowhole lies well back and slightly to left of center
Flippers can be pressed tight against body

SIZE
L: ♂ 2.1–2.4m, ♀ 2.1–2.4m;
WT: 135–270kg; MAX: 2.7m, 303kg
Calf – L: c. 0.9–1.1m; WT: 40–50kg

**DIVE SEQUENCE** Inconspicuous blow; floats motionless in same position on surface (with front of melon to dorsal fin visible); usually sinks vertically out of sight but (particularly if startled) may roll forward with little arching of back; does not show flukes. • **DEPTH** Unknown, but believed to forage at shallower depths than pygmy sperm whale (600–1,200m suggested). • **DURATION** 7–15 minutes (possibly up to 30 minutes); brief 1–3 minutes at surface in between.

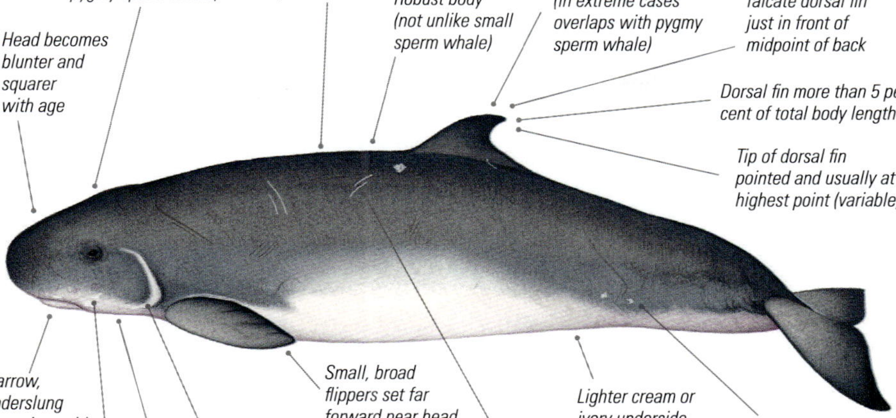

**ADULT**

- Blunt, squarish head (proportionately smaller and squarer than on pygmy sperm whale)
- Dark bluish-gray to brownish-black upperside (color quite variable)
- Robust body (not unlike small sperm whale)
- Height, position and shape of dorsal fin highly variable (in extreme cases overlaps with pygmy sperm whale)
- Relatively large, falcate dorsal fin just in front of midpoint of back
- Head becomes blunter and squarer with age
- Dorsal fin more than 5 per cent of total body length
- Tip of dorsal fin pointed and usually at highest point (variable)
- Narrow, underslung lower jaw with long, sharp teeth (fit into sockets in upper jaw)
- Small, broad flippers set far forward near head
- Lighter cream or ivory underside (sometimes with pinkish tinge)
- May have scarring from cookiecutter shark bites (quickly re-pigment to background color)
- May be circular scars around mouth (caused by squid bites)
- Crescent-shaped, light-colored marks usually present on sides of head (dubbed 'false gills')
- Two or more short, longitudinal throat grooves (similar to those on beaked whales)
- Scarring on both sexes (attributed to fighting during mating season or shark attacks)

**COMPARISON OF SILHOUETTES**

**Pygmy sperm whale**

- Distinctive rounded bulge on back (between blowhole and dorsal fin) visible when logging.
- Distinct 'neck' behind relatively large head.
- Smaller, hooked dorsal fin, further back.
- Sometimes dorsal fin invisible until animal rolls out of sight.

**Dwarf sperm whale**

- Flatter profile in water (usually without a prominent bulge), reminiscent of an upside-down surfboard.
- Tends to float lower in water.
- Larger, pointed dorsal fin, further forward.
- Dorsal fin not unlike that of bottlenose dolphin.

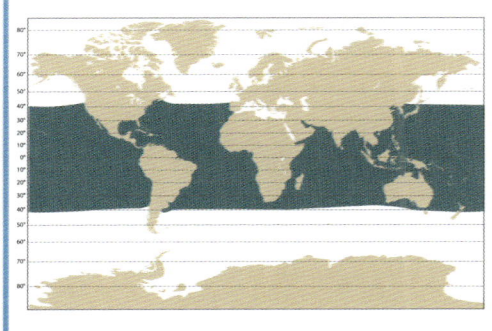

**AT A GLANCE** Deep tropical to warm temperate waters worldwide • Small size • Generally appears dark gray at sea • Blunt, squarish head • Tall, pointed, falcate dorsal fin slightly further forward than on pygmy sperm whale • Floats motionless on surface between dives • Back appears flat when logging

# NARWHAL
*Monodon monoceros*  Linnaeus, 1758

The narwhal can be difficult to see: living in remote regions of the High Arctic, it spends half the year in dense pack ice under continuous darkness. However, it does have predictable migratory patterns, and the male is unmistakable, with its extraordinary long, spiralling tusk.

**IUCN status** Least Concern (2017).
**Population** *c.* 170,000 (123,000 mature individuals), excluding the Russian Arctic (for which no estimates are available). Trend unknown.
**Classification** Odontoceti, family Monodontidae.
**Taxonomy** No recognized forms or subspecies.
**Other names** Narwhale, unicorn whale, sea unicorn.

**DISTRIBUTION** Mainly in the Atlantic sector of the Arctic (60–85°N, most commonly 70–80°N). Occasional stragglers in the Pacific sector. Relatively rare in Svalbard and south of the Arctic Circle. Discontinuous range, separated by Greenland. Winter and summer ranges up to 2,000km apart. Vagrants have been recorded in Newfoundland, UK, Germany, Belgium and the Netherlands (in the North Atlantic); and as far south as the Alaska Peninsula and the Commander Islands (in the North Pacific).
**Winter distribution** Wintering areas tend to be deep, offshore, ice-covered habitats along the continental slope. Two-thirds of all narwhals winter in deep water offshore under dense pack ice in Baffin Bay and Davis Strait (between Baffin Island and Greenland), in two distinct 'grounds' (northern and southern); the east Greenland population winters offshore in deep waters of the Greenland Sea. On the wintering grounds, they spend six months in continuous darkness, in air temperatures as low as -40°C, where there is often less than 3 per cent open water.
**Summer distribution** Spends about two months each summer in ice-free bays, fjords and island passages in the Canadian Arctic, western and eastern Greenland, Svalbard, and the northwestern Russian High Arctic. In some parts of the range, glacial fronts are an important summer habitat. It prefers deep water but readily enters shallow water to hide from hunting killer whales.
**Migrations** There are predictable migrations between the summer and winter grounds, as the ice retreats in spring and refreezes in fall, and in many areas narwhals pass certain promontories, bays and fjords at precisely the same time each year. Migrations last about two months each way.

**ADULT MALE**

Right tusk (measuring *c.* 3cm) usually remains embedded in skull

Dark brown dorsal surface (strongest on top and front of head, along dorsal ridge and along borders of flippers and flukes)

Erupted tusk appears on left side of upper jaw and angles slightly to left

**DIVE SEQUENCE** Visible but inconspicuous blow (clearly audible on calm days and at close range); male's tusk sometimes (but not always) appears above surface (usually briefly); alternatively, may see impression of tusk just below surface; may fluke before deep dive (rarely before shallow dive). • **DEPTH** Summer 13–850m, though usually less than 50m; winter much deeper (typically spending over three hours per day at depths of at least 800m, during 18–25 dives); over half winter dives reach 1,500m; maximum recorded 1,800m. • **DURATION** Mostly 7–20 minutes, maximum documented 25 minutes.

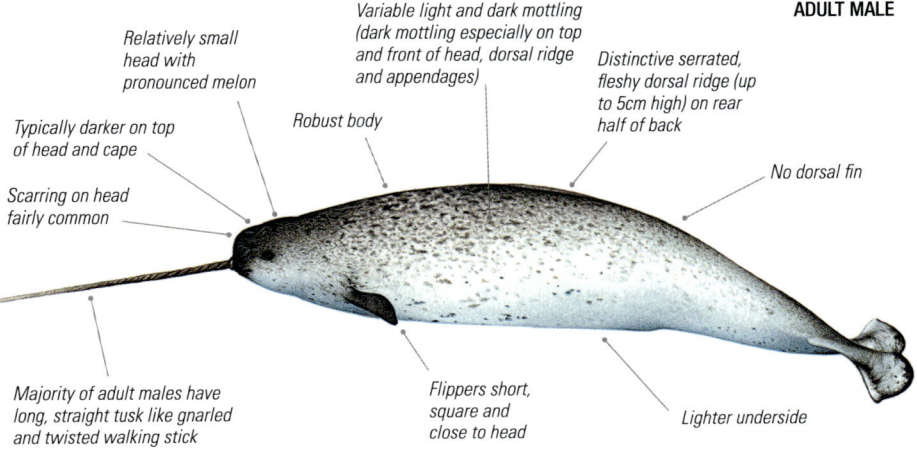

**ADULT MALE**

- Relatively small head with pronounced melon
- Variable light and dark mottling (dark mottling especially on top and front of head, dorsal ridge and appendages)
- Distinctive serrated, fleshy dorsal ridge (up to 5cm high) on rear half of back
- Typically darker on top of head and cape
- Robust body
- No dorsal fin
- Scarring on head fairly common
- Majority of adult males have long, straight tusk like gnarled and twisted walking stick
- Flippers short, square and close to head
- Lighter underside

**ADULT FEMALE**

- Female typically smaller than male
- Tusks normally remain in tooth sockets within upper jaw

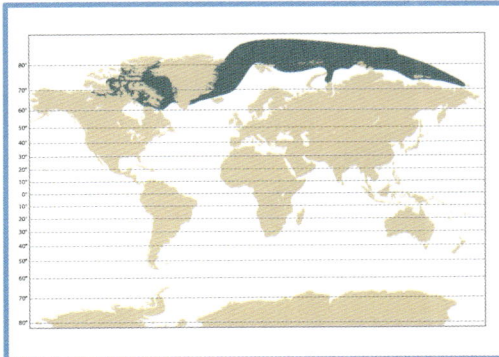

**SIZE**
**L:** ♂ 4.3–4.8m (excluding tusk of up to 3m), ♀ 3.7–4.2m;
**WT:** 700–1,650kg; **MAX:** 5m, 1,800kg
**Calf – L:** 1.5–1.7m; **WT:** c. 80kg

**AT A GLANCE** High Arctic • Small to medium size • Long tusk of male • Relatively small bulbous head • Little or no beak • No dorsal fin (but slight dorsal ridge) • Variable light and dark mottling

**BEHAVIOR** Not given to spontaneous exuberance such as breaching or speed-swimming, but occasionally spyhops and lunges. Frequently logs (or rafts), with the top of its head and back visible, and may roll around at the surface while socializing. Spends less time at the surface in choppy or rough seas. Male may wave its tusk in the air or rest it on the back of another individual. Before a deep dive, a narwhal often swims directly towards the ice edge, then flukes about 5–30m away. Navigates easily under ice and can travel several kilometers between breathing opportunities, or uses head or back to break through ice several centimeters thick; the only barrier to its movements is fast ice without cracks. Often associates with bowhead whales; rarely forms mixed herds with belugas. Often shy and wary of boats (especially where hunted), but in Canada, at least, can be less nervous of people standing on the floe edge or shore.

**FOOD AND FEEDING** Fish (especially Greenland halibut, Arctic cod; also Arctic eelpout, polar cod, roughhead grenadier, capelin), squid (especially small deepwater *Gonatus* spp.), shrimp (especially deep-sea prawn); precise diet varies with region and season. Most feeding in winter (November–March), very little or no feeding in summer (July–September), will feed under sea ice in spring and fall; probably sucks prey into mouth and swallows it whole; no evidence of cooperative hunting.

**TEETH** Upper jaw 2 (usually only 1 erupts, typically only in the male); lower jaw 0.

**GROUP SIZE AND STRUCTURE** Most narwhal pods or 'clusters' typically contain 2–10 individuals (range 1–50). Mostly these comprise a single sex: all males, or all females with young. Groups often combine to form large dispersed herds, each with up to 600 clusters, containing hundreds or even thousands of individuals and a mix of sexes and ages.

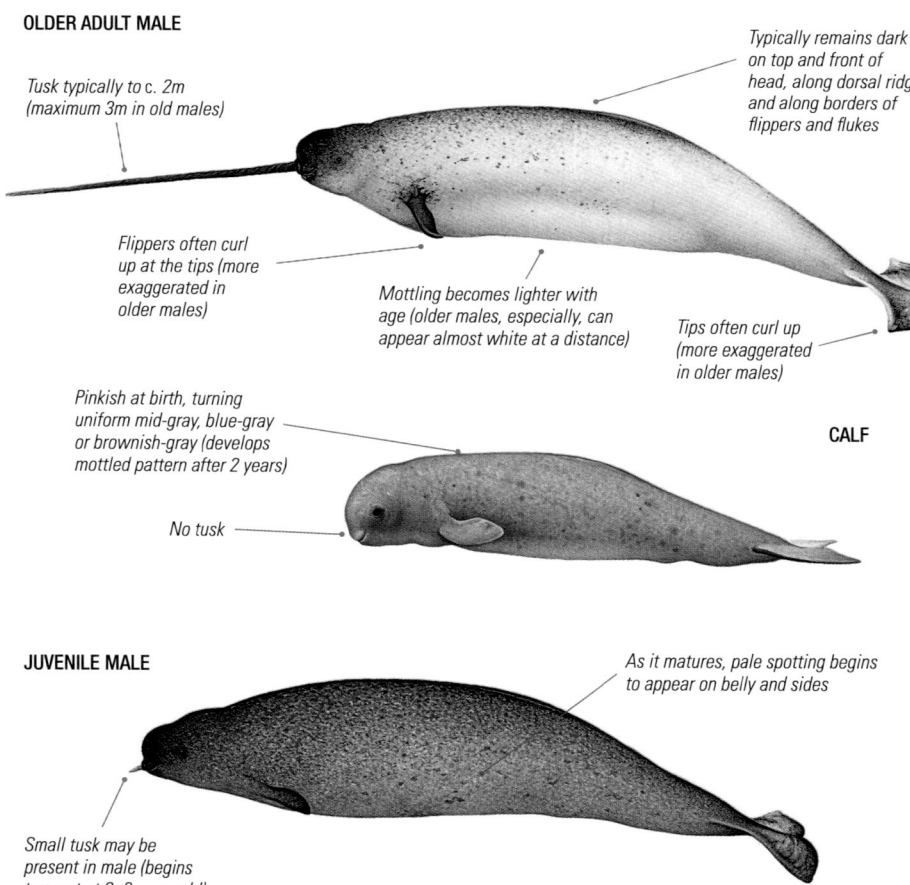

**OLDER ADULT MALE**

Tusk typically to c. 2m (maximum 3m in old males)

Typically remains dark on top and front of head, along dorsal ridge and along borders of flippers and flukes

Flippers often curl up at the tips (more exaggerated in older males)

Mottling becomes lighter with age (older males, especially, can appear almost white at a distance)

Tips often curl up (more exaggerated in older males)

**CALF**

Pinkish at birth, turning uniform mid-gray, blue-gray or brownish-gray (develops mottled pattern after 2 years)

No tusk

**JUVENILE MALE**

As it matures, pale spotting begins to appear on belly and sides

Small tusk may be present in male (begins to erupt at 2–3 years old)

**ADULT FEMALE WITH TUSK**

c. 1.5 per cent of females have 1 tusk (though estimates vary since tusked females are easy to confuse with juvenile males)

Female tusks shorter (maximum 1.5m), less robust and whiter (do not collect as much algae on surface as male tusks)

**ADULT MALE**

Tusk usually ends in shiny white point

Tusk typically 1.8–2.7m long

Little or no beak

Tusk length, girth, morphology, wear and color highly variable

Short, upturned mouthline

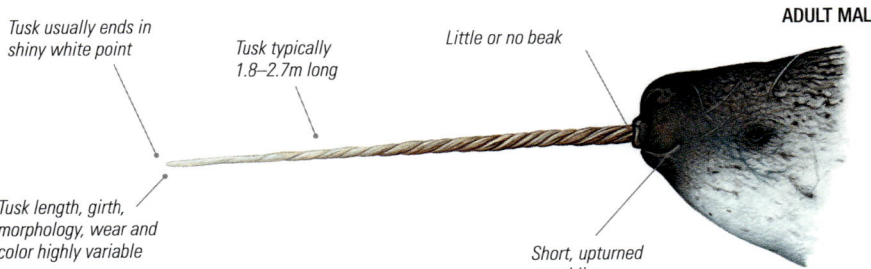

Right tusk usually shorter and less robust

c. 0.9 per cent of males have two tusks

One record of female with two tusks (collected in 1684)

**ADULT MALE WITH TWO TUSKS**

The tusk always spirals anticlockwise from the whale's-eye view

**ADULT MALE FLUKES UPPERSIDE**

Strongly convex trailing edges (giving 'back-to-front' appearance), especially in older males

Deep median notch

Tips may curl upwards (especially in older males)

**ADULT FEMALE FLUKES UPPERSIDE**

Trailing edges of flukes less convex, even straight (more dolphin-like)

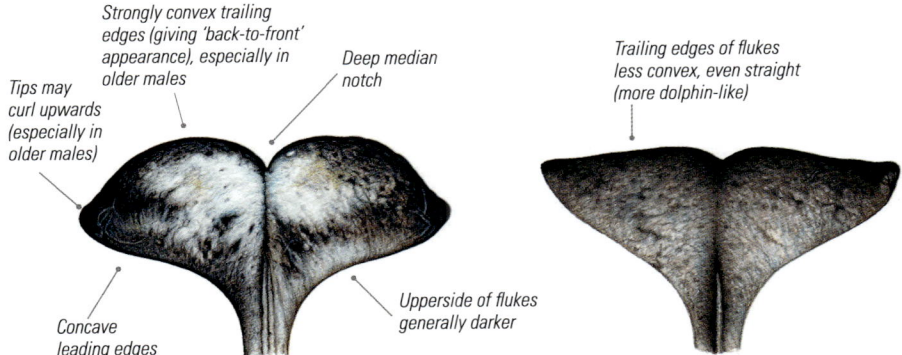

Concave leading edges

Upperside of flukes generally darker

# BELUGA
*Delphinapterus leucas* (Pallas, 1776)

Ancient mariners used to call the pale white beluga the 'sea canary' because of its great repertoire of groans, roars, whistles, squawks, moos, buzzes and trills. When seen from above the surface, its ghost-like glow is hard to mistake for any other species.

**IUCN status** Least Concern (2017). Cook Inlet sub-population Critically Endangered (2018).
**Population** c. 150,000–200,000 including 136,000 mature individuals (though many parts of the range remain unsurveyed). Trend unknown.
**Classification** Odontoceti, family Monodontidae.
**Taxonomy** No recognized forms or subspecies.
**Other names** Beluga whale, white whale; historically – sea canary.
**DISTRIBUTION** Cold waters of the Arctic and sub-Arctic, ranging from 47–82°N. Wide choice of habitat, including estuaries, coastal waters (as shallow as 1–3m), continental shelves and deep ocean basins, in open water and loose ice (it generally avoids dense pack ice, though commonly overwinters in polynyas). Will swim up rivers. There is an isolated population in the St Lawrence River estuary, Canada, which is resident year-round. Dense concentrations are common in shallow coastal waters in summer.
**BEHAVIOR** Rarely given to aerial displays, though it can be more demonstrative (spyhopping, tail-waving and lobtailing) in nearshore concentrations. Able to turn its head sideways (the cervical vertebrae are unfused, making the neck more flexible), which is unusual in cetaceans. Little fear of shallow water (if stranded, it is often able to wait and refloat on the next tide – assuming it is not found by a polar bear first). Sometimes shows curiosity towards boats and frequently towards snorkelers and divers. Often associates with bowhead whales, though rarely forms mixed herds with narwhals. The annual moult, which involves rubbing along the seafloor to remove sloughed skin, is rare in cetaceans.
**FOOD AND FEEDING** Mainly fish such as salmon, herring, Greenland halibut, smelt, Arctic and polar cod and capelin, but also squid, octopuses, shrimps, crabs, clams, mussels and even marine worms and large zooplankton. Sucks prey into mouth with flexible 'lips'; some evidence of cooperative hunting (e.g. in groups of 3–5, hunting smelt in Russia's Sea of Okhotsk), but usually hunts alone (even within a group).
**TEETH** Upper jaw 16–20; lower jaw 16–20. Teeth are often heavily worn – even down to the gums in older animals.
**GROUP SIZE AND STRUCTURE** Usually in groups of 5–20 (although large adults are occasionally seen alone). Many groups together may form herds of hundreds, or more than 1,000, which can be mixed or segregated by age and sex. Group structure tends to be fluid.

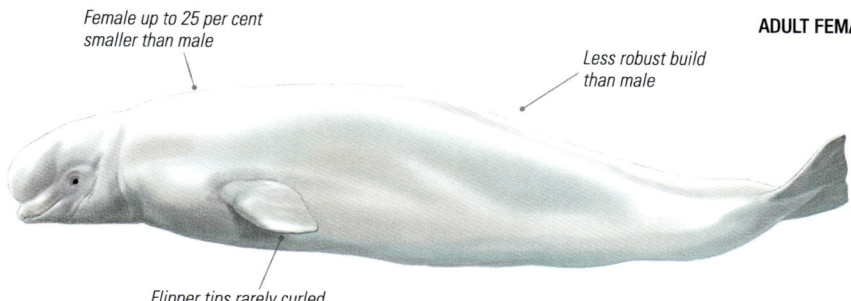

**ADULT FEMALE**
Female up to 25 per cent smaller than male
Less robust build than male
Flipper tips rarely curled

**DIVE SEQUENCE** Swims in slow rolling motion (look for white arc that appears, grows, shrinks, then disappears); flukes occasionally appear at low angle above water; indistinct low, steamy blow (often not visible). • **DEPTH** Regularly 300–600m, sometimes beyond 800m (maximum recorded 956m). • **DURATION** 9–18 minutes (feeding dives typically 18–20 minutes); maximum documented 25 minutes.

## ADULT MALE

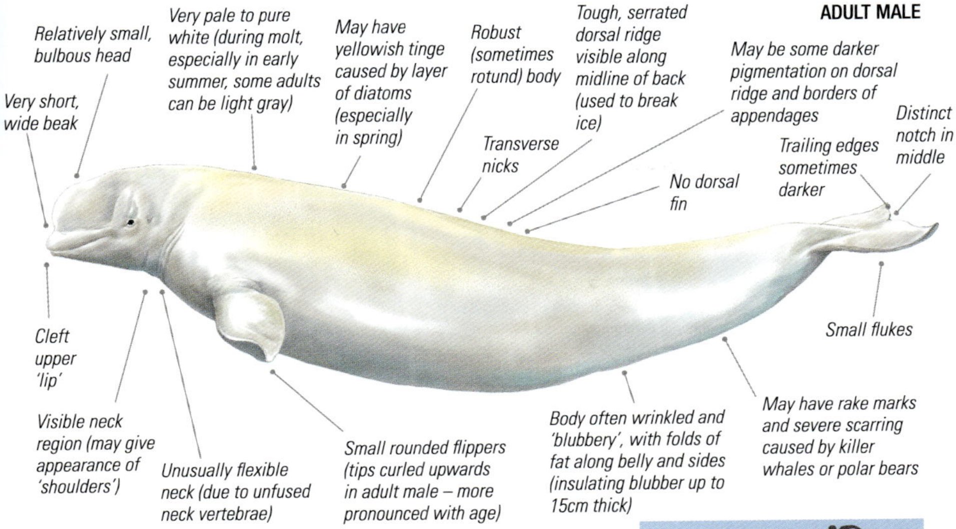

- Relatively small, bulbous head
- Very short, wide beak
- Very pale to pure white (during molt, especially in early summer, some adults can be light gray)
- May have yellowish tinge caused by layer of diatoms (especially in spring)
- Robust (sometimes rotund) body
- Transverse nicks
- Tough, serrated dorsal ridge visible along midline of back (used to break ice)
- No dorsal fin
- May be some darker pigmentation on dorsal ridge and borders of appendages
- Trailing edges sometimes darker
- Distinct notch in middle
- Small flukes
- Cleft upper 'lip'
- Visible neck region (may give appearance of 'shoulders')
- Unusually flexible neck (due to unfused neck vertebrae)
- Small rounded flippers (tips curled upwards in adult male – more pronounced with age)
- Body often wrinkled and 'blubbery', with folds of fat along belly and sides (insulating blubber up to 15cm thick)
- May have rake marks and severe scarring caused by killer whales or polar bears

## CALF

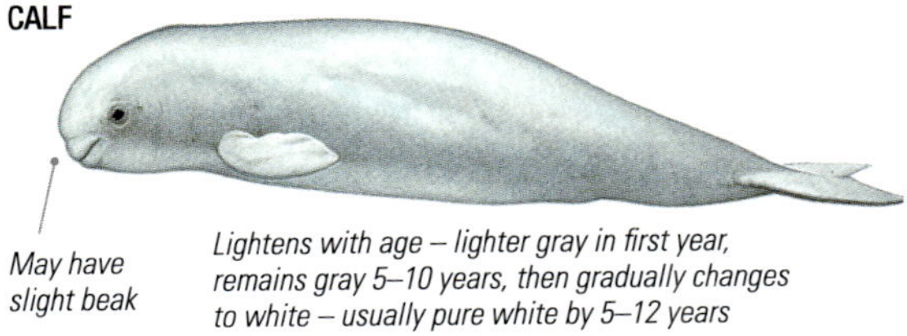

- May have slight beak
- Lightens with age – lighter gray in first year, remains gray 5–10 years, then gradually changes to white – usually pure white by 5–12 years

**SIZE**
L: ♂ 3.7–4.8m; ♀ 3.0–3.9m;
WT: 500–1,300kg; MAX: 5.5m, 1.9t
Calf – L: 1.5–1.6m; WT: 80–100kg
Body size varies considerably between sub-populations.

## ADULT FACIAL EXPRESSIONS

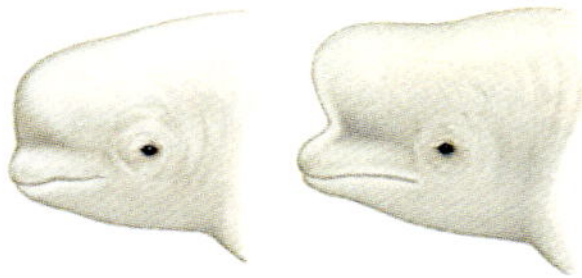

Can alter shape of 'lips' and melon (impressive array of facial expressions)

## FLUKES ADULT

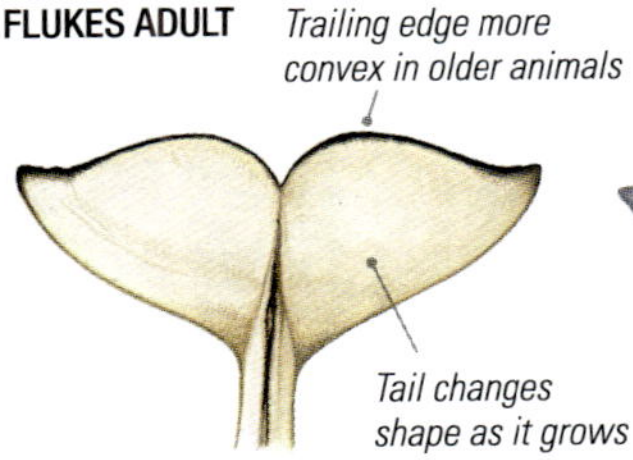

Trailing edge more convex in older animals

Tail changes shape as it grows

## FLUKES IMMATURE

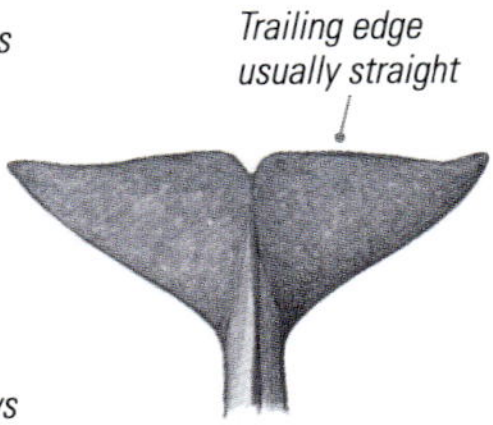

Trailing edge usually straight

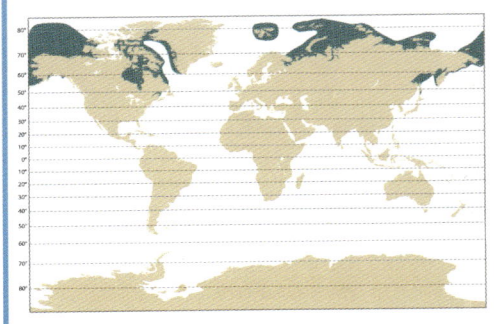

**AT A GLANCE** Arctic and sub-Arctic • Small to medium size • Very pale to pure white, pale gray or yellowish with no mottling • Robust body • Small, bulbous head • No dorsal fin • Surfaces often with distinctive slow rolling motion

# BAIRD'S BEAKED WHALE
*Berardius bairdii*　　　　　　　　　　　　　　　　　　　　　　　　　　Stejneger, 1883

The largest of all the beaked whales, Baird's beaked whale is one of the easiest members of the family to identify in the North Pacific. It is strikingly similar to Arnoux's beaked whale (though the two species are widely separated geographically). Baird's is one of few beaked whale species to be commercially hunted; Japanese hunting began in the early 1600s, reached a peak after the Second World War (322 were killed in 1952) and continues to this day (with an annual quota of *c.* 68).

**IUCN status** Least Concern (2020).
**Population** Unknown. Regional estimates: *c.* 7,100 in Japanese waters, *c.* 7,960 in the California Current. Trend unknown.
**Classification** Odontoceti, family Ziphiidae.
**Taxonomy** No recognized forms or subspecies.
**Other names** Giant bottlenose whale, North Pacific bottlenose whale, four-toothed whale, northern fourtooth whale.
**DISTRIBUTION** In summer, deep, offshore, cool temperate to sub-polar waters in the northern North Pacific and the adjacent Sea of Japan, Sea of Okhotsk and Bering Sea, at least from *c.* 30–62°N. Appears to be migratory in most areas, with seasonal peaks in abundance. In winter, it is believed to move into deeper waters away from the continental slope, but little is known about its wintering grounds. Prefers continental slope waters, 1,000–3,000m deep, and areas with complex topography such as submarine canyons, seamounts and ridges.
**BEHAVIOR** Moderately aerially active. Will sometimes breach repeatedly, leap in a low arc or, especially in mixed groups, jump on top of one another. Will also spyhop, flipper-slap and lobtail. May swim belly up or sideways or roll at the surface. Reaction to vessels varies.
**FOOD AND FEEDING** Pelagic and benthic fish, squid, octopuses, some crustaceans. Often (but not exclusively) down to seafloor; pebbles in stomachs may reflect bottom-feeding; suction feeder.
**TEETH** Upper jaw 0; lower jaw 4.
**GROUP SIZE AND STRUCTURE** Typically 5–20, with some regional variation; occasionally up to 50. There may be some sexual segregation. Lone individuals are rare. Groups can be so tightly packed that individuals may be in physical contact with each other.

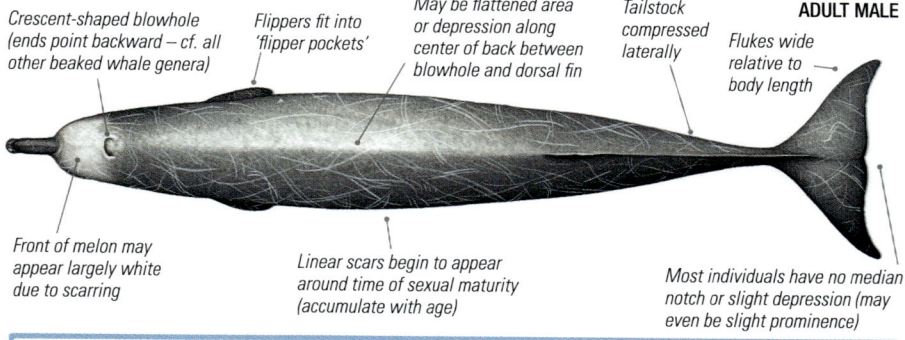

**DIVE SEQUENCE** Typically in tight school, all surfacing in unison; body appears very long; blows continuously while swimming slowly (easily identifiable from distance); shallow roll with little arching of back (more arching before deep dive); rarely raises flukes before deep dive. • **DEPTH** Deep dives routinely to 1,000m or more; maximum recorded 1,777m (Japan) but probably capable of deeper. • **DURATION** Typically 11–30 minutes; maximum recorded 82 minutes; unsubstantiated reports of up to two hours.

**BLOW** Strong, low, puffy or rather shapeless blow (up to *c.* 2m), quite conspicuous in calm weather (but tends to dissipate quickly). • Blow sometimes angled slightly forward.

# BAIRD'S BEAKED WHALE

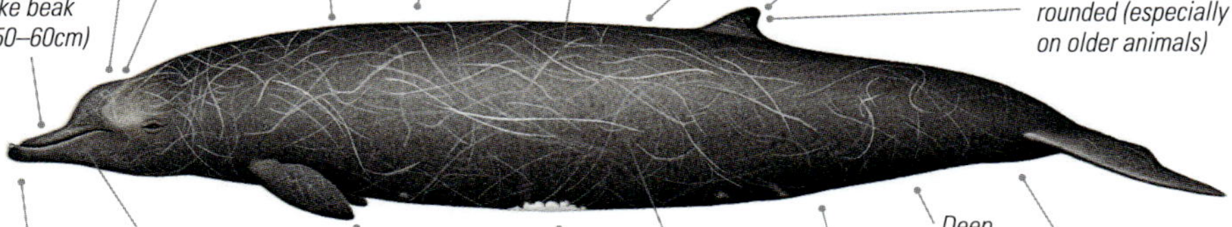

**Bulbous melon** with moderately steep forehead (more bulbous with age)

**Long, fairly slender, slightly spindle-shaped body**

**Extensive single and closely paired linear scarring** (tooth rake marks made mainly by conspecifics) especially on upperside

**Upperside of older males may appear whitish** (due to heavy scarring)

**ADULT MALE**

**Small triangular to slightly falcate dorsal fin** (25–32cm high) two-thirds of the way along back

**Long, well-defined tube-like beak** (50–60cm)

**Relatively small head**

**Predominantly slate-gray to black**

**Tip can be very rounded** (especially on older animals)

**Long, sinuous mouthline**

**Small flippers set far forward**

**More extensive scarring in older animals**

**Lighter underside** (may be irregular white area along ventral midline)

**Some animals have brownish or greenish-brown tinge** (due to diatoms)

**Deep tailstock**

**May be extensive light oval scarring from cookiecutter shark bites** (possibly also from Pacific lampreys)

**Two pairs of teeth near tip of lower jaw** (only larger pair at front visible outside closed mouth)

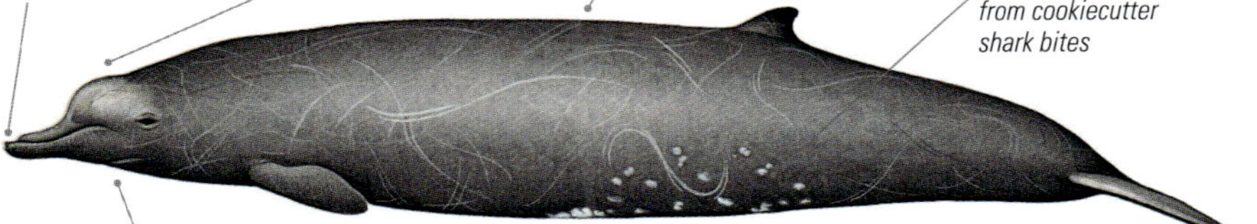

**One of the least sexually dimorphic beaked whales** (both sexes have erupted teeth and look remarkably similar)

**Melon slightly less bulbous**

**Far less linear scarring**

**May be more extensive scarring from cookiecutter shark bites**

**ADULT FEMALE**

**Pair of V-shaped throat grooves** (additional smaller pairs may be present)

**Lower jaw extends c. 10cm further forward than upper jaw** (exposing front teeth)

**OLD ADULT MALE**

**Teeth may appear white and 'flash' brightly in sunlight**

**Larger pair of teeth may be worn down** (sometimes invisible) and heavily infested with stalked (goose) barnacles in older individuals

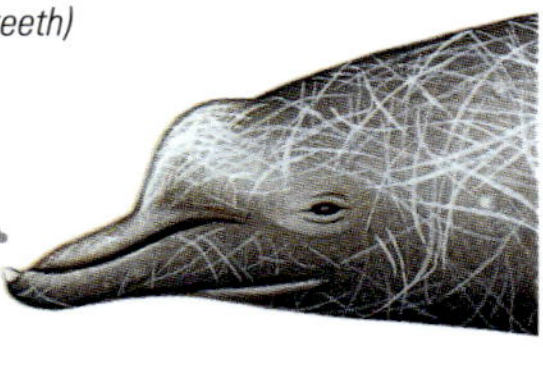

**SIZE**
**L:** ♂ 9.1–10.7m, ♀ 9.8–11.1m; **WT:** 8–12t; **MAX:** 13m, 12.8t
**Calf – L:** 4.5–4.9m; **WT:** unknown

**AT A GLANCE** Cool offshore waters of northern North Pacific • Predominantly dark with heavy scarring • Medium to large size • Bulbous melon • Long, slender beak • Two visible teeth at tip of lower jaw • Small, rounded fin two-thirds of the way along back • Tightly packed groups surface in unison

# SATO'S BEAKED WHALE
*Berardius minimus*  Yamada, Kitamuru and Matsuishi, 2019

Hokkaido whalers in Japan have traditionally recognized two different kinds of Baird's beaked whale: the relatively common 'slate-gray' form (p. 92) and a rarer, smaller 'black' form. The existence of the rarer form as a new species has been debated for decades. It was officially described as a new species, Sato's beaked whale, in 2019.

**IUCN status**  Near Threatened (2020).
**Population**  Unknown, but probably rare given the small number of records. Decreasing.
**Classification**  Odontoceti, family Ziphiidae.
**Taxonomy**  No recognized forms or subspecies.
**Other names**  Whalers from the Japanese island of Hokkaido call it kuro-tsuchi (black Baird's beaked whale); another name they use is karasu (crow or raven – after its dark color), though it is unclear whether this refers to the same species, yet another new species, or one of the *Mesoplodon* species found in Hokkaido.
**DISTRIBUTION**  Cool temperate and sub-Arctic waters in the North Pacific, between 40°N and 60°N, and 140°E and 160°W. Confirmed records reveal a disjunct distribution with two main centers (as well as a small number of records in the Sea of Okhotsk, Russia): Nemuro Strait north of Hokkaido, in Japan; and around the Aleutian Islands, in Alaska. However, the distribution may be continuous. The only known location to have 'regular' sightings is the Nemuro Strait (late April to June). Bite scars from cookiecutter sharks (*Isistius brasiliensis*) – with a northern limit of c. 40°N – suggest that at least some individuals spend at least some time at lower latitudes. Little is known about habitat preferences, but limited evidence suggests a preferred depth of 400–900m.
**BEHAVIOR**  Little known. Quite skittish and difficult to observe. Appears to spend c. 2 minutes on the surface, followed by a prolonged dive of several tens of minutes. Inconspicuous with a low profile when lying flat on the surface, but typically arches its back before diving. Lack of a visible blow may help to differentiate it from Baird's beaked whale (though Sato's sometimes shows a faint blow when seen against the sun). When searching from a boat, with the engine turned off, groups can be detected by the sounds of exhalation (it's probably best to search by moving the boat 2–3km, then drifting and listening for 10–20 mins).
**A NEW SPECIES**  The existence of a new species of beaked whale – closely related to Baird's beaked whale – had been speculated for more than 75 years. There was anecdotal evidence from whalers operating from the island of Hokkaido, in northern Japan, but until recently it was unknown to scientists. The species was officially described in 2019, after researchers analyzed the DNA and physical features of several dead specimens. (There is a greater genetic difference between Sato's beaked whale and Baird's beaked whale than between Baird's and Arnoux's beaked whales.) There were several possible sightings of live Sato's beaked whales at sea, but the first indisputable live sighting was not until 2022 (in the Nemuro Strait – confirmed by a skin sample taken from one of at least 14 individuals observed). There have been several other scientific observations since and Japanese whale-watching tour operators in the Nemuro Strait are now reporting sightings several times a year.
**GROUP SIZE AND STRUCTURE**  Little information, but possibly 2–5. There is limited evidence of sexual segregation, with groups of males or females with calves and juveniles.

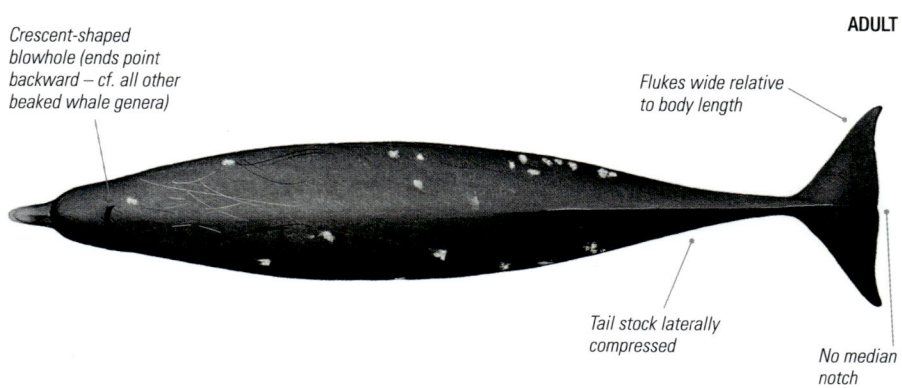

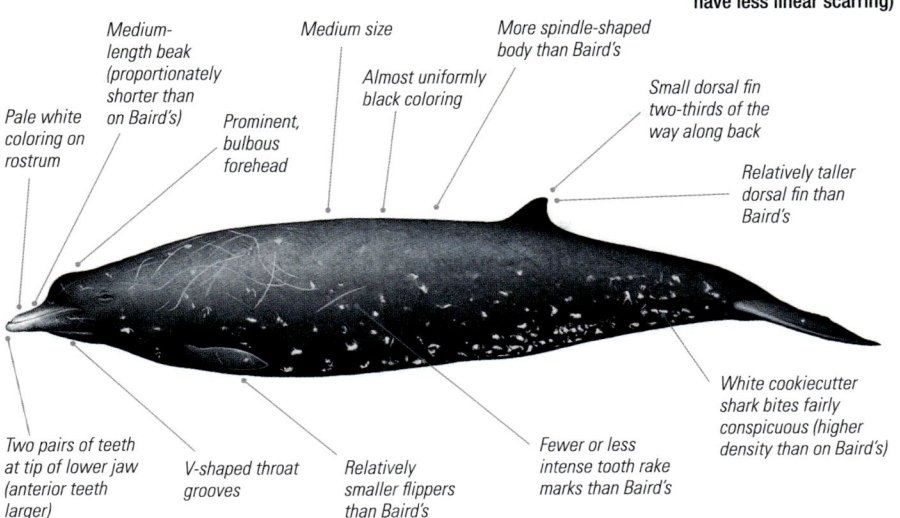

**ADULT MALE** (females similar but may have less linear scarring)

- Medium-length beak (proportionately shorter than on Baird's)
- Medium size
- More spindle-shaped body than Baird's
- Pale white coloring on rostrum
- Prominent, bulbous forehead
- Almost uniformly black coloring
- Small dorsal fin two-thirds of the way along back
- Relatively taller dorsal fin than Baird's
- White cookiecutter shark bites fairly conspicuous (higher density than on Baird's)
- Two pairs of teeth at tip of lower jaw (anterior teeth larger)
- V-shaped throat grooves
- Relatively smaller flippers than Baird's
- Fewer or less intense tooth rake marks than Baird's

**DORSAL FIN VARIATIONS**

**SIZE**
**L:** 6.2–6.9m; **MAX:** 6.9m;
**WT:** unknown, possibly c. 2–3t;
**Calf – L:** unknown; **WT:** unknown

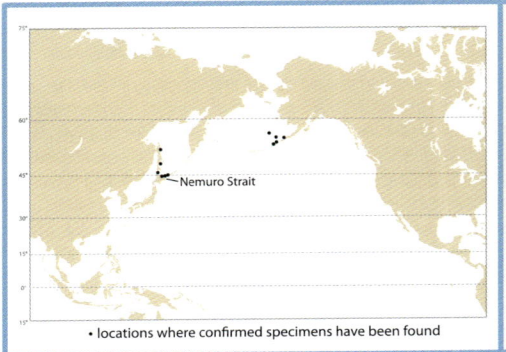

• locations where confirmed specimens have been found

**AT A GLANCE** Cool temperate and sub-Arctic waters of North Pacific • Medium size (60–70 per cent of length of Baird's beaked whale) • Predominantly black body color • Pale white coloring on rostrum • Less scarring than Baird's beaked whale • Dorsal fin relatively taller than Baird's beaked whale and two-thirds of the way along back

SATO'S BEAKED WHALE

# CUVIER'S BEAKED WHALE
*Ziphius cavirostris*  G. Cuvier, 1823

Cuvier's beaked whale is one of the most frequently seen, easily recognizable and widely distributed species of beaked whale, though it is still poorly understood. It holds the records for the deepest and longest dives of any mammal.

**IUCN status** Least Concern (2018). Mediterranean sub-population Vulnerable (2018).
**Population** Likely to be well over 100,000 (including 5,800 in the Mediterranean). Global trend unknown (Mediterranean sub-population decreasing).
**Classification** Odontoceti, family Ziphiidae.
**Taxonomy** No recognized forms or subspecies; genetically distinct population in the Mediterranean Sea.
**Other names** Goose-beaked whale, goosebeak whale.

**DISTRIBUTION** Widely distributed in cool polar to warm tropical waters worldwide, though generally absent from very high latitudes (uncommon south of the Antarctic Convergence around 60°S). Present in many enclosed seas, including the Gulf of Mexico, Caribbean Sea, Sea of Okhotsk, Gulf of California and the Mediterranean Sea (where it is the only species of beaked whale commonly occurring). Seems to prefer deeper waters over and near the continental slope, or oceanic waters, with a complex seabed topography; it is most common in canyons along shelf margins or around oceanic islands or seamounts. Usually occurs in water deeper than 1,000m. Evidence of seasonal movements in some parts of the range, but other populations appear to be resident.

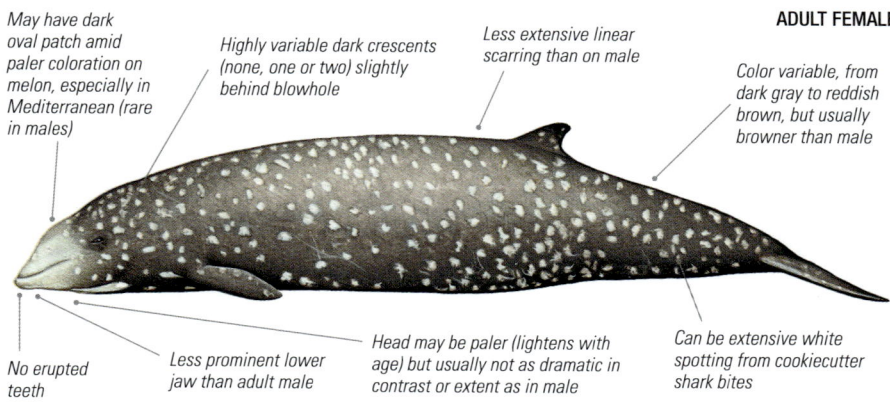

**ADULT FEMALE**

- May have dark oval patch amid paler coloration on melon, especially in Mediterranean (rare in males)
- Highly variable dark crescents (none, one or two) slightly behind blowhole
- Less extensive linear scarring than on male
- Color variable, from dark gray to reddish brown, but usually browner than male
- No erupted teeth
- Less prominent lower jaw than adult male
- Head may be paler (lightens with age) but usually not as dramatic in contrast or extent as in male
- Can be extensive white spotting from cookiecutter shark bites

**DIVE SEQUENCE** Often exposes beak on surfacing; entire head and part of body may be exposed when swimming fast or just before long dive (when often lunges out of water); occasionally flukes before a deep dive in some regions (not north-west Atlantic). • **DEPTH** Holds mammalian record for dive depth (2,992m off southern California, USA); probably physiologically capable of diving much deeper (possibly to 5,000m); believed to dive to more than 1,000m for 60+ minutes both day and night, non-stop, year-round (indeed, estimated to spend two-thirds of its life at depths greater than 1,000m). • **DURATION** Holds mammalian record for longest dive duration (3 hours 42 minutes off Cape Hatteras, North Carolina, USA); same study analyzed 3,680 dives by 26 individuals over 5 years: 5 per cent of dives exceeded 77.7 minutes; foraging dives average 58 minutes (followed by average 1.9 minutes on the surface), often interspersed with shorter 'bounce' dives; non-feeding dives average *c.* 12 minutes.

**BLOW** Bushy blow usually about 1m high, projecting slightly forward and to left (but usually inconspicuous or invisible).

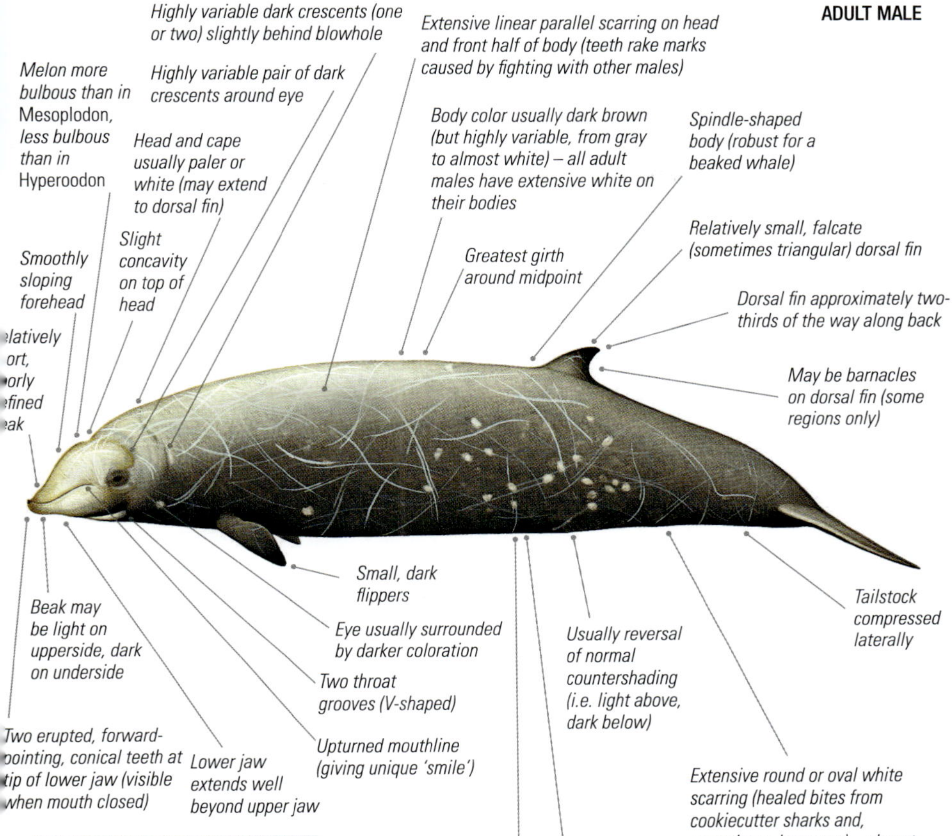

**ADULT MALE**

- Highly variable dark crescents (one or two) slightly behind blowhole
- Extensive linear parallel scarring on head and front half of body (teeth rake marks caused by fighting with other males)
- Melon more bulbous than in Mesoplodon, less bulbous than in Hyperoodon
- Highly variable pair of dark crescents around eye
- Head and cape usually paler or white (may extend to dorsal fin)
- Body color usually dark brown (but highly variable, from gray to almost white) – all adult males have extensive white on their bodies
- Spindle-shaped body (robust for a beaked whale)
- Smoothly sloping forehead
- Slight concavity on top of head
- Greatest girth around midpoint
- Relatively small, falcate (sometimes triangular) dorsal fin
- Relatively short, poorly defined beak
- Dorsal fin approximately two-thirds of the way along back
- May be barnacles on dorsal fin (some regions only)
- Beak may be light on upperside, dark on underside
- Small, dark flippers
- Eye usually surrounded by darker coloration
- Two throat grooves (V-shaped)
- Usually reversal of normal countershading (i.e. light above, dark below)
- Tailstock compressed laterally
- Two erupted, forward-pointing, conical teeth at tip of lower jaw (visible when mouth closed)
- Lower jaw extends well beyond upper jaw
- Upturned mouthline (giving unique 'smile')
- Extensive round or oval white scarring (healed bites from cookiecutter sharks and, sometimes, lampreys) – absent in some parts of range
- Orange-brown or greenish-brown patches caused by films of diatoms or algae (some individuals completely or partially covered)
- Mature males have seven times more scarring than females (1–6 new scars appear per year)

**SIZE**
L: ♂ 5.3–6m, ♀ 5.5–6m;
WT: 2.2–2.9t; **MAX:** 8.4m, 3t
Calf – L: 2.3–2.8m; **WT:** 250–300kg

**AT A GLANCE** Worldwide except High Arctic and Antarctic • Color highly variable, from slate-gray to brown to white • May be covered in round or oval white scars from cookiecutter shark bites (some regions only – e.g. rare in north-west Atlantic) • Numerous linear scars (especially on male) • Medium size • Smoothly sloping forehead with relatively short beak • Upturned mouthline (giving unique 'smile') • Two conical, forward-pointing teeth at tip of lower jaw of male (visible when mouth closed) • Dorsal fin approximately two-thirds of the way along back

CUVIER'S BEAKED WHALE

**BEHAVIOR** Occasionally breaches. Lobtailing often observed in some regions, especially when there is more than one male in a group. Tends to be very sensitive to low levels of human-made noise, typically escaping quickly, and generally avoids boats; but it can be inquisitive in some areas. Significant scarring suggests males fight for access to females.

**FOOD AND FEEDING** Primarily deepwater squid; may consume deepwater fish and crustaceans in some parts of range. Most feeding occurs at or near seabed (though also sometimes in water column). Likely to be a suction feeder.

**TEETH** Upper jaw 0; lower jaw 2. Teeth erupt in male only; 8cm tall (including portion buried in jawbone).

**GROUP SIZE AND STRUCTURE** Typically in small fluid groups of 1–4 animals; groups of up to 25 reported, but more than 10 is rare. Larger groups typically contain at least two adult males and two or three adult females and juveniles. Lone individuals are usually older males. Rarely interacts with other cetacean species.

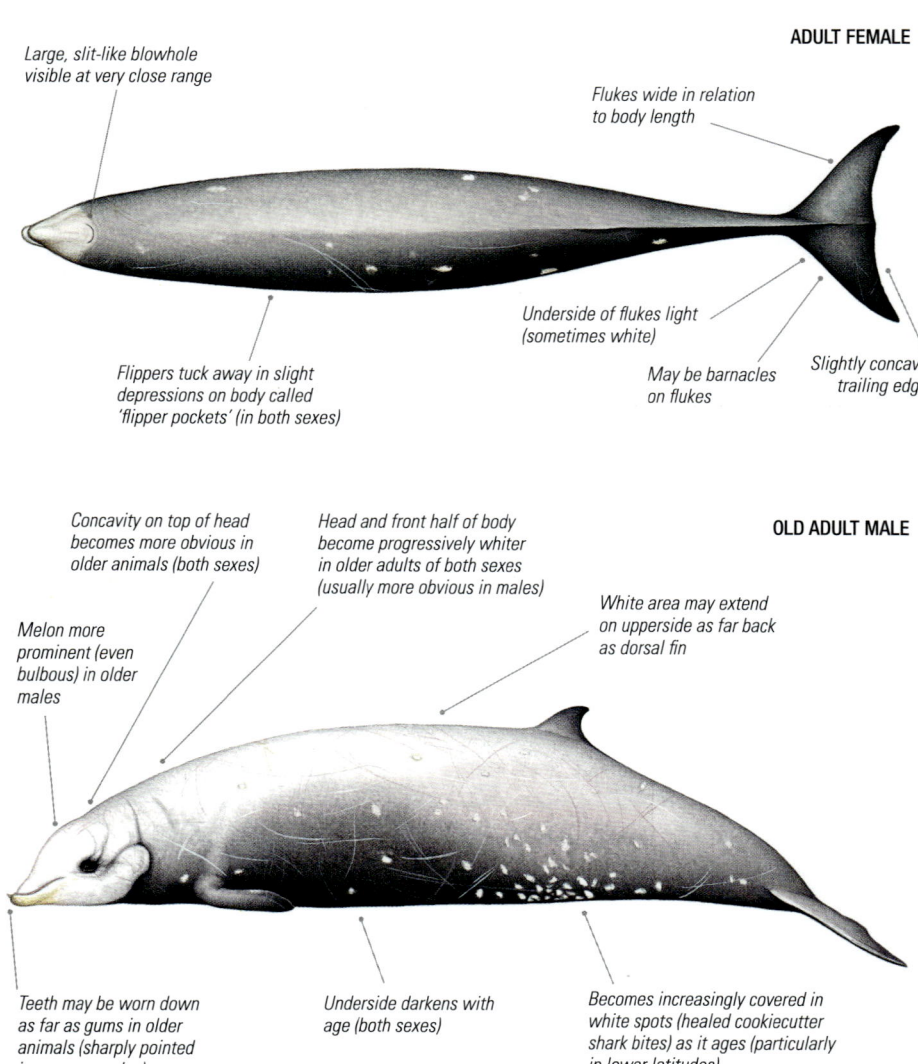

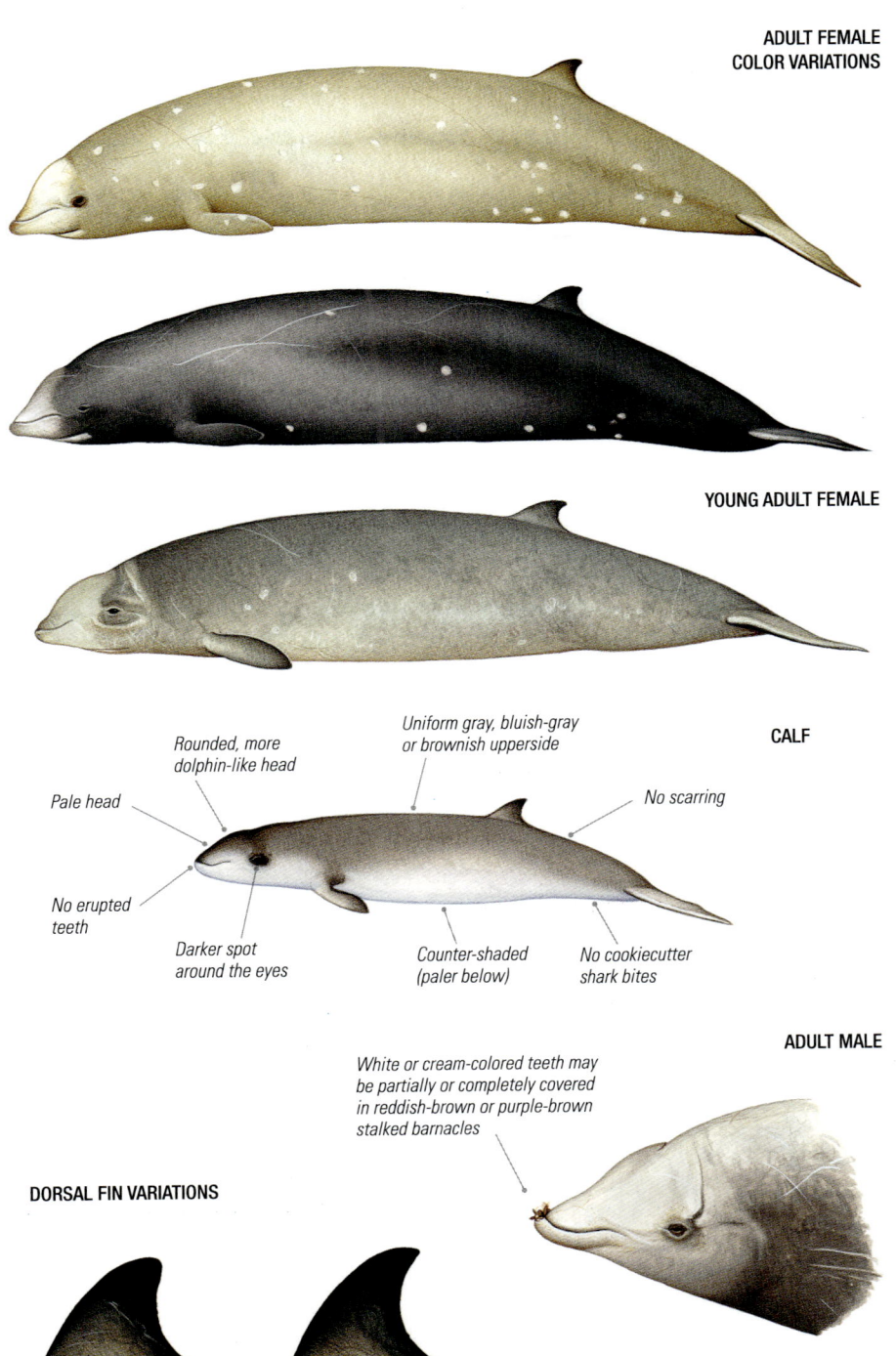

# NORTHERN BOTTLENOSE WHALE
*Hyperoodon ampullatus* (Forster, 1770)

The northern bottlenose whale is the largest beaked whale in the North Atlantic, one of the best-known members of the family and one of the few to have been targeted by whalers on a large scale. The heads of males and females look so different that early anatomists believed them to be two separate species. References to 'bottlenose whales' in the northern North Pacific refer to Baird's beaked whales.

**IUCN status** Near Threatened (2020).
**Population** Probably in the order of 20,000+. Possibly up to 100,000 before intense whaling began in the 1880s, but this was reduced to a few tens of thousands by the time whaling ceased in the 1970s. Trend unknown.
**Classification** Odontoceti, family Ziphiidae.
**Taxonomy** No recognized forms or subspecies.
**Other names** Northern bottle-nosed whale, bottlehead, flathead, steephead, common bottlenose whale.
**DISTRIBUTION** Cold temperate to Arctic waters in the North Atlantic. Ranges from the ice edge to at least *c.* 37°N. However, rarely seen south of *c.* 55°N – with one notable exception: a small but well-studied population in a submarine canyon called The Gully, 200km south-east of the Atlantic coast of Halifax, Nova Scotia, Canada. Usually frequents waters with steep and deep topography – deeper than 500m (with a preference for 800–1,800m) over the continental slope. Rarely strays over the continental shelf, except in submarine canyons, and prefers areas with complex seabed topography, such as the shelf edge, oceanic islands and seamounts. Generally occurs in open water.
**BEHAVIOR** Males may use their large, bulbous heads to headbutt one other. When resting at the surface, both sexes and all ages may hang in the water at a 45° angle, with the entire melon and beak above the surface. Breaching and lobtailing are not uncommon. Can be very curious towards boats and even large ships, and will often approach closely.
**FOOD AND FEEDING** Mainly deepwater squid; sometimes fish; rarely prawns, sea cucumbers and starfish. Most feeding appears to be on or near seabed in deep water; probably suction-feeder.
**TEETH** Upper jaw 0; lower jaw 2. Teeth erupt in male only (up to 5cm tall).
**GROUP SIZE AND STRUCTURE** Typically 1–10, rarely more than 20. Group size varies according to region (in The Gully average 3, rarely more than 6). There is some segregation by age and sex. Females appear to form loose, fluid associations, but pairs of males form long-term relationships that can last anything from days to 1–2 years.

**ADULT FEMALE**
- Melon bulbous but relatively smaller and less square than adult male
- Much of melon and face slightly paler
- No linear scarring
- No erupted teeth
- Beak may be thicker than on male

**DIVE SEQUENCE** Rolls forward, sometimes with head, back and dorsal fin visible simultaneously; rarely flukes.
• **DEPTH** Routinely dives to 800+m (average in one study 1,065m); maximum recorded 2,339m. • **DURATION** Routinely 30–40 minutes; maximum recorded 94 minutes; possibly capable of 2 hours.
**BLOW** Low, puffy blow (1–2m), often clearly visible and canted forward.

**ADULT MALE**

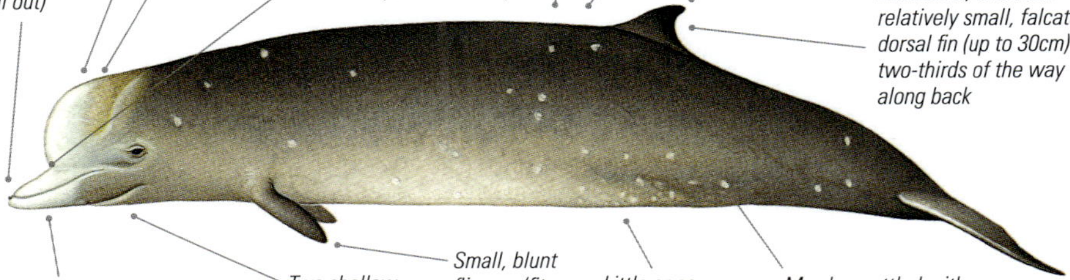

- Two conical teeth at tip of lower jaw (rarely visible outside closed mouth and often wear down or fall out)
- Huge, rounded melon (often overhanging rostrum) with very steep (almost square-shaped) forehead as it matures
- Much of melon and face (back to eyes) white or cream-colored
- Distinct junction between beak and melon (but no crease)
- Robust body
- Uniform medium to dark gray, chocolate- to olive-brown or tan coloration (brownish or yellowish tinge may be enhanced by film or patches of diatoms)
- Tip usually pointed
- Prominent, erect but relatively small, falcate dorsal fin (up to 30cm) two-thirds of the way along back
- Medium-length, thick, well-defined beak (highly variable appearance – beak looks smaller relative to melon growing larger)
- Two shallow, V-shaped throat grooves
- Small, blunt flippers (fit into 'flipper pockets' on sides)
- Little or no linear scarring
- May be mottled with round or oval white scarring (healed bites from cookiecutter sharks)

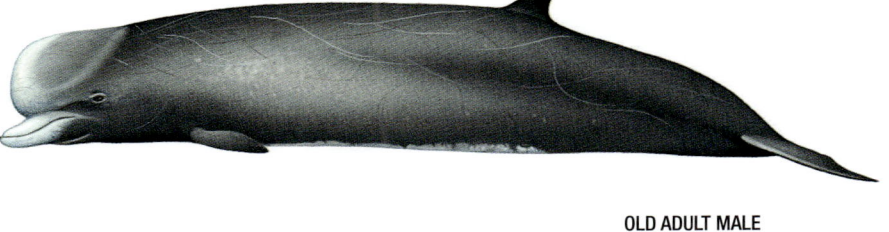

**ADULT MALE GRAY FORM**

**OLD ADULT MALE**

- Huge melon with very steep, flattened forehead (giving distinctly squarish profile)
- Much of melon and face very pale (white or cream-colored)

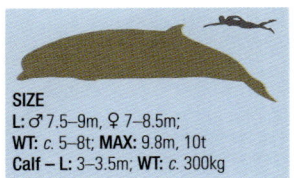

**SIZE**
L: ♂ 7.5–9m, ♀ 7–8.5m;
**WT:** c. 5–8t; **MAX:** 9.8m, 10t
Calf – L: 3–3.5m; **WT:** c. 300kg

**AT A GLANCE** Cold, deep waters of North Atlantic • Medium size (larger than other beaked whales in region) • Gray, tan or brownish coloration • Huge, squared-off, bulbous white or cream-colored melon (especially in male) • Medium-length, thick, well-defined beak • Prominent falcate dorsal fin two-thirds of the way along back • Little or no linear scarring • Male's teeth not clearly visible • Often inquisitive and may approach stationary vessels

NORTHERN BOTTLENOSE WHALE

# LONGMAN'S BEAKED WHALE
*Indopacetus pacificus* (Longman, 1926)

Longman's beaked whale was one of cetology's great long-standing mysteries. Until 2003, the only evidence for its existence came from two weathered skulls: one found on an Australian beach in 1882, and the other on the floor of a Somalian fertiliser factory in 1955. However, live animals are now being seen with some regularity at scattered locations in the tropical Indo-Pacific and there have been about 20 strandings.

**IUCN status** Least Concern (2020).
**Population** Unknown. Despite the paucity of sightings, its wide distribution suggests that it is not rare. Trend unknown.
**Classification** Odontoceti, family Ziphiidae.
**Taxonomy** Originally placed in the genus *Mesoplodon*, but morphological and genetic studies prove that it should be in its own genus; no recognized forms or subspecies.
**Other names** Indo-Pacific beaked whale, tropical bottlenose whale.

**DISTRIBUTION** Distribution is poorly known, but it appears to be widespread and fairly continuous in the subtropical and tropical Indo-Pacific. Thought to be more abundant in the western part of its distribution. Sightings tend to be in areas with surface water temperatures of 21–31°C (mostly in water warmer than 26°C). May push further south or north with warm currents. Most sightings are oceanic, over or near areas with steep bottom topography, in depths of more than 1,000m. Most sightings in US waters have been offshore of the Hawaiian Islands.

**BEHAVIOR** Larger groups tend to be more active at the surface and may ignore or approach boats.

**FOOD AND FEEDING** Presumed to feed mainly on deepwater squid, with some fish. Feeding techniques unknown.

**TEETH** Upper jaw 0; lower jaw 2. Teeth erupt in male only.

**GROUP SIZE AND STRUCTURE** Tight groups tend to be large (up to 110 recorded, with typical size 10–20 and overall average of 18.5); may vary regionally.

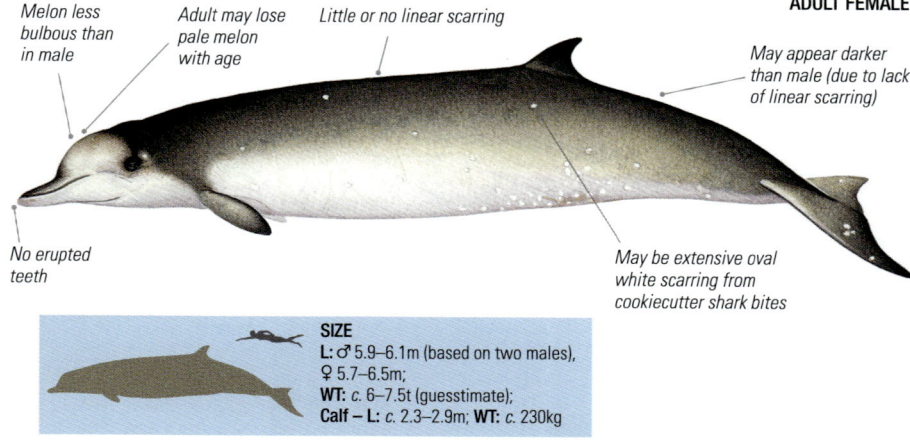

**ADULT FEMALE**
- Melon less bulbous than in male
- Adult may lose pale melon with age
- Little or no linear scarring
- May appear darker than male (due to lack of linear scarring)
- No erupted teeth
- May be extensive oval white scarring from cookiecutter shark bites

**SIZE**
**L:** ♂ 5.9–6.1m (based on two males); ♀ 5.7–6.5m;
**WT:** *c.* 6–7.5t (guesstimate);
**Calf – L:** *c.* 2.3–2.9m; **WT:** *c.* 230kg

**DIVE SEQUENCE** Swims faster and more 'aggressively' than most other beaked whales; when surfacing quickly, head and beak appear quite high out of water as it throws up rooster-tail of spray; little arching of back (cf. Cuvier's beaked whale) before long dive. • **DEPTH** Unknown, but probably a deep diver. • **DURATION** 11–33 minutes; one individual tracked underwater acoustically for 45 minutes (contact lost before it surfaced).

**BLOW** Low, bushy blow fairly conspicuous and angled slightly forwards.

**ADULT MALE**

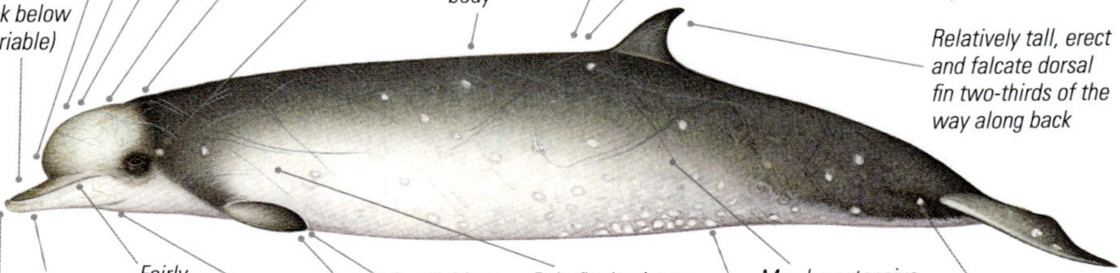

- Melon drops steeply to beak (average c. 75°)
- Pale tan or lighter-colored head (extends as far back as blowhole)
- Crease between melon and beak
- Dark area behind melon forms part of dark patch around eye
- Bluish-gray or brownish-gray to olive-brown or bronze-brown upperside and tailstock (highly variable between individuals and according to light conditions)
- Bulging melon
- May have paler 'ear patch' in dark area behind eye
- Appears grayer in overcast conditions, browner in bright conditions
- Beak usually dark above, and white, light gray, pale tan or pink below (variable)
- Spindle-shaped body
- Dorsal fin may have darker leading edge and pale center
- Relatively tall, erect and falcate dorsal fin two-thirds of the way along back
- Moderately long, well-defined dolphin-like beak
- Fairly straight mouthline
- Two shallow, V-shaped throat grooves
- Small, blunt flippers
- Pale flanks rise to shoulder (can be conspicuous at sea)
- Flippers mostly dark on upperside, light on underside
- May be extensive linear scarring
- Much lighter sides and underside (shading to white around genital region)
- May be extensive oval-white scarring from cookiecutter shark bites (often looks spotty)
- Single pair of pear-shaped teeth at tip of lower jaw (male only), barely visible outside closed mouth except at close range

**ADULT MALE VARIATION**

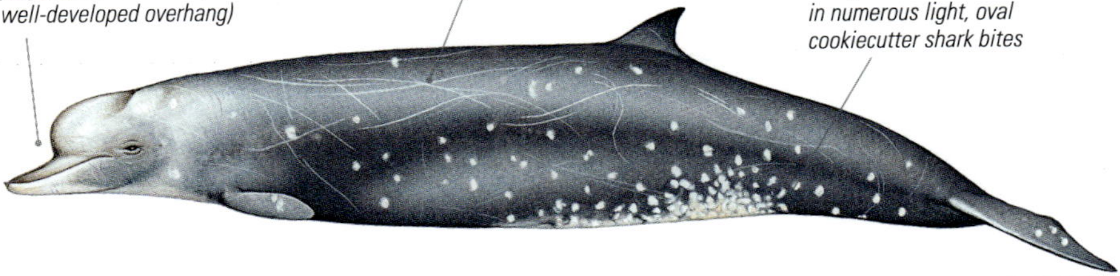

- In older males, melon may drop almost perpendicular to beak (sometimes with well-developed overhang)
- Coloration variable
- Many individuals covered in numerous light, oval cookiecutter shark bites

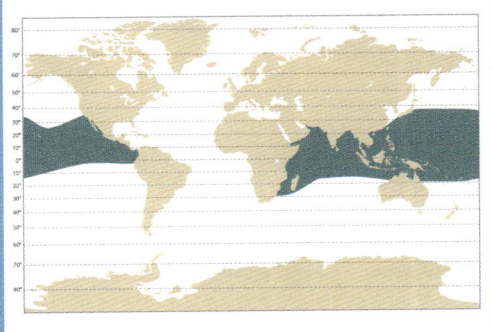

**AT A GLANCE** Warm waters of the Indo-Pacific • Medium size • Spindle-shaped body • Low bushy blow often angled forward • Conspicuous, bulging pale melon (especially in male) • Distinct, sharply demarcated beak • Falcate, dolphin-like dorsal fin • Fin two-thirds of the way along back • Apparent color varies according to weather conditions

**LONGMAN'S BEAKED WHALE**

# PERRIN'S BEAKED WHALE
*Mesoplodon perrini* — Dalebout, Mead, Baker & van Helden, 2002

Officially named in 2002, this is one of the least-known beaked whales (all information is based on just six stranded specimens, from southern California in 1975–1997). Its external appearance closely resembles Hector's beaked whale (though the two species are not closely related). Its nearest relative is the pygmy beaked whale.

**IUCN status** Endangered (2020).
**Population** Possibly fewer than 1,000 mature individuals. Decreasing.
**Classification** Odontoceti, family Ziphiidae.
**Taxonomy** No recognized forms or subspecies.
**Other names** California beaked whale.

**DISTRIBUTION** Currently known only from southern California, USA. Strandings range from Torrey Pines State Reserve (32°55′N – just north of San Diego) north to Fisherman's Wharf, Monterey (36°37′N). It might be expected in waters with depths of more than 1,000m (mainly beyond the continental shelf but possibly also close to shore where the water is sufficiently deep). Researchers recorded unique echolocation pulses 350 nautical miles off the coast of California, believed to have been made by Perrin's. There have also been probable acoustic detections off Baja California, Mexico.

**BEHAVIOR** This species has never been reliably identified at sea, so there is no information on behavior. Probably unobtrusive and difficult to spot in anything but calm conditions.

**FOOD AND FEEDING** Virtually no information, but very limited evidence from stomach contents suggests primarily mid- and deepwater squid; may also include deepwater fish and shrimps.

**TEETH** Upper jaw 0; lower jaw 2. Laterally compressed teeth erupt in males only (females have similar teeth, but they do not erupt); the exposed portion is roughly the shape of an isosceles triangle; up to 64mm long.

**GROUP SIZE AND STRUCTURE** No information.

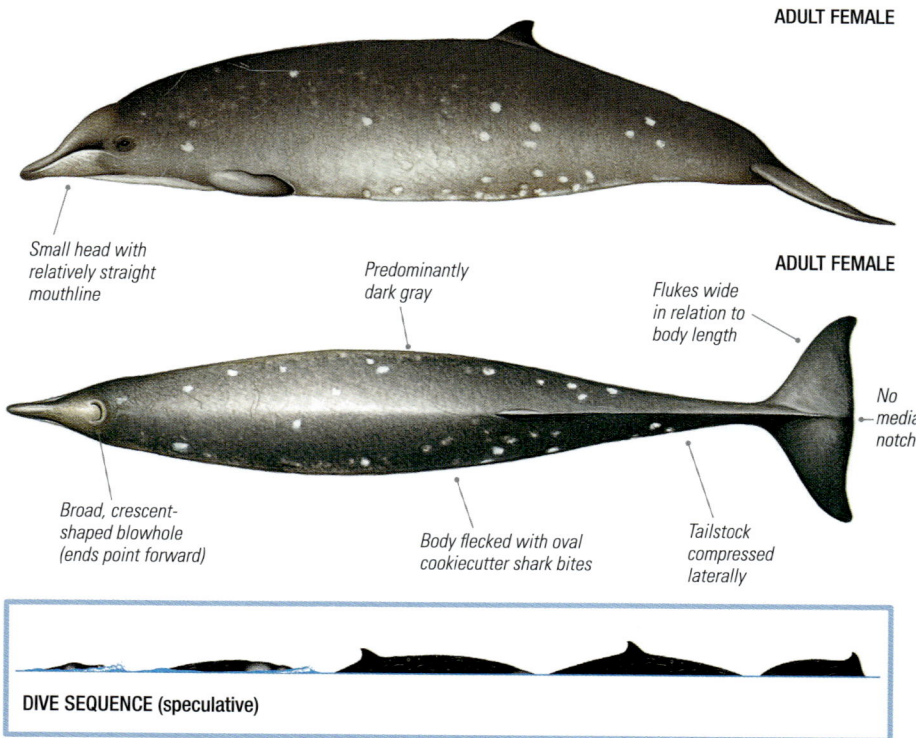

# PERRIN'S BEAKED WHALE

**ADULT MALE** (based on little information)

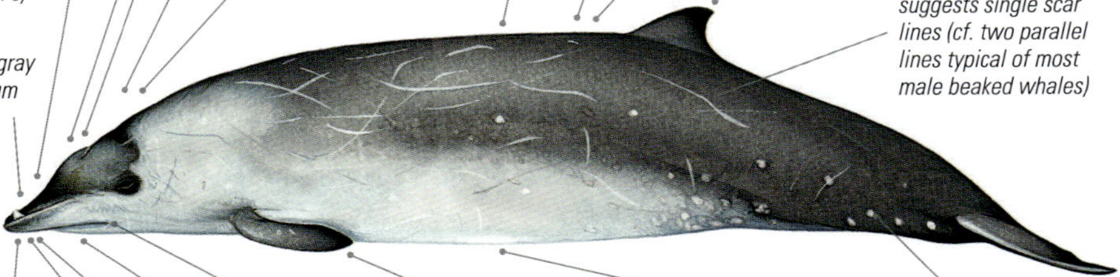

- Smoothly sloping, non-bulbous melon
- Appearance of darker face mask (dark gray color extending from near gape to eye and to just behind blowhole)
- Dark gray upperside grading to lighter gray or white underside
- Melon may be paler and/or browner
- Small head with relatively straight mouthline (does not rise towards rear)
- Spindle-shaped body
- Moderate to heavy linear scarring (single, rather than parallel, tooth rake marks)
- Relatively short beak (shorter than other *Mesoplodon* beaked whales except pygmy and Hector's)
- Small, triangular or slightly falcate dorsal fin two-thirds of the way along back
- Dark gray rostrum
- Limited evidence suggests single scar lines (cf. two parallel lines typical of most male beaked whales)
- Two relatively large, laterally compressed triangular teeth c. 1–2cm behind tip of lower jaw
- Two shallow throat grooves
- Small, narrow flippers
- White patch around umbilicus
- Body flecked with oval cookiecutter shark bites
- Light gray to white throat and lower jaw
- Teeth may be colonized by stalked barnacles
- Teeth exposed above gumline by c. 3cm (visible outside closed mouth)

**FLUKES (UNDERSIDE)**

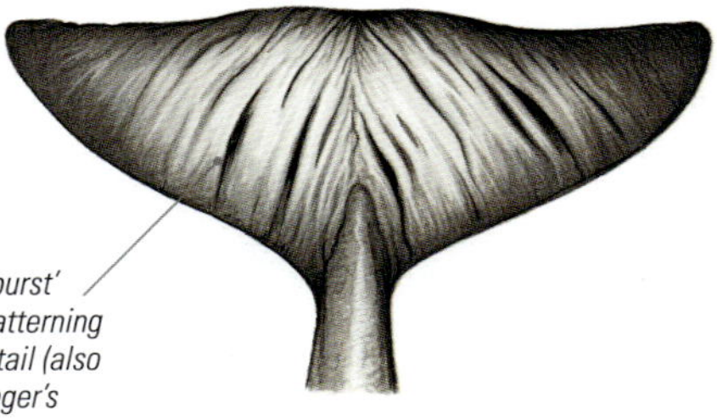

- Distinctive 'starburst' light-and-dark patterning on underside of tail (also occurs on Stejneger's beaked whale)

**SIZE**
L: ♂ 3.9m, ♀ 4.3–4.4m;
**WT:** c. 900kg; **MAX:** 4.53m
Calf – L: c. 2–2.1m; **WT:** unknown
(based on very few specimens)

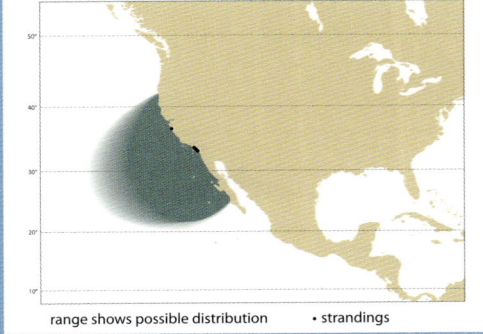

range shows possible distribution • strandings

**AT A GLANCE** Eastern North Pacific • Small to medium size • Generally nondescript counter-shaded coloration • Small triangular or slightly falcate dorsal fin two-thirds of the way along back • Appearance of darker face mask • Two large triangular teeth near tip of lower jaw may be visible at close range (male only)

# PYGMY BEAKED WHALE
*Mesoplodon peruvianus*　　　　　　　　　　　　　　　　Reyes, Mead & Van Waerebeek, 1991

Formally described in 1991, the pygmy or Peruvian beaked whale is the smallest species of beaked whale. It is very poorly known: for many years the majority of records were curated specimens from Peru, but recently there have been more live sightings in the Gulf of California, Mexico, and elsewhere.

**IUCN status** Least Concern (2020).
**Population** Unknown, but likely tens of thousands (reasonably common within its limited range). Trend unknown.
**Classification** Odontoceti, family Ziphiidae.
**Taxonomy** No recognized forms or subspecies. This is the previously unnamed species formerly known as '*Mesoplodon* sp. A'.
**Other names** Peruvian beaked whale (used interchangeably with 'pygmy'), lesser beaked whale.
**DISTRIBUTION** Originally described primarily from freshly captured specimens landed in Peruvian fishing ports. However, in recent years it has become the most frequently sighted *Mesoplodon* in deep waters of the sub-tropical and tropical eastern Pacific Ocean (although this is relative); Mexico's Gulf of California is a sightings hotspot. A single male stranded near Kaikoura, South Island, New Zealand (42°31'S), in 1991, is considered to be a vagrant.
**BEHAVIOR** Generally unobtrusive and difficult to spot in anything but calm conditions. Groups typically dive for 15–30 minutes, surface again some distance away, breathe half a dozen times and then dive again. Has been known to approach small boats very closely (albeit briefly) but usually keeps its distance. Breaching, lobtailing and other surface behaviors have been recorded, but appear to be rare.
**FOOD AND FEEDING** Limited evidence suggests primarily mid- and deepwater fish, but also probably deepwater squid and shrimps. Feeding techniques unknown.
**TEETH** Upper jaw 0; lower jaw 2. Teeth erupt in male only; 31–65mm long.
**GROUP SIZE AND STRUCTURE** Typically in groups of 2–5 (sometimes 1–8). Groups usually include mixed sex and age classes.

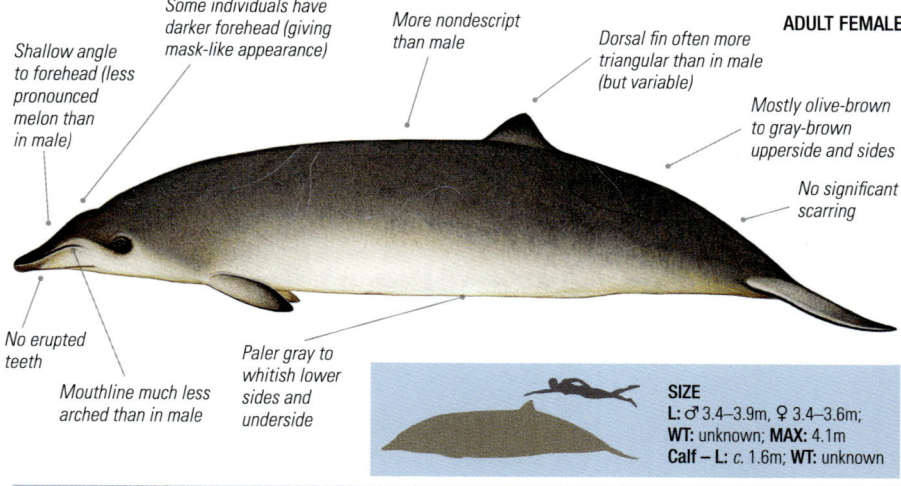

SIZE
L: ♂ 3.4–3.9m, ♀ 3.4–3.6m;
WT: unknown; MAX: 4.1m
Calf – L: *c.* 1.6m; WT: unknown

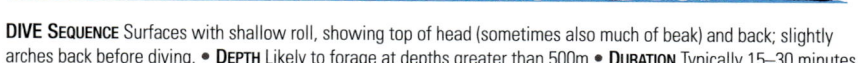

**DIVE SEQUENCE** Surfaces with shallow roll, showing top of head (sometimes also much of beak) and back; slightly arches back before diving. • **DEPTH** Likely to forage at depths greater than 500m. • **DURATION** Typically 15–30 minutes (based on limited observations).
**BLOW** Indistinct and rarely visible.

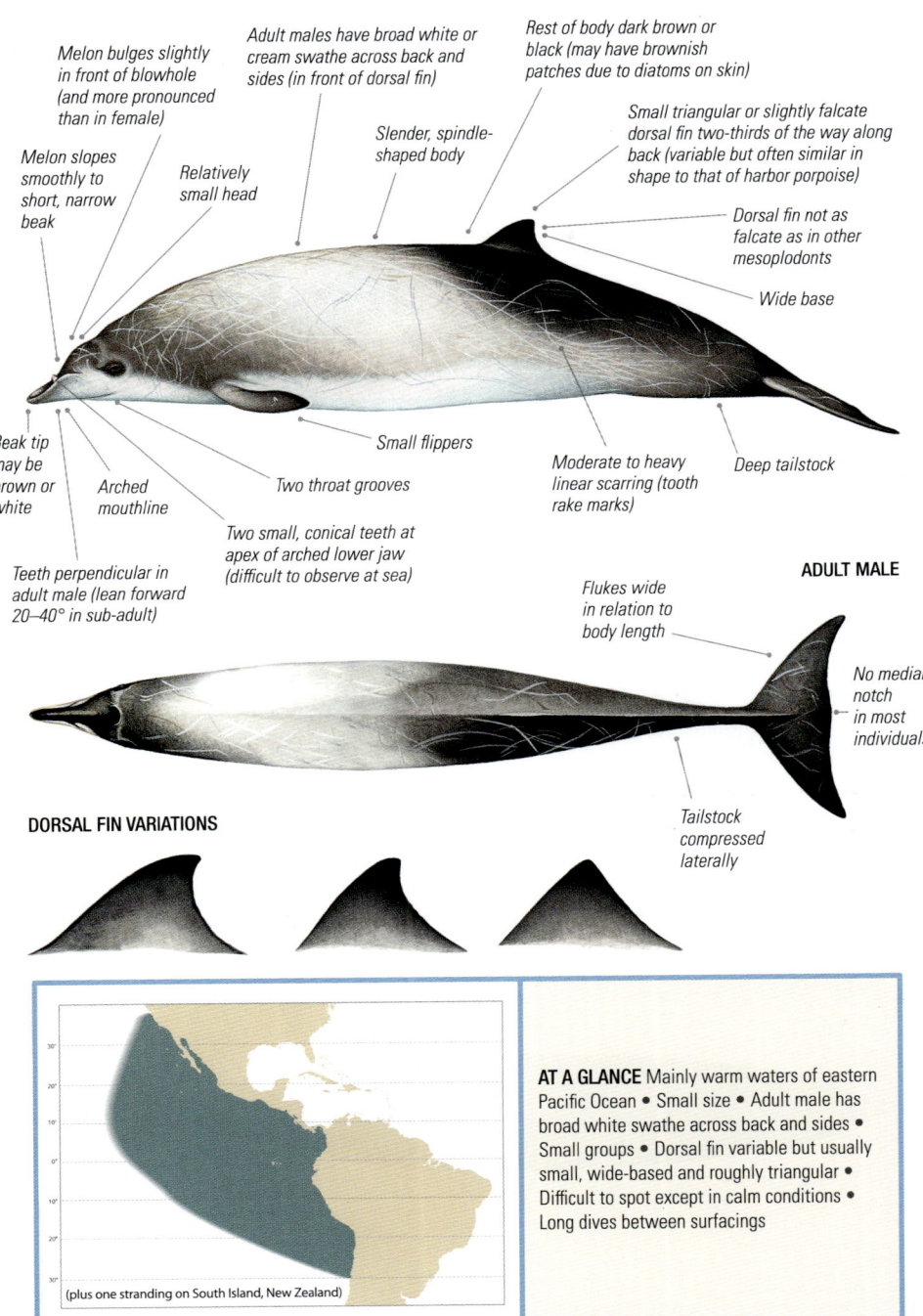

# GINKGO-TOOTHED BEAKED WHALE
*Mesoplodon ginkgodens*  Nishiwaki and Kamiya, 1958

There has been one confirmed sighting of ginkgo-toothed beaked whales at sea (off north-west Baja California, Mexico, in 2024) and several possible ones. There have also been fewer than 100 widely scattered strandings and captures of this poorly known species, spread across the Pacific and Indian Oceans.

**IUCN status**  Data Deficient (2020).
**Population**  Unknown. Probably uncommon, given the small number of records. Trend unknown.
**Classification**  Odontoceti, family Ziphiidae.
**Taxonomy**  No recognized forms or subspecies. For many years Deraniyagala's beaked whale was considered synonymous with ginkgo-toothed, but Deraniyagala's was formally accepted as a separate species in 2014. The two species may be indistinguishable at sea (though Deraniyagala's does not occur in North America).
**Other names**  Japanese beaked whale, ginkgo-toothed whale.
**DISTRIBUTION**  Exact distribution is unclear, due to the small number of records. Records are widely scattered – including Japan, the US West Coast, Mexico, the Galapagos Islands, Thailand, South Korea, China, Taiwan, Australia, New Zealand, Micronesia and the Marshall Islands. Most concentrated in the western Pacific (Indian Ocean records are more likely to represent Deraniyagala's). Mainly deep tropical to temperate waters.
**BEHAVIOR**  No information on behavior.
**FOOD AND FEEDING**  Like other beaked whales, it is presumed to eat mainly deepwater squid, and some fish.
**TEETH**  Upper jaw 0; lower jaw 2. Teeth erupt in male only (6.5cm tall).
**GROUP SIZE AND STRUCTURE**  Nothing known.

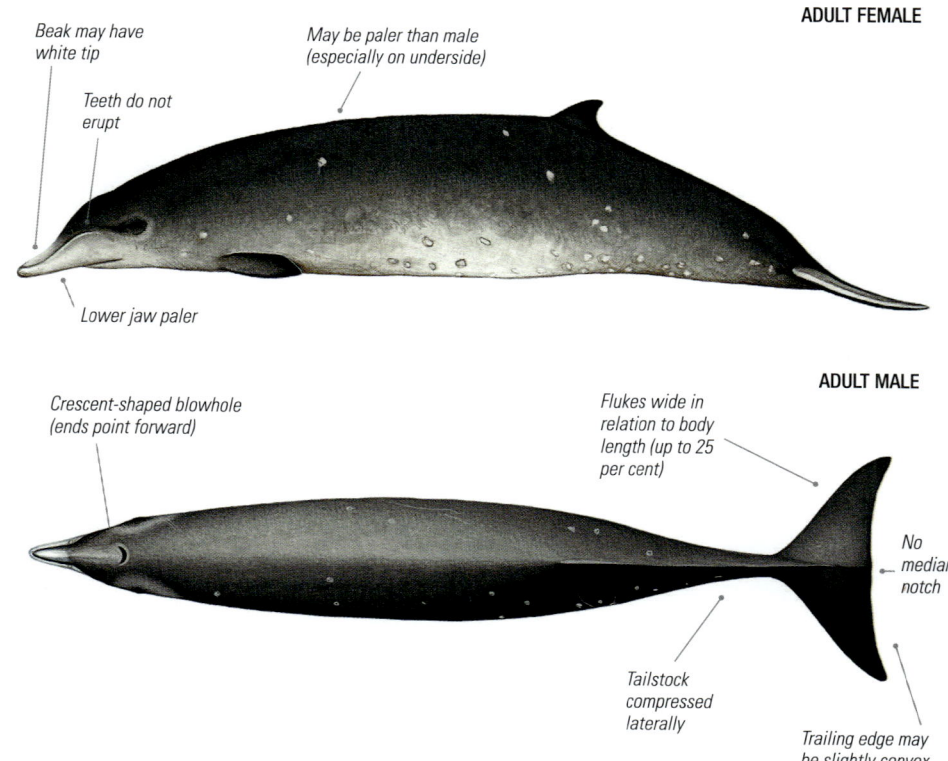

**ADULT MALE**

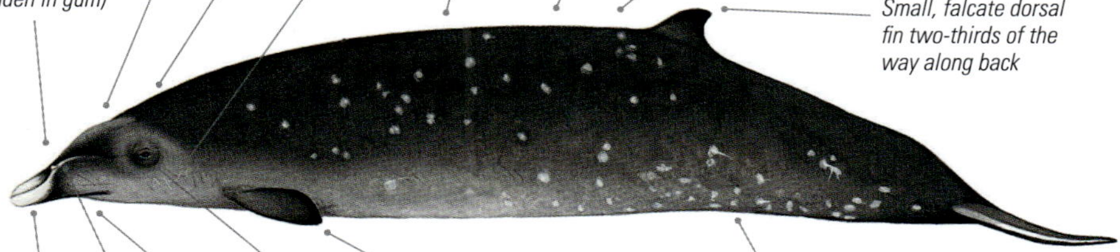

- Smoothly sloping, slightly bulbous melon merges with rostrum (no crease)
- Basic dark gray (or brownish) countershading coloration (poorly known from few fresh specimens)
- May be lighter mottled gray on cheek
- Spindle-shaped body
- Little or no pale linear scarring
- Flattened tooth near apex of arch on each side of lower jaw (mostly hidden in gum)
- Small head
- Small, falcate dorsal fin two-thirds of the way along back
- Moderately long beak with pale tip (upper and lower jaws)
- Dark eye patch
- Small, narrow flippers
- Extensive round or oval white scars from cookiecutter shark bites (especially in urogenital area)
- Prominent arch in mouthline slightly behind middle of lower jaw
- Two shallow, V-shaped throat grooves
- Tooth 10cm wide
- Teeth resemble leaves of ginkgo tree
- Teeth similar (in size and shape) to those of Deraniyagala's beaked whale and wider-toothed variants of Gray's beaked whale

**SIZE**
L: ♂ 4.7–5.3m, ♀ 4.7–5.3m;
WT: c. 1–1.5t; MAX: 5.3m, 2t
Calf – L: 2–2.5m; WT: unknown

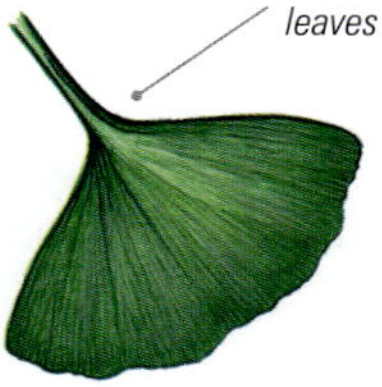

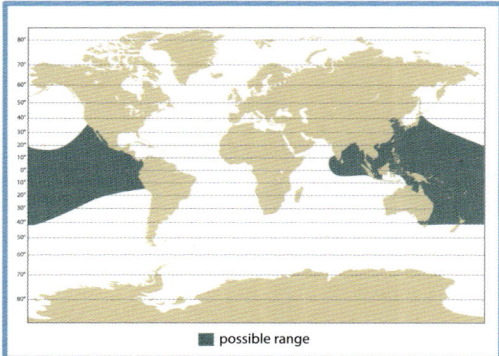

possible range

**AT A GLANCE** Tropical to temperate waters of Pacific and Indian Oceans • Medium size • Little or no pale linear scarring but extensive round or oval scars • Moderately long beak with white tip • Smoothly sloping, slightly bulbous melon • Prominent arch in mouthline slightly behind middle of lower jaw • Wide tooth near apex of arch on each side of lower jaw (male only) • Small, falcate dorsal fin two-thirds of the way along back • May be indistinguishable from Deraniyagala's beaked whale at sea

**GINKGO-TOOTHED BEAKED WHALE**

# HUBBS' BEAKED WHALE
*Mesoplodon carlhubbsi* — Moore, 1963

Known from just over 60 records in the North Pacific — mostly strandings and fishery entanglements, with just a few reliable sightings at sea — Hubbs' beaked whale is very poorly understood. It is strikingly similar to Andrews' beaked whale, which lives far away in the cold waters of the Southern Ocean.

**IUCN status** Data Deficient (2020).
**Population** Unknown. Paucity of sightings suggests that it may be rare but, like all mesoplodonts, it is inconspicuous at sea and may simply be missed. Trend unknown.
**Classification** Odontoceti, family Ziphiidae.
**Taxonomy** No recognized forms or subspecies.
**Other names** Arch-beaked whale.

**DISTRIBUTION** Distribution is known mainly from strandings. Endemic to deep offshore cold temperate waters of the North Pacific. The majority of records are from western North America, largely along the path of the south-flowing, cold-water California Current, and Japan. The lack of land masses where strandings could be recorded may account for the paucity of records in the central North Pacific, so it is possible that the distribution is continuous.

**BEHAVIOR** There have been only a few confirmed sightings, so virtually nothing is known about behavior at sea.

**FOOD AND FEEDING** Limited evidence suggests mainly deepwater squid, and some deepwater fish. Probably suction-feeder (sucks prey into mouth and swallows it whole).

**TEETH** Upper jaw 0; lower jaw 2. Teeth erupt in male only.

**GROUP SIZE AND STRUCTURE** Very little information suggests groups of 1–5.

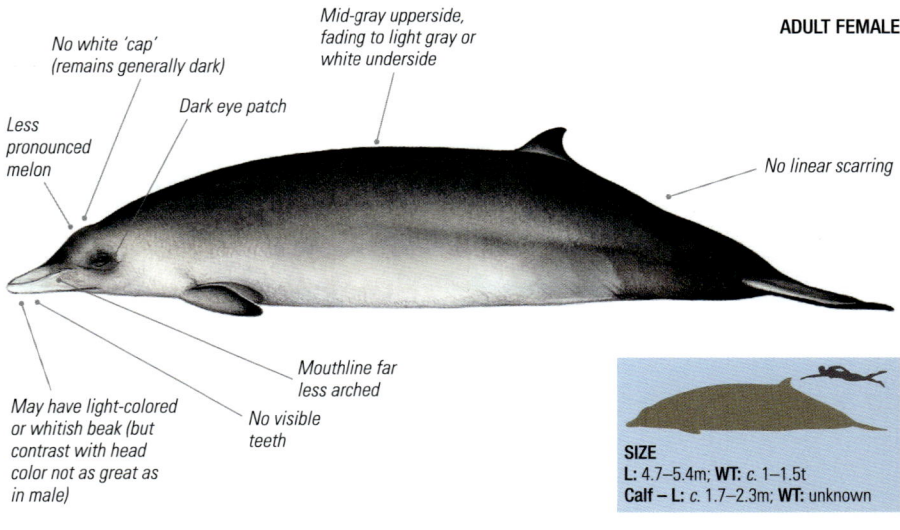

**ADULT FEMALE**
- No white 'cap' (remains generally dark)
- Mid-gray upperside, fading to light gray or white underside
- Dark eye patch
- Less pronounced melon
- No linear scarring
- Mouthline far less arched
- No visible teeth
- May have light-colored or whitish beak (but contrast with head color not as great as in male)

**SIZE**
L: 4.7–5.4m; **WT:** c. 1–1.5t
Calf – L: c. 1.7–2.3m; **WT:** unknown

**DIVE SEQUENCE (SPECULATIVE)** Beak and white cap on head visible (reports of lifting head or beak out of water on surfacing); probably does not show flukes. • **DEPTH** Possibly 500–3,000m. • **DURATION** Unknown, but likely up to one hour.

**BLOW** Blow indistinct.

**ADULT MALE**

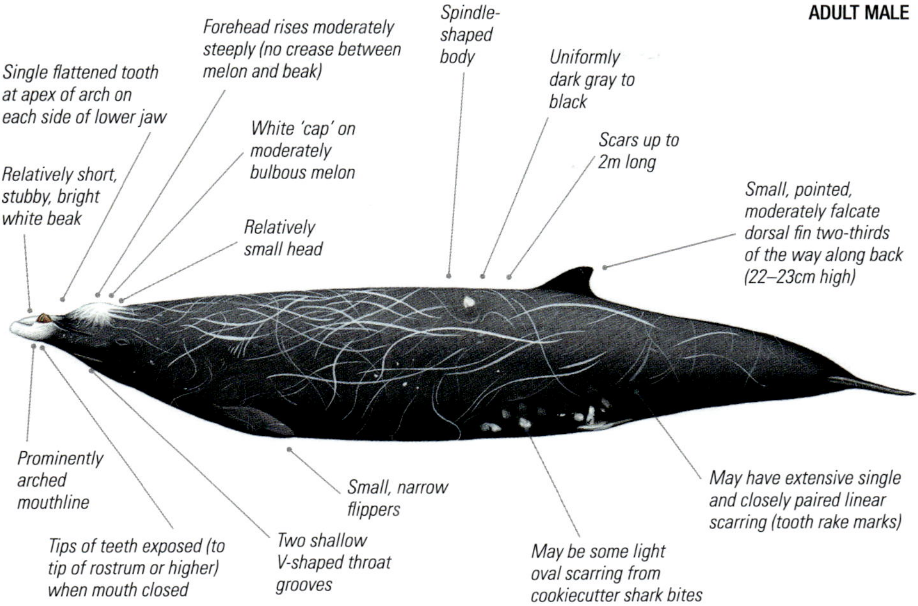

- Single flattened tooth at apex of arch on each side of lower jaw
- Forehead rises moderately steeply (no crease between melon and beak)
- Spindle-shaped body
- Uniformly dark gray to black
- White 'cap' on moderately bulbous melon
- Scars up to 2m long
- Small, pointed, moderately falcate dorsal fin two-thirds of the way along back (22–23cm high)
- Relatively short, stubby, bright white beak
- Relatively small head
- Prominently arched mouthline
- Small, narrow flippers
- May have extensive single and closely paired linear scarring (tooth rake marks)
- Tips of teeth exposed (to tip of rostrum or higher) when mouth closed
- Two shallow V-shaped throat grooves
- May be some light oval scarring from cookiecutter shark bites

**ADULT MALE**

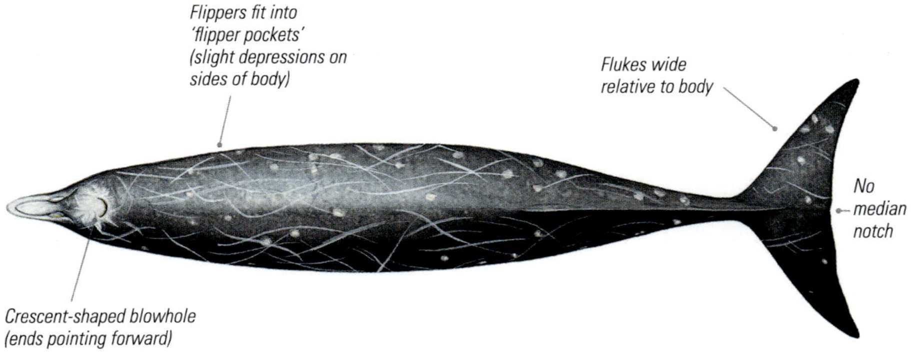

- Flippers fit into 'flipper pockets' (slight depressions on sides of body)
- Flukes wide relative to body
- No median notch
- Crescent-shaped blowhole (ends pointing forward)

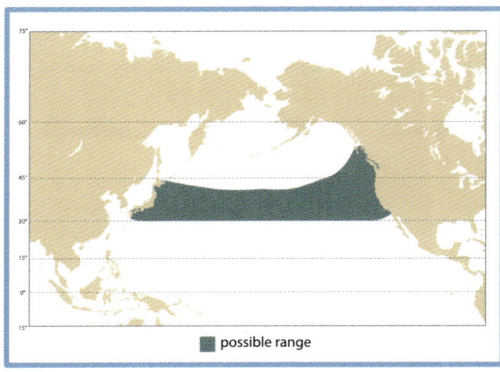

possible range

**AT A GLANCE (male)** Cool temperate waters of North Pacific • Medium size • Uniformly dark • Bright white 'cap' and beak • Heavily scarred • Small falcate dorsal fin two-thirds of the way along back • Two tusks on arched lower jaw

HUBBS' BEAKED WHALE

# BLAINVILLE'S BEAKED WHALE
## *Mesoplodon densirostris* (Desmarest, 1817)

Blainville's beaked whale is the most commonly observed *Mesoplodon* in tropical waters worldwide (though this is relative) and is the most widely distributed member of the genus. It has a strongly arched lower jaw, with teeth that protrude like a pair of horns, and its rostrum is formed of the densest bone of any animal.

**IUCN status** Least Concern (2020).
**Population** Unknown, but appears to be relatively common in most tropical seas. Trend unknown.
**Classification** Odontoceti, family Ziphiidae.
**Taxonomy** No recognized forms or subspecies.
**Other names** Dense-beaked whale, tropical beaked whale.
**DISTRIBUTION** Tropical to warm temperate waters in both hemispheres. There have been nearly 400 known strandings worldwide, and Blainville's beaked whale is seen fairly frequently in a few key hotspots, including Hawai'i, the Bahamas and the Canary Islands. It is one of the most tropical of the mesoplodonts; higher-latitude records are usually associated with warm-water currents. Occurs in many enclosed seas, including the Gulf of Mexico, the Caribbean Sea and the Sea of Japan, but is considered a vagrant in the Mediterranean. Seems to prefer waters of intermediate depth (500–1,500m in Hawai'i and the Bahamas) over continental shelf waters, deep submarine canyons and steeply sloping regions around seamounts. However, it is also known in much deeper waters (at least 5,000m) in the open ocean and has been reported in waters as shallow as 320m (the mean depth for seven sightings in the Canary Islands). In the few areas where it has been studied, it shows a high degree of site fidelity (known individuals have been seen repeatedly in the same area over one or two decades).
**BEHAVIOR** Behavior better known than for any other species of *Mesoplodon*. It rarely breaches or performs other aerial behaviors. Not known to occur in mixed aggregations with other cetaceans. Behavior around boats varies enormously.
**FOOD AND FEEDING** Mainly deepwater squid and fish, some crustaceans, with regional differences. Foraging dives occur day and night; believed to forage along seabed, at least sometimes; suction-feeder.

**DIVE SEQUENCE** Briefly lifts beak out of water at angle of *c*. 45° on surfacing (entire head may clear surface); may be slight pause as it levels and blows; tailstock appears as it arches slightly and rolls forward to dive (arches higher on terminal dives); rarely, if ever, shows flukes. • **DEPTH** Three types of dive: shallow (2–4m); long, deep, foraging (regularly more than 1,000m); and intermediate (30–300m) to avoid detection by predators. Record in Hawai'i of adult female diving to 800m accompanied all the way with calf; maximum recorded 1,599m.
• **DURATION** Typically 45–60 minutes; maximum recorded 83 minutes.
**BLOW** Inconspicuous blow usually low and canted forward.

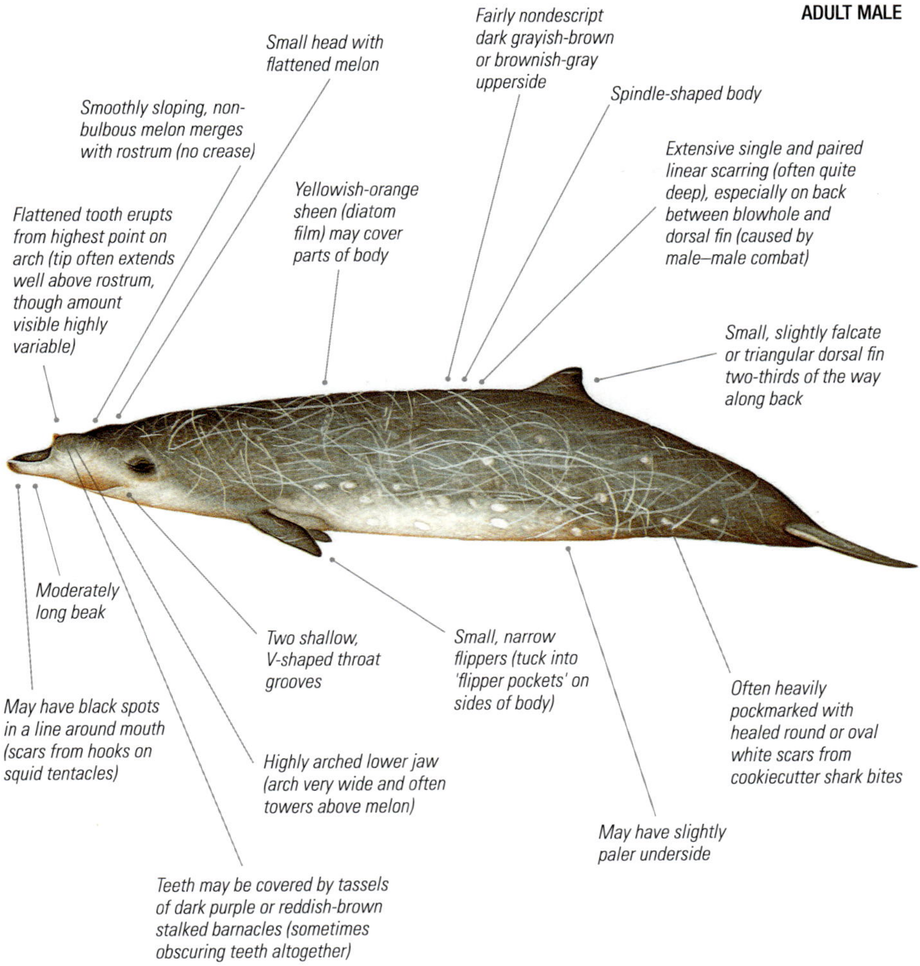

**ADULT MALE**

- Small head with flattened melon
- Fairly nondescript dark grayish-brown or brownish-gray upperside
- Spindle-shaped body
- Smoothly sloping, non-bulbous melon merges with rostrum (no crease)
- Extensive single and paired linear scarring (often quite deep), especially on back between blowhole and dorsal fin (caused by male–male combat)
- Flattened tooth erupts from highest point on arch (tip often extends well above rostrum, though amount visible highly variable)
- Yellowish-orange sheen (diatom film) may cover parts of body
- Small, slightly falcate or triangular dorsal fin two-thirds of the way along back
- Moderately long beak
- Two shallow, V-shaped throat grooves
- Small, narrow flippers (tuck into 'flipper pockets' on sides of body)
- Often heavily pockmarked with healed round or oval white scars from cookiecutter shark bites
- May have black spots in a line around mouth (scars from hooks on squid tentacles)
- Highly arched lower jaw (arch very wide and often towers above melon)
- May have slightly paler underside
- Teeth may be covered by tassels of dark purple or reddish-brown stalked barnacles (sometimes obscuring teeth altogether)

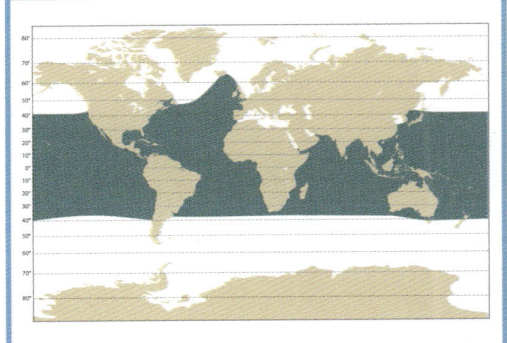

**AT A GLANCE (male)** Tropical to warm temperate waters worldwide • Medium size • Mostly nondescript gray-brown coloring • Pockmarked with healed cookiecutter shark bites • Tangled web of mainly parallel linear scarring • Very strongly arched lower jaw • Flattened, forward-tilting teeth on jaw arches • Stalked barnacles on teeth look like pompoms • Small head with flattened melon • Small, slightly falcate or triangular dorsal fin two-thirds of the way along back

**BLAINVILLE'S BEAKED WHALE**

**TEETH** Upper jaw 0; lower jaw 2. Teeth erupt in male only.

**GROUP SIZE AND STRUCTURE** Varies geographically, but typically alone, in pairs or in groups of 3–7 (mean 4.1); the largest groups observed in Hawai'i and the Bahamas each contained 11 individuals. Groups are usually harems, with a single adult male accompanying several adult females with their calves and/or juveniles. Sub-adults appear to stay in separate groups, and tend to occur in less productive waters. Occasional sightings of larger aggregations, including more than one adult male, are probably temporary aggregations of two or more groups.

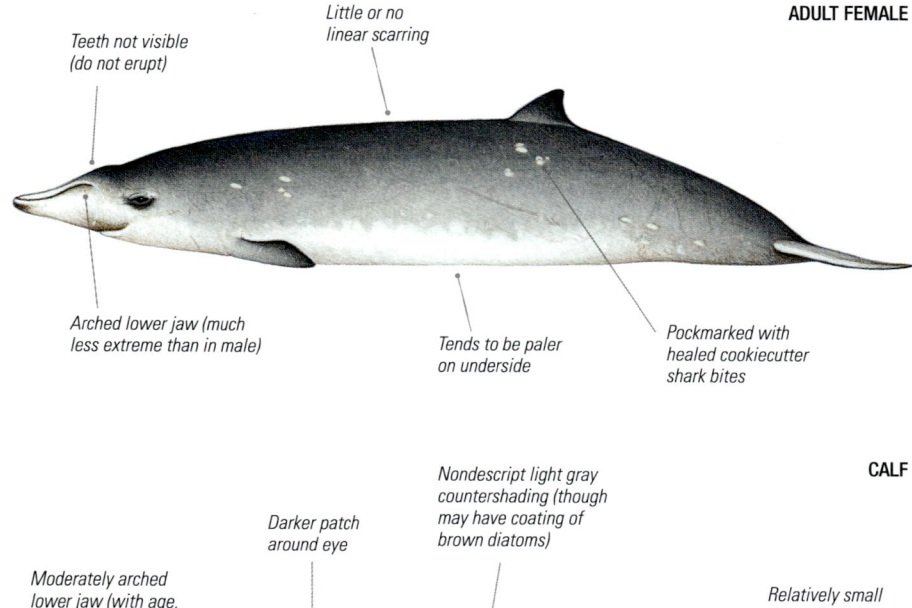

## DENSE BEAK

The rostrum of Blainville's beaked whale becomes secondarily ossified as individuals mature, especially in males, forming the densest bone currently known. Three possible functions have been proposed: to act as ballast (reducing the energetic cost of deep diving); as an adaptation for transmitting sound, during echolocation; or, most likely, as a mechanical reinforcement to prevent impact damage to the skull during male–male combat.

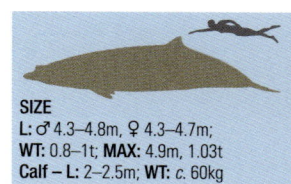

**SIZE**
**L:** ♂ 4.3–4.8m, ♀ 4.3–4.7m;
**WT:** 0.8–1t; **MAX:** 4.9m, 1.03t
**Calf – L:** 2–2.5m; **WT:** *c.* 60kg

**ADULT TWO-TONED MALE**

*Ochre to gold color patches caused by diatom infestations*

**ADULT TWO-TONED FEMALE**

**ADULT MALE VARIATION**

**ADULT FEMALE VARIATION**

BLAINVILLE'S BEAKED WHALE

# SOWERBY'S BEAKED WHALE
*Mesoplodon bidens* (Sowerby, 1804)

Sowerby's beaked whale was the first *Mesoplodon* beaked whale to be described: a male stranded in 1800, in the Moray Firth, northeastern Scotland (UK), and the skull was preserved. A few years later, James Sowerby, an English watercolor artist and naturalist, painted a picture of it and how he imagined the whole animal might have looked.

**IUCN status** Least Concern (2020).
**Population** Unknown, but probably fairly common in parts of its range. One (likely very low) estimate of 7,200. Trend unknown.
**Classification** Odontoceti, family Ziphiidae.
**Taxonomy** No recognized forms or subspecies, though recent evidence suggests two spatially distinct populations on either side of the Atlantic.
**Other names** North Sea beaked whale, North Atlantic beaked whale.
**DISTRIBUTION** Deep, cold offshore waters of the northern North Atlantic (among the most northerly species of *Mesoplodon*). Appears to be considerably more common in the eastern North Atlantic, and the center of abundance appears to be northern Europe. The vast majority of strandings have been between 50°N and 60°N. Occurs mainly in deep waters beyond the continental shelf edge and often associated with areas of complex seabed topography. May be seen close to shore where deep water approaches the coast.
**BEHAVIOR** Small groups typically surface within a couple of body lengths from each other, diving and resurfacing synchronously. Breaching, spyhopping and tail-slapping have been observed but rarely.
**FOOD AND FEEDING** Unusual among beaked whales in taking mainly small mid- and deepwater fish; some squid.
**TEETH** Upper jaw 0; lower jaw 2. Teeth erupt in male only; both sexes also possess small vestigial teeth, which do not normally erupt.
**GROUP SIZE AND STRUCTURE** Very little information, but seems to be 3–10 in mixed groups of males, females and young (8–10 have been recorded on a number of occasions).

**ADULT MALE**

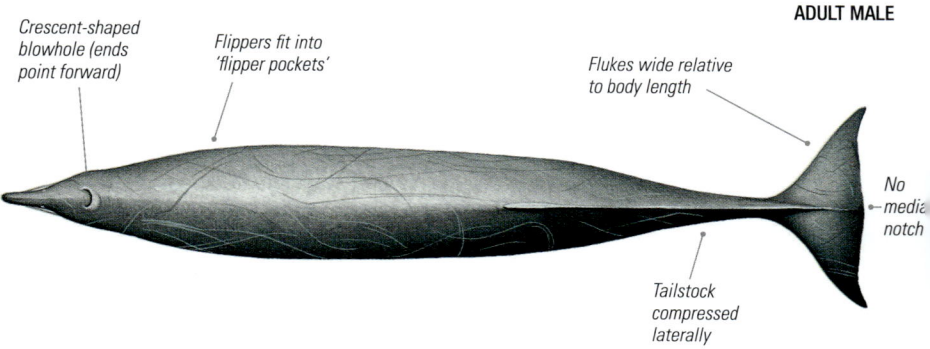

**DIVE SEQUENCE** Usually surfaces at 30–45° angle, clearly showing beak; melon and much of head may also be visible; swimming behavior often described as 'calm and unhurried'. • **DEPTH** Main prey typically occurs at 400–750m but probably dives to at least 1,000m. • **DURATION** Typically 12–30 minutes.
**BLOW** Invisible or inconspicuous (small and diffuse) blow angled slightly forward.

**ADULT MALE**

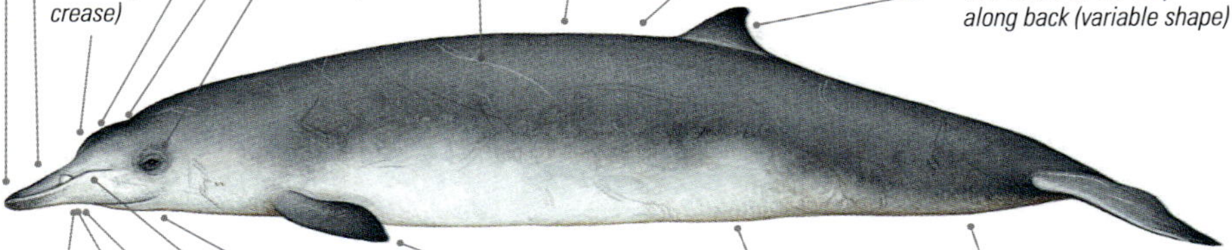

- Beak can be dark gray or very light gray
- Long, relatively slender beak (length variable)
- Melon slopes gently onto rostrum (no crease)
- Distinctive bulge on forehead (in front of blowhole)
- Relatively small head
- May be dark around eyes
- Linear scarring on many individuals (variable – less extensive than many other mesoplodonts – more with age)
- Fairly streamlined, spindle-shaped body
- Dark bluish-gray to lighter slate-gray upperside (may have brownish tinge due to diatoms)
- Small, falcate dorsal fin two-thirds of the way along back (variable shape)
- Small, narrow flippers
- Slightly paler sides and underside
- May have white or light gray spots (probably cookiecutter shark bites)
- Two shallow, V-shaped throat grooves
- Two small teeth two-thirds of the way along lower jaw (visible outside closed mouth)
- Mouthline mostly straight with slight arch at rear
- Stalked barnacles may attach to teeth
- Teeth do not rise above level of upper rostrum, so rake marks mostly single (not paired)

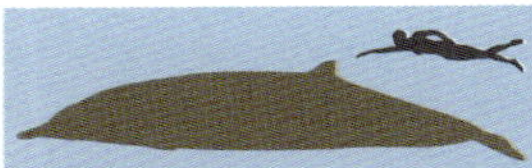

**SIZE**
L: ♂ 4.4–5.5m, ♀ 4.4–5.5m;
WT: 1–1.3t; **MAX:** 5.5m, 1.5t
Calf – L: 2.1–2.4m; WT: 170–185kg

**ADULT FEMALE**

- No erupted teeth
- Little or no linear scarring

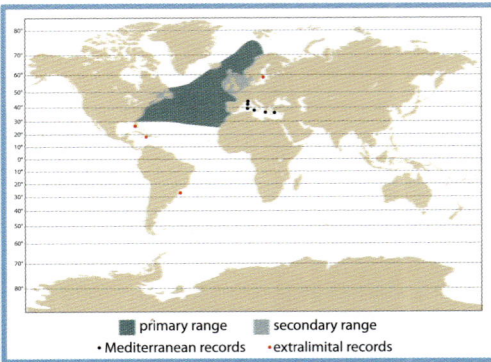

- primary range
- secondary range
- Mediterranean records
- extralimital records

**AT A GLANCE** Cool waters of North Atlantic • Nondescript light to dark gray above, lighter below • White linear scars may be present (male) • Medium size • Long, slender beak visible on surfacing • Two teeth two-thirds of the way along beak (male only) • Distinctive bulge on forehead • Usually unobtrusive/elusive behavior

# TRUE'S BEAKED WHALE
*Mesoplodon mirus* — True, 1913

The taxonomic status of True's beaked whale has long been a puzzle. Until recently, there were believed to be two disjunct populations — in the North Atlantic and the southern hemisphere — widely separated by the tropics. However, it is now accepted that these represent two different species: True's beaked whale in the North Atlantic and Ramari's beaked whale in the southern hemisphere.

**IUCN status** Least Concern (2022).
**Population** Unknown (largely because it is cryptic and usually skittish in the field and difficult to separate from several other *Mesoplodon* species). One 2016 estimate of 2,562 in US waters of the North Atlantic. Trend unknown.
**Classification** Odontoceti, family Ziphiidae.
**Taxonomy** See 'Anti-tropical species' box opposite.
**Other names** Wonderful beaked whale.

**DISTRIBUTION** Prefers deep, offshore, relatively warm temperate waters (though it does occur in deep waters near the coast in some areas). Sightings during surveys conducted in US waters occurred over an average seafloor depth of 2,425m. May favor areas of complex seabed topography. There has been a relative abundance of live sightings in deep coastal waters off the Azores, and to some extent off the Canary Islands.

**BEHAVIOR** Energetic breaching has been recorded on several occasions. Tail-slapping has also been observed.

**FOOD AND FEEDING** Very limited evidence suggests primarily small fish, with occasional deepwater squid. Feeding techniques unknown.

**TEETH** Upper jaw 0; lower jaw 2. Teeth erupt in males only.

**GROUP SIZE AND STRUCTURE** Typically alone or in small, closely associated groups of 3–6 (based on a few rare sightings).

**ADULT FEMALE**

Teeth do not erupt

Some individuals may have lighter tailstock (not unlike Ramari's)

**SIZE**
**L:** ♂ 4.5–5.3m, ♀ 4.5–5.4m;
**WT:** 1–1.4t; **MAX:** 5.4m, 1.4t
**Calf – L:** 2–2.5m; **WT:** unknown
Females may be slightly larger than males.

**DIVE SEQUENCE** Surfaces at angle, possibly showing entire beak and head (to just below eye level). • **DEPTH** Likely to forage deeper than 500m (one study suggests main prey found in waters 200–800m deep); maximum recorded 1,037m. • **DURATION** Unknown.

**ADULT MALE**

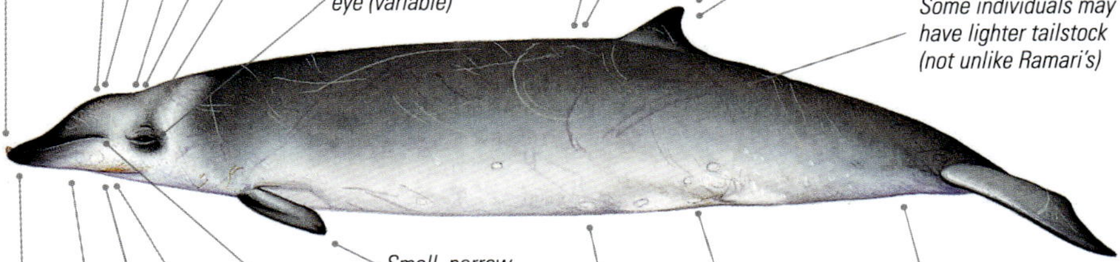

- Rounded melon slopes steeply onto rostrum (no crease)
- Relatively small head
- May be distinct indentation behind blowhole
- Slender, spindle-shaped body
- Medium gray to brownish-gray upperside and tailstock (posterior region of tailstock may be darker)
- Two small teeth at tip of lower jaw visible when mouth closed (do not erupt in female)
- May be pale or whitish blaze covering melon from behind blowhole
- Dorsal fin may be darker than back
- May be narrow, dark line between eye and top of head
- Small, moderately falcate to triangular dorsal fin two-thirds of the way along back
- Dark patch around eye (variable)
- Some individuals may have lighter tailstock (not unlike Ramari's)
- Front half of beak dark
- Small, narrow flippers
- Mid-length, stubby, dolphin-like beak
- Fairly straight or slightly curved mouthline
- May be white or light pink urogenital patch in some animals
- May be closely spaced parallel scars on some mature males – and cookiecutter shark bites
- Two well-defined V-shaped throat grooves
- Lighter gray to slate-gray underside
- May be dark flecking on throat and lower jaw (more in older animals)

### ANTI-TROPICAL SPECIES

Recent genetic and morphological analysis of museum and archival specimens of True's beaked whales in the North Atlantic and in the southern hemisphere revealed very different genetics and skull shape (there is believed to have been no gene flow between the populations for at least 0.35 million years). They were separated formally in 2021: the North Atlantic animals remain True's beaked whales (*Mesoplodon mirus*) and the southern hemisphere animals became Ramari's beaked whales (*Mesoplodon eueu*). The external appearances of the two species are not known to be consistently distinguishable.

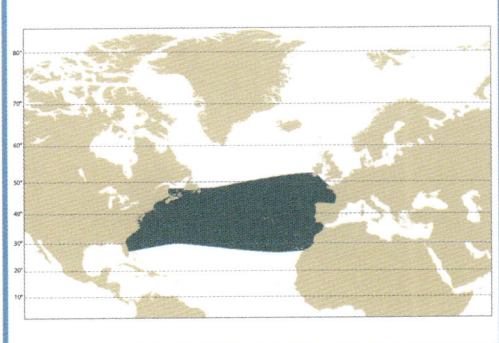

**AT A GLANCE** North Atlantic • Temperate offshore waters • Medium size • Rounded melon • Mid-length beak with two small teeth at tip (male only) • Closely spaced parallel scars • Small dorsal fin two-thirds of the way along back

**TRUE'S BEAKED WHALE**

# STEJNEGER'S BEAKED WHALE
*Mesoplodon stejnegeri* — True, 1885

Known mostly from strandings — predominantly along the west coast of Honshu, Japan, and in the Aleutian Islands, Alaska — Stejneger's beaked whale is rarely seen alive at sea. Sometimes called the sabre-toothed beaked whale, the male has two particularly large teeth, like tusks, that are used for fighting.

**IUCN status** Near Threatened (2020).
**Population** Unknown, but given the paucity of records it appears to be rare in most parts of its range. Decreasing.
**Classification** Odontoceti, family Ziphiidae.
**Taxonomy** No recognized forms or subspecies.
**Other names** Bering Sea beaked whale, North Pacific beaked whale, sabre-toothed beaked whale.
**DISTRIBUTION** Primarily cold temperate and sub-Arctic waters of the North Pacific. Seems to prefer deep areas of complex seabed topography. The central Aleutian Islands are considered the strandings hotspot. May move to warmer southern waters on an unknown seasonal or temporal basis.
**BEHAVIOR** Very little information. Known to breach. Appears to be shy and difficult to approach. Reports of roaring and groaning sounds made at the surface.
**FOOD AND FEEDING** Mostly deepwater squid, possibly some fish. Suction is main feeding method.
**TEETH** Upper jaw 0; lower jaw 2. Teeth erupt in male only.
**GROUP SIZE AND STRUCTURE** Typically 5–15 but smaller groups have been observed. Groups may contain animals of mixed sexes and ages, or can be segregated. May be tightly bunched together at the surface — sometimes touching or nearly touching — and typically swim and dive in unison.

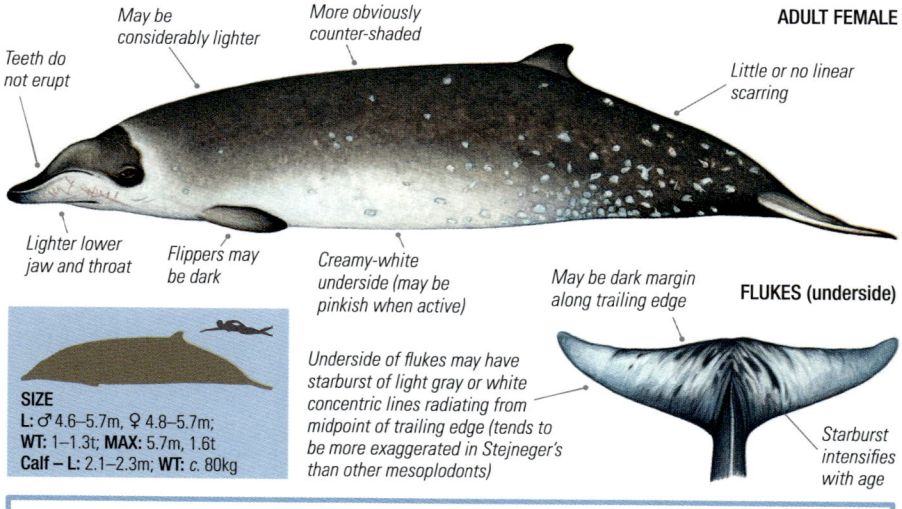

**SIZE**
**L:** ♂ 4.6–5.7m, ♀ 4.8–5.7m;
**WT:** 1–1.3t; **MAX:** 5.7m, 1.6t
**Calf – L:** 2.1–2.3m; **WT:** c. 80kg

**DIVE SEQUENCE** Tip of beak breaks surface first; blowhole and upperside of head appear briefly; low profile as head quickly disappears and dorsal fin rolls forward. • **DEPTH** Preferred prey suggests at least 200m; presumably capable of diving much deeper (possibly to 1,500m). • **DURATION** At least 15 minutes, probably much longer.
**BLOW** Blow indistinct.

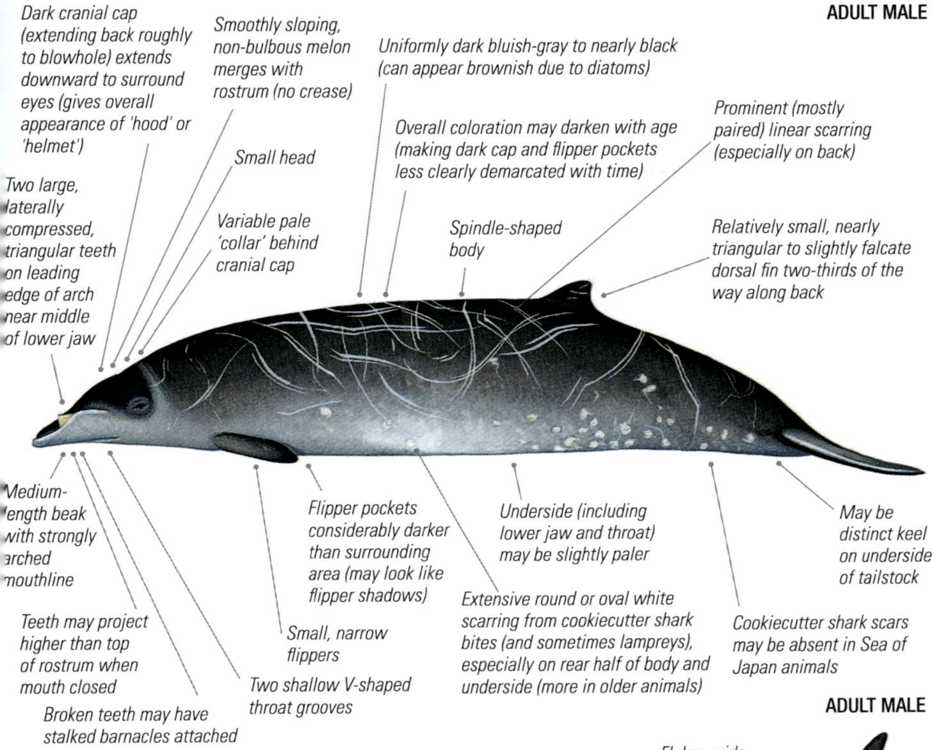

# STEJNEGER'S BEAKED WHALE

**ADULT MALE**

- Dark cranial cap (extending back roughly to blowhole) extends downward to surround eyes (gives overall appearance of 'hood' or 'helmet')
- Smoothly sloping, non-bulbous melon merges with rostrum (no crease)
- Uniformly dark bluish-gray to nearly black (can appear brownish due to diatoms)
- Small head
- Overall coloration may darken with age (making dark cap and flipper pockets less clearly demarcated with time)
- Prominent (mostly paired) linear scarring (especially on back)
- Two large, laterally compressed, triangular teeth on leading edge of arch near middle of lower jaw
- Variable pale 'collar' behind cranial cap
- Spindle-shaped body
- Relatively small, nearly triangular to slightly falcate dorsal fin two-thirds of the way along back
- Medium-length beak with strongly arched mouthline
- Flipper pockets considerably darker than surrounding area (may look like flipper shadows)
- Underside (including lower jaw and throat) may be slightly paler
- May be distinct keel on underside of tailstock
- Teeth may project higher than top of rostrum when mouth closed
- Small, narrow flippers
- Extensive round or oval white scarring from cookiecutter shark bites (and sometimes lampreys), especially on rear half of body and underside (more in older animals)
- Cookiecutter shark scars may be absent in Sea of Japan animals
- Broken teeth may have stalked barnacles attached
- Two shallow V-shaped throat grooves

**ADULT MALE**

- Crescent-shaped blowhole (ends point forward)
- Flukes wide relative to body length
- No median notch
- Teeth point forward and slightly inward (may constrict opening of jaw)
- Flippers fit into darkly pigmented 'flipper pockets' (slight depressions on sides of body)
- Tailstock compressed laterally

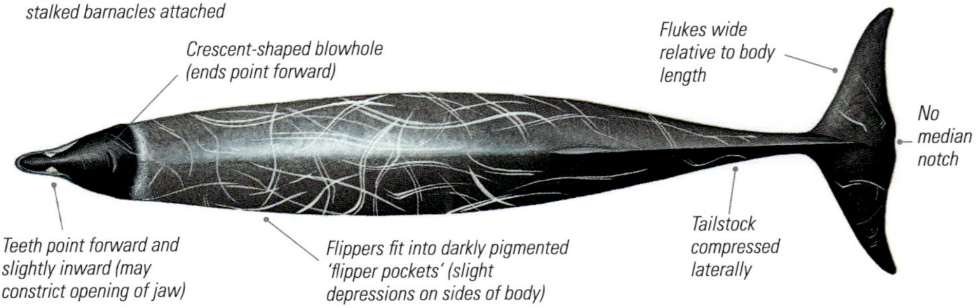

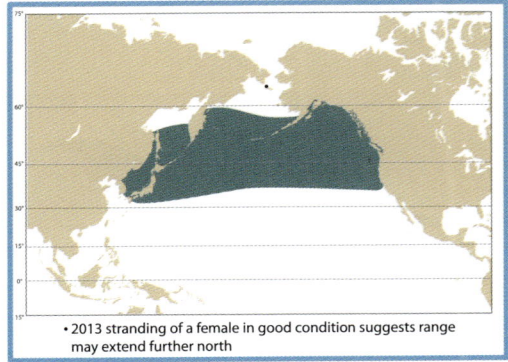

- 2013 stranding of a female in good condition suggests range may extend further north

**AT A GLANCE** Cold offshore waters of northern North Pacific • Medium size • Spindle-shaped body • Dark cranial 'cap' • Gently-sloping forehead • Strongly arched mouthline • Two large, exposed, flattened teeth (male only) • Small groups bunched together

# GERVAIS' BEAKED WHALE
*Mesoplodon europaeus* (Gervais, 1855)

Gervais' beaked whale is known from more than 300 records in the North Atlantic, and just a handful in the South Atlantic. Most of these are strandings – there have been few reliable sightings at sea – so information on its life and habits is sparse.

**IUCN status** Least Concern (2020).
**Population** Unknown, but believed to be relatively common. Trend unknown.
**Classification** Odontoceti, family Ziphiidae.
**Taxonomy** No recognized forms or subspecies.
**Other names** European beaked whale, Gulf Stream beaked whale, Antillean beaked whale.
**DISTRIBUTION** Most records are from the western North Atlantic – Florida and North Carolina account for more than 40 per cent of all worldwide records combined – and it is the most commonly sighted mesoplodont off the US Atlantic coast and in the Gulf of Mexico. Nearly half of all records in the eastern North Atlantic (21 strandings involving 24 individuals) are from the Canary Islands. There are scattered records in the South Atlantic. Seems to prefer deep waters in the tropics and subtropics, but there are records from warm temperate and even cold temperate waters. Likely to be more common in areas of complex seabed topography.
**BEHAVIOR** Virtually nothing known. Breaching has been observed.
**FOOD AND FEEDING** Little known. Feeds mainly on deepwater squid, possibly some deepwater fish and shrimp.
**TEETH** Upper jaw 0; lower jaw 2. Teeth erupt in male only.
**GROUP SIZE AND STRUCTURE** Limited information suggests that Gervais' beaked whales are usually found alone or in small, close-knit groups of up to five individuals.

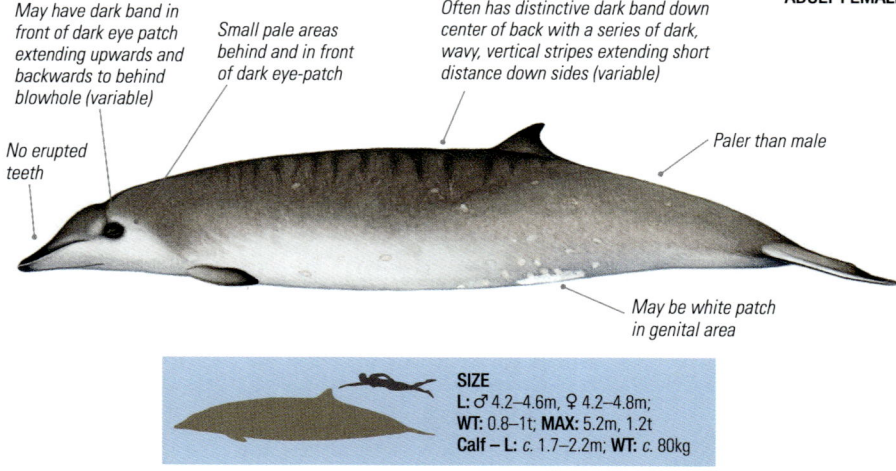

**ADULT FEMALE**

**SIZE**
**L:** ♂ 4.2–4.6m, ♀ 4.2–4.8m; **WT:** 0.8–1t; **MAX:** 5.2m, 1.2t
**Calf – L:** *c.* 1.7–2.2m; **WT:** *c.* 80kg

**DIVE SEQUENCE** May surface at angle of about 45°; briefly shows beak and much of head; slight pause before rolls forward; tends to sink below surface rather than arching back; does not show flukes on sounding dive. •
**DEPTH** In one encounter dives lasted around 1 hour.
**BLOW** Inconspicuous blow.

# GERVAIS' BEAKED WHALE

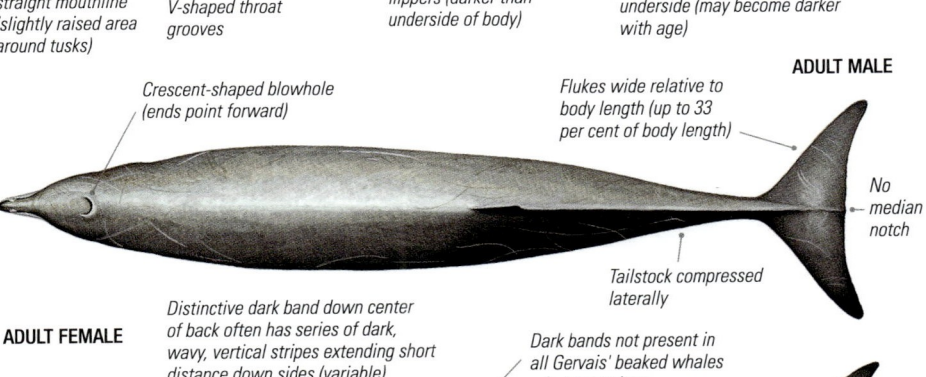

**ADULT MALE**

- Single tooth on small arch about one-third of the way along each side of lower jaw (7–10cm from tip), visible outside closed mouth
- Dark band down center of back may be partially obscured by darker upperside (often more conspicuous in female and calf)
- Relatively little pale linear scarring (usually single lines when present)
- Dark patch around eye often more pronounced than in other mesoplodonts (variable)
- Smoothly sloping, slightly bulbous melon merges with rostrum (no crease)
- Very small head
- Spindle-shaped body
- Small, wide-based, slightly falcate dorsal fin two-thirds of the way along back (variable – shark-like to falcate)
- Mid-length beak with relatively straight mouthline (slightly raised area around tusks)
- Two shallow, V-shaped throat grooves
- Small, narrow flippers (darker than underside of body)
- Medium to dark gray (sometimes brownish) upperside, paler underside (may become darker with age)

**ADULT MALE**

- Crescent-shaped blowhole (ends point forward)
- Flukes wide relative to body length (up to 33 per cent of body length)
- No median notch
- Tailstock compressed laterally

**ADULT FEMALE**

- Distinctive dark band down center of back often has series of dark, wavy, vertical stripes extending short distance down sides (variable)
- Dark bands not present in all Gervais' beaked whales – but are unique

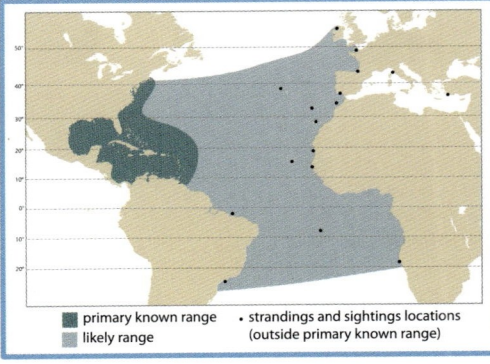

primary known range
likely range
• strandings and sightings locations (outside primary known range)

**AT A GLANCE** Tropical to warm temperate waters of Atlantic Ocean • Medium size • Medium to dark gray above, paler below • Little or no linear scarring • Very small head with slightly bulbous melon • Dark patch around eye • Medium-length beak • Two teeth on small arch one-third of the way along lower jaw (male only) • Small dorsal fin two-thirds of the way along back • Females and juveniles may have tiger-like stripes

# KILLER WHALE or ORCA
*Orcinus orca* (Linnaeus, 1758)

Two thousand years ago, Roman scholar Pliny the Elder described the killer whale as 'an enormous mass of flesh armed with savage teeth'. Even as recently as the early 1970s, US Navy diving manuals described it as 'extremely ferocious', warning that it 'will attack human beings at every opportunity'. But the killer whale does not deserve its killer reputation any more than any other apex predator.

**IUCN status** Data Deficient (2017). Strait of Gibraltar sub-population Critically Endangered (2019).
**Population** No reliable global estimate, but likely minimum 50,000 (probably considerably more, especially given the lack of information from large oceanic areas and the Arctic, and likely underestimates in the Antarctic). The latest estimate for the Antarctic south of 60°S is *c.* 25,000–27,000. There are estimated to be 15,000 in the North Atlantic (2022) and 2,500 in the eastern North Pacific (2024). Trend unknown.
**Classification** Odontoceti, family Delphinidae.
**Taxonomy** Three subspecies are currently recognized: resident killer whale (*O. o. ater*), Bigg's (aka 'transient') killer whale (*O. o. rectipinnus*) and common killer whale (*O. o. orca*). There is a strong case (given the high level of genetic, morphological and ecological differentiation) for separating resident and Bigg's as separate species (*O. ater* and *O. rectipinnus* respectively) but a more complete global review and revision of killer whales is required. In the meantime, the term 'ecotype' is used for ecologically distinct populations that do not interbreed (even if they inhabit the same waters) while recognizing scientific uncertainty about killer whale taxonomy. There could be as many as 40 ecotypes altogether; as well as those illustrated in this guide, they might include New Zealand coastal, New Zealand pelagic, Eastern Tropical Pacific and ecotypes in Argentina, Papua New Guinea and others.
**Other names** Blackfish (non-taxonomic group of six dark-colored members of Delphinidae with 'whale' in their name); historically, grampus.

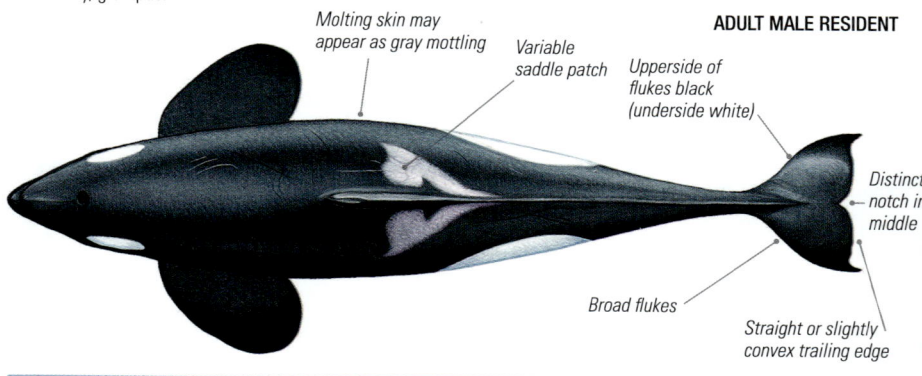

**ADULT MALE RESIDENT**
- Molting skin may appear as gray mottling
- Variable saddle patch
- Upperside of flukes black (underside white)
- Distinct notch in middle
- Broad flukes
- Straight or slightly convex trailing edge

**DIVE SEQUENCE** Outline of adult male's tall dorsal fin unmistakable; tip of dorsal fin typically breaks surface first (followed by top of head). • **DEPTH** Varies with prey and location; maximum recorded more than 1,000m (near South Georgia), but potentially even deeper (especially males); foraging residents usually dive less than 100m. • **DURATION** Varies with prey and location; maximum recorded 16 minutes.

**BLOW** Fairly tall, columnar blow (up to 5m), bushy at the top and projects slightly forward.

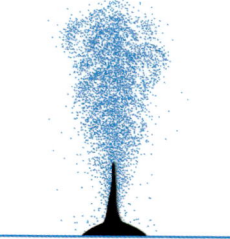

**ADULT MALE RESIDENT**

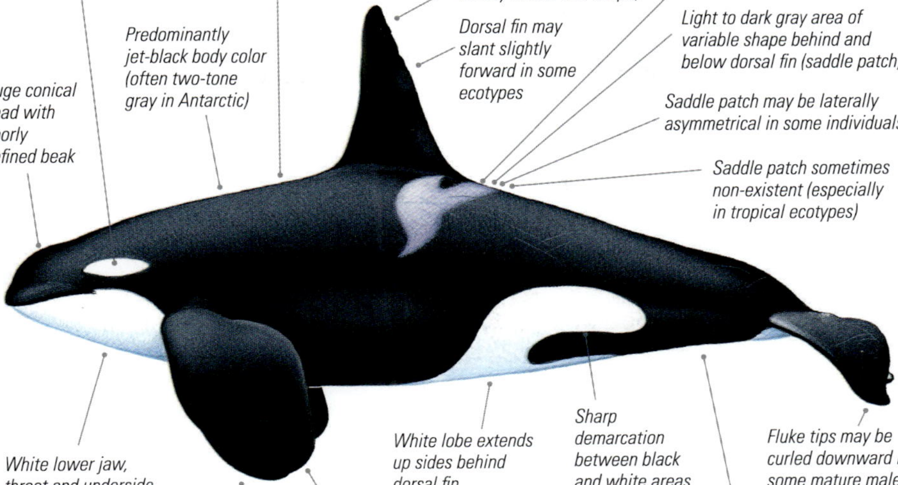

- Conspicuous elliptical white patch above and behind each eye
- Robust, spindle-shaped body
- Exceptionally tall, erect dorsal fin (up to 1.8m)
- Fin triangular (varies widely in size and shape)
- Saddle patch often 'open' (i.e. with dark incursion)
- Predominantly jet-black body color (often two-tone gray in Antarctic)
- Dorsal fin may slant slightly forward in some ecotypes
- Light to dark gray area of variable shape behind and below dorsal fin (saddle patch)
- Huge conical head with poorly defined beak
- Saddle patch may be laterally asymmetrical in some individuals
- Saddle patch sometimes non-existent (especially in tropical ecotypes)
- White lower jaw, throat and underside
- White lobe extends up sides behind dorsal fin
- Sharp demarcation between black and white areas
- Fluke tips may be curled downward in some mature males (cf. female)
- Disproportionately large, oval flippers (grow with age – up to 2m)
- Dorsal fin, flippers and tail flukes all substantially larger in proportion to body size (cf. female)
- Scarring usually made by killer whale tooth marks during bouts of play (though sometimes more serious interactions)

**FLUKES**

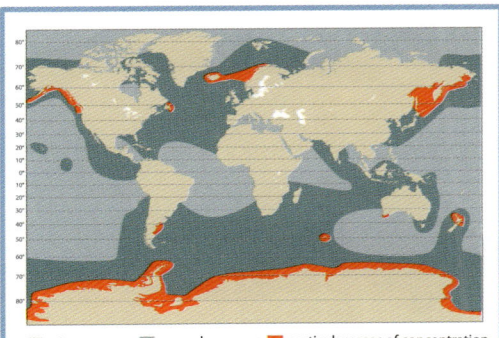

**SIZE**
**L:** ♂ 5.6–9m, ♀ 4.5–7.7m;
**WT:** 1.3–6.6t; **MAX:** 9.8m, 10t
**Calf – L:** 2–2.8m; **WT:** 160–200kg
Body size varies considerably among ecotypes. Highly sexually dimorphic – mature males up to 17 per cent longer and 40 per cent heavier than mature females.

**AT A GLANCE** Worldwide distribution
• Medium size • Two-tone coloring, predominantly jet black (or gray) and white
• Exceptionally tall dorsal fin of male • Pronounced sexual dimorphism • White patch above and behind each eye • Usually in family groups

■ primary range ■ secondary range ■ particular areas of concentration

**KILLER WHALE**

**DISTRIBUTION** The most cosmopolitan cetacean, with worldwide (though patchy) distribution. Occurs in all oceans and many enclosed seas. Found in all temperatures and depths, from tropical to polar waters and from the surf zone to the open sea, though the highest densities are in cold temperate to polar coastal waters with high productivity. Most abundant in the Southern Ocean south of 60°S. Commonly enters heavy, consolidated ice in the Antarctic, but rarely does so in the Arctic (though it is becoming seasonally more abundant further north as pack-ice extent and duration decline with climate change). Widespread but rarer in tropical and offshore waters.

**BEHAVIOR** Can be very active at the surface, especially when socializing or after a successful hunt. Often breaches, flipper-slaps and lobtails. Frequently spyhops; several animals may do so together. Occasionally bow-rides or (more frequently) wake-rides. Northern residents beach-rub on smooth pebbles in shallow water.

**FOOD AND FEEDING** Apex predator with extremely diverse diet (c. 150 prey species known globally) but high level of specialization depending on ecotype and location. Includes 32 species of cetaceans, 19 species of pinnipeds, 44 species of bony fishes, 22 species of sharks and rays (including great white), 20 species of seabirds, five species of squid and octopus, two species of sea turtle and two species of terrestrial mammals (sitka black-tailed deer and moose, swimming between islands). Huge range of feeding strategies, including 'beaching' in Punta Norte, Argentina, and Crozet Island, Indian Ocean, to catch seals on shore; 'wave-washing' in Antarctic, to wash Weddell seals from ice floes; 'endurance-exhaustion' in Strait of Gibraltar, to catch bluefin tuna; 'carousel feeding' in Iceland and Norway, to catch schooling herring. Often hunts cooperatively. Will take fish from longline fishing operations.

**TEETH** Upper jaw 20–28; lower jaw 20–28. Teeth may be worn flat in some ecotypes.

**GROUP SIZE AND STRUCTURE** Fewer than 20 typical for most ecotypes. Occasionally gather in larger, temporary groups of up to 150 individuals, which are probably to reinforce social bonds between matrilines and pods, or for mating. Individuals seen alone are almost always males.

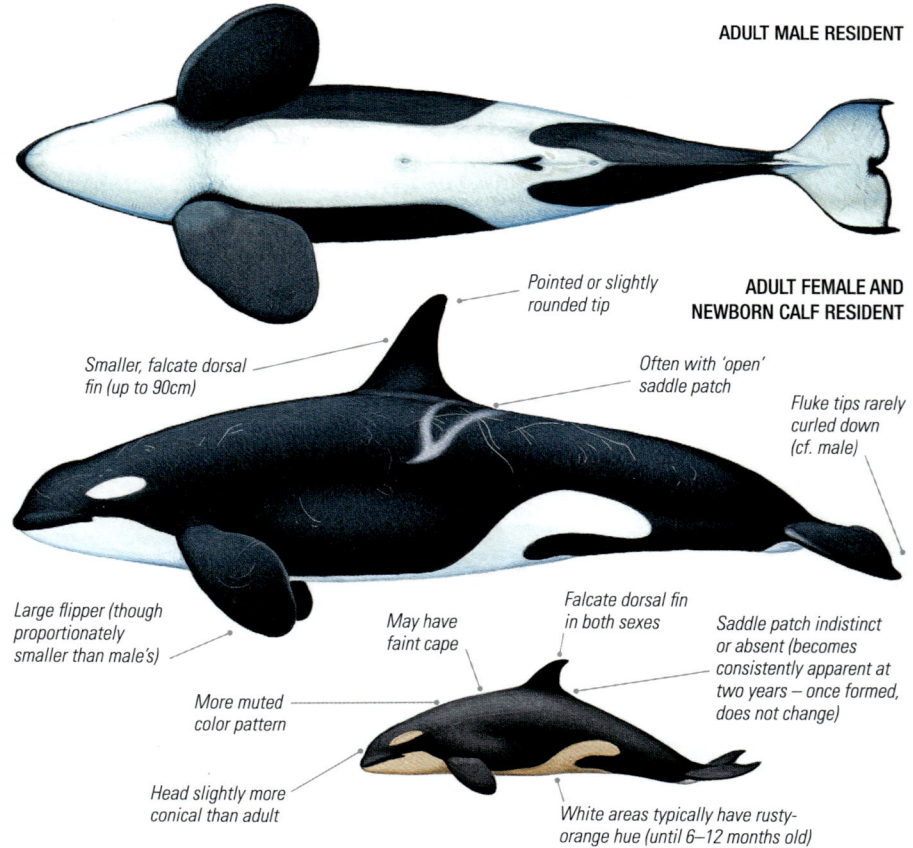

# KNOWN KILLER WHALE SUB-SPECIES AND ECOTYPES IN THE NORTH PACIFIC

## MALE DORSAL FIN AND SADDLE PATCH COMPARISONS

- Open saddle patch common (either uniform gray or contains varying amounts of black)
- Dorsal fin has rounded tip (but usually with sharp angle at rear corner)
- Dorsal fin has pointed tip
- Saddle patch often larger than in residents and offshores
- Saddle patch never open
- Dorsal fin continuously rounded over tip (without sharp angle at rear corner seen in residents)
- Saddle patch usually quite faint and roughly same size as in residents
- Saddle patch typically closed (but some open)

**RESIDENT**     **BIGG'S**     **OFFSHORE**

## RESIDENT (OR FISH-EATING) KILLER WHALE

- 'Typical' black-and-white killer whale
- Male dorsal fin tip usually more rounded than Bigg's and ends in pointed trailing tip
- Typically a few nicks and scars on trailing edge of the dorsal fin
- Male dorsal fin may lean forward to varying degrees
- Fin tip tends to be positioned over front end of base
- Dorsal fin often has wavy trailing edge (especially in older males)
- Leading edge of dorsal fin tends to be straight or slightly concave
- Saddle patch usually very open (considerable black incursion in otherwise pale gray) and rarely closed
- Middle of saddle patch rarely extends further forward than midpoint of dorsal fin base
- No obvious dorsal cape
- White eye patch a medium-sized oval (parallel to body axis)

**L:** ♂ 6.9m, ♀ 6m; **Max:** 7.2m

**Distribution** Known mainly from the north-east Pacific, ranging from the Aleutian Islands, Alaska (USA), through British Columbia (Canada) and Washington to Monterey Bay, California (USA). May also occur elsewhere in the North Pacific: a large population of fish-eating killer whales in the Russian Far East, for example, corresponds in appearance, behavior, acoustic activity and genetics with this ecotype (its range includes the central Sea of Okhotsk, southern and central Kamchatka Peninsula, the Commander and Kuril Islands, and the southern Bering Sea); it is known as Type R. Often inhabits sheltered coastal waterways and not known to venture far beyond the continental shelf.

**Food and feeding** Primarily fish. Usually ignores marine mammals, which rarely show avoidance behavior (though southern residents have been observed harassing porpoises without killing them).

**Group size** Pod typically composed of 1–3 matrilines (ranges from 1–11) with average of 18 whales (typically 5–50).

**Remarks** The term 'resident' is rather misleading when describing the site-fidelity and movement patterns of these whales – so they are often referred to as fish-eating killer whales.

**JUVENILE RESIDENT**

Falcate, female-like dorsal fin in both sexes (difficult to tell apart)

Dorsal fin starts to grow quickly in male when c. 15 years old

# BIGG'S (OR TRANSIENT) KILLER WHALE

- 'Typical' black-and-white killer whale
- Largest of three North Pacific ecotypes
- Male dorsal fin tip straighter and more pointed than resident
- Typically many nicks and scars on trailing edge of fin
- Fin tip tends to be positioned over center of base
- Large saddle patch uniformly gray
- Saddle patch always closed (no black incursion into gray)
- Middle of saddle patch typically extends further forward (cf. residents) – often past midline of dorsal fin
- White eye-patch medium-sized oval (slanted very slightly downwards towards rear)
- No obvious dorsal cape

**L:** ♂ 8m, ♀ 7m; **Max:** 9.8m

**Distribution** Known mainly from the Bering Sea through British Columbia (Canada) and Washington (USA) to Baja California (Mexico). May also occur elsewhere in the North Pacific: a very poorly known population of mammal-eating killer whales in the Russian Far East, known as Type T, corresponds in appearance, behavior and acoustic activity with this ecotype. Movement patterns coincide with seasonally available prey species. Coastal and offshore waters. Travels over a wider range than residents and rarely keeps to predictable routes or stays in the same place for long.

**Food and feeding** Primarily mammals, especially cetaceans, pinnipeds and sea otters. Preference varies with location. Will kill swimming seabirds (usually abandoning the carcasses) and takes some squid. Not known to take fish.

**Group size** Typically 3–7 animals, consisting of a mother and her offspring (but more fluid than residents). In recent years, large temporary aggregations of 30 or more have been observed. Lone individuals (usually males) are sometimes seen.

**Remarks** The consensus among researchers is that the ecotype formerly known as the 'transient killer whale' should be called 'Bigg's killer whale', in honour of the late pioneering killer whale researcher Dr Michael Bigg (and because the term 'transient' is rather misleading when describing the site-fidelity and movement patterns of these whales).

ADULT MALE BIGG'S

ADULT FEMALE AND CALF BIGG'S

# OFFSHORE

- 'Typical' black-and-white killer whale
- Overall appearance very similar to resident
- Smallest of three known North Pacific ecotypes
- Relatively smaller dorsal fin
- Male dorsal fin tip continuously rounded over tip (more rounded than Bigg's, without sharp corner of resident)
- Usually more nicks and scars on trailing edge of dorsal fin than resident
- May have oval scars from cookiecutter shark bites
- Gray saddle patch similar in size to that of resident
- Saddle patch usually quite faint (normally closed – no black incursion into gray – though open in some individuals)
- Regularly slaps tail when swimming
- Less sexual dimorphism than in resident or Bigg's (tendency for sexes to be more similar in size)
- Extreme tooth wear normal even in sub-adults (often worn flat to gum line – likely from eating sharks that have abrasive skin); similar tooth wear unknown in resident or Bigg's
- White eye-patch a medium-sized oval (parallel to body axis)

**L:** ♂ 6.5m, ♀ 5.5m; **Max:** 7.2m

**Distribution** The least known of the three North Pacific ecotypes. Ranging between southern California and the eastern Aleutian Islands, Alaska, offshores travel extensively throughout their range. Primarily associated with the outer continental shelf and inshore, with a peak of sightings just inshore of the 200m depth contour; they will sporadically visit coastal (and occasionally protected inshore) waters.
**Food and feeding** Bony and cartilaginous fish, especially sharks (including great white, blue, Pacific sleeper, Pacific spiny dogfish); also known to take Chinook salmon and Pacific halibut. There is no evidence of mammal-eating.
**Group size** Typically 20+ (50–100 animals is not unusual); total population only *c.* 300.

ADULT MALE OFFSHORE

ADULT FEMALE AND CALF OFFSHORE

KILLER WHALE

# KNOWN KILLER WHALE POPULATIONS IN THE NORTH ATLANTIC

**Note** Possible ecotype delineation is less clear in the North Atlantic, but the best-known populations are described here. Previously, they were dubbed Type 1 or Type 2 killer whales (based on museum and stranded specimens) according to the absence and presence of tooth wear.

## ICELANDIC SUMMER-SPAWNING HERRING-FEEDERS

- 'Typical' black-and-white killer whale
- Smaller than those of West Coast Community
- More closely resembles north-east Pacific residents
- Medium- to large-sized oval eye patch (parallel to body axis)
- Front end of eye patch in front of blowhole
- Significant tooth wear (teeth often worn smooth to the gum line)
- Conspicuous saddle patch
- Worn teeth produce wide rake marks

**L:** ♂ 6.3m, ♀ 5.9m; **Max:** 6.6m

**Distribution** Iceland, but *c.* 5 per cent move to north-east Scotland (UK) in the spring and summer (especially Shetland). A very small number have been detected moving between Iceland and Norway.

**Food and feeding** Primarily schooling herring, moving between the herring wintering, spawning and feeding grounds around Iceland. Carousel feeding frequently observed during winter and summer: the whales work in groups of 3–9 to round up herring; they split a group of fish from the larger school, swim in fast circles (blowing bubbles, flashing their undersides and lobtailing) to herd them into a tighter ball, whip their tails into the ball to stun or kill as many fish as possible, then pick them off one-by-one. Some Icelandic killer whales appear to specialize on herring and follow it year-round, while others feed on it only seasonally or opportunistically. At least 12 other prey species have been recorded in Icelandic waters, including minke whales, dolphins, porpoises, seals, eider ducks, fish and squid. In Scotland, killer whales feed on herring (offshore) and harbor and gray seals (inshore), and are known to take porpoises, otters and seabirds. Scottish individuals return to Iceland for the winter, where they feed on the same herring stocks as those that remain year-round.

**Group size** 4–6 when feeding on marine mammals; 6–30 when feeding on fish close to shore; up to 300 feeding on fish along the continental shelf edge.

ADULT MALE NORTH-EAST ATLANTIC
(ICELANDIC HERRING-FEEDERS, NORWEGIAN HERRING-FEEDERS AND NORTH-EAST ATLANTIC MACKEREL-FEEDERS)

ADULT FEMALE AND CALF NORTH-EAST ATLANTIC

## NORWEGIAN SPRING-SPAWNING HERRING-FEEDERS
- No discernible differences from Icelandic herring-feeders

**L:** ♂ 6.2m, ♀ 5.5m; **Max:** 6.6m

**Distribution** Most of Norway's killer whales follow the movements of its spring-spawning herring. No contemporary movement has been detected between Norway and Iceland.
**Food and feeding** Primarily schooling herring; recent evidence suggests that, in addition to feeding on overwintering herring inside the Norwegian fjords, at least some of the whales follow the herring to their offshore spawning grounds. Other prey items documented include mackerel, cod, salmon, squid, harbor porpoise and harbor and gray seals. Many whales in this population may include seals in their diet on a regular basis.
**Group size** Herring-feeding groups range from 6–30 individuals (median 15), seal-feeding groups from 3–11 (median 5).

## NORTH-EAST ATLANTIC MACKEREL-FEEDERS
- Worn teeth (due to suction feeding)
- Subtly different features to Icelandic and Norwegian herring-feeding killer whales, e.g. eye patch often smaller and male dorsal fin often more rounded at tip (butter knife shape)

**L:** ♂ 6.3m, ♀ 5.9m; **Max:** 6.6m

**Distribution** Ranges throughout the northern North Sea, the Irish Sea and the Norwegian Sea, and into the Arctic. Mostly offshore, but also coastal. Known mainly between the Northern Isles (UK) and southern Norway in mid- to late fall; west of the Hebrides (UK) in winter; and in the Norwegian Sea (including Iceland) up to 72°N during late summer.
**Food and feeding** Mackerel (for at least part of the year – whether it targets mackerel year-round or switches to other seasonally available prey is unknown). Frequently feeds around fishing trawlers.
**Group size** Those feeding in the Norwegian Sea have been observed in groups ranging from 1–40 (average 8); groups feeding around fishing trawlers range from 1–70 (average 13). Maximum group size 200.

## WEST COAST COMMUNITY
- 'Typical' black-and-white killer whale
- Larger than Icelandic or Norwegian herring specialists
- More closely resembles NE Pacific Bigg's and Type A
- Medium- to large-sized oval eye patch (slanting down towards rear – a key distinguishing feature)
- Faint saddle patch

**L:** ♂ unknown, ♀ 6.1m; **Max:** unknown

**Distribution** The UK and Ireland, primarily around Scotland and especially the Outer Hebrides (where they are seen most often).
**Food and feeding** Little understood, but known to have taken harbor seals and harbor porpoises.
**Group size** Currently only two males remain (John Coe and Aquarius); there were *c.* 20 individuals in the 1980s.

**ADULT MALE WEST COAST COMMUNITY**
(most distinct individual – adult male called John Coe)

## STRAIT OF GIBRALTAR BLUEFIN TUNA-FEEDERS

- 'Typical' black-and-white killer whale
- Medium- to large-sized oval eye patch (parallel to body axis)
- Conspicuous saddle patch uniformly gray and closed
- Male dorsal fin tip usually rounded and ends in pointed trailing tip (similar to NE Pacific resident)

**L:** ♂ 6m, ♀ 5.3m; **Max:** 7.3m

**Distribution** In spring and summer, known mainly from the Strait of Gibraltar, with sporadic sightings in the Gulf of Cadiz and Alboran Sea plus along the west and north coasts of the Iberian peninsula. They have been sighted in the Strait in fall and winter, but it is also possible that they follow bluefin tuna into the eastern Atlantic Ocean.

**Food and feeding** Atlantic bluefin tuna, which is chased at high speed and for up to 30 minutes at a time during spring and summer. Approximately half the population (pods A1 and A2) frequently depredates tuna from baited hooks in a drop longline fishery during summer. It is unknown whether they are dependent on tuna year-round; however, at least one pod (D) may opportunistically feed on coastal fish species.

**Group size** The small population of fewer than 40 is subdivided into five pods (A1, A2, B, C and D) each comprising 7–15 individuals.

**Remarks** In recent years there have been hundreds of apparently aggressive interactions between some individuals in this population and boats (mainly yachts). The cause is likely to be playful social behavior.

ADULT MALE STRAIT OF GIBRALTAR BLUEFIN TUNA-FEEDER

ADULT FEMALE STRAIT OF GIBRALTAR BLUEFIN TUNA-FEEDER

## NORTH-WEST ATLANTIC

- 'Typical' black-and-white killer whale
- Medium- to large-sized oval eye patch (parallel to body axis)
- Dorsal fin tip usually rounded and ends in pointed trailing tip (similar to NE Pacific resident)
- Conspicuous saddle patch uniformly gray and mostly closed

**L:** ♂ 6.7m, ♀ 5.5–6.5m; **Max:** unknown

**Distribution** Known mainly from Newfoundland and Labrador (Canada), with some sightings further south into Maine (USA), especially in summer (though there is considerably less observer effort in winter, the seasonal arrival of pack ice likely limits distribution). Fishermen are also reporting killer whales far offshore, on the Grand Banks of Newfoundland.

**Food and feeding** Takes a variety of seals, other cetaceans, fish, cephalopods and occasionally seabirds. There appears to be some group-specific prey specialization (e.g. other whales in eastern and Arctic Canada, fish and cephalopods in west Greenland, marine mammals in east Greenland).

**Group size** Typically 2–6 (average five); rarely more than 15, occasionally as many as 30; one-quarter of all sightings are lone individuals.

# KNOWN KILLER WHALE ECOTYPES IN THE ANTARCTIC AND ADJACENT WATERS

## TYPE A (ANTARCTIC KILLER WHALE)

- 'Typical' black-and-white killer whale
- Possibly largest of Antarctic ecotypes (Large Type B may be equal in size)
- Medium-sized oval eye patch (oriented parallel to body axis)
- Male saddle patch usually closed, female's can be slightly open
- Usually no visible dorsal cape
- Saddle patch can be brownish
- White patches occasionally tinted slightly yellowish (with diatoms)
- Black body color occasionally tinted slightly brownish (with diatoms)

**L:** ♂ 7.3m, ♀ 6.4m; **Max:** 9.2m

**Distribution** During the southern summer, it is circumpolar in Antarctic waters, mostly in offshore, ice-free, open water; frequently seen around the Antarctic Peninsula. Seasonal movements are poorly understood, but it is known to migrate away from Antarctica to lower, warmer latitudes, at least for short periods. Killer whales that look like Type A have been observed in New Zealand, Australia, South Africa, West Africa and Chile, as well as the Crozet Archipelago, the Kerguelen Islands and Macquarie Island, but whether any of these other whales migrate to Antarctic waters is unknown. Currently, Type A is a 'catch-all' ecotype that includes anything that is not Type B, C or D – and, ultimately, it could include more than one ecotype.

**Food and feeding** In Antarctic waters, predominantly Antarctic minke whales and elephant seals, though it may take calves of other baleen whale species and other seals; it has also been observed chasing (though not catching) penguins. It is not known what this ecotype feeds on when away from Antarctic waters, though it is likely Type A in Chile has taken sea lions and dolphins.

**Group size** 10–15 (ranges from 1–38).

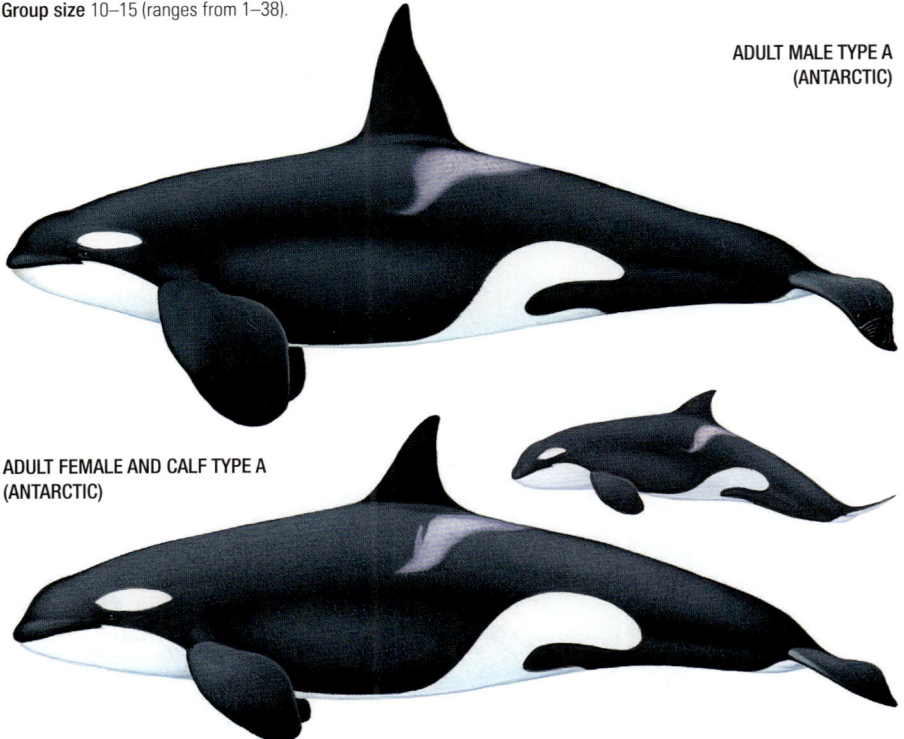

**ADULT MALE TYPE A (ANTARCTIC)**

**ADULT FEMALE AND CALF TYPE A (ANTARCTIC)**

## LARGE TYPE B (PACK ICE KILLER WHALE)

- Two-tone gray and white (not black and white)
- Larger and more robust body than Small Type B (Gerlache) killer whale
- Often covered in diatoms (turns white areas yellowish, gray areas brownish)
- Eye patch variable but always larger than any other killer whale (at least twice as large as in Type A)
- Eye patch oriented parallel to body axis
- Small oval scars from cookiecutter shark bites common
- Saddle patch almost always closed
- Dark gray dorsal cape (often demarcated by narrow white border originating as an extension of lower saddle) stretching from forehead to just behind dorsal fin
- Very similar to Small Type B (Gerlache) killer whales but twice the size
- Often seen spyhopping around ice floes (looking for seals)

**L:** ♂ unknown, ♀ unknown. **Max:** 9m; estimated to be at least twice the bulk of the Gerlache killer whale.

**Distribution** Circumpolar in Antarctic waters during summer, mostly inshore around dense pack ice and, especially, floe ice. It is common around the northwestern half of the Antarctic Peninsula and retreats south with the summer break-up of fast ice. Winter distribution is unknown but, while it apparently spends most of the year in Antarctic waters, it periodically undertakes rapid round-trip migrations to tropical and sub-tropical waters (30–37°S). These are known as 'maintenance migrations', believed to allow skin regeneration without the high cost of heat loss. The Antarctic Peninsula population migrates north, east of the Falkland Islands and Argentina, to Uruguay and Brazil. It is likely that 'clean' gray-and-white individuals have recently returned from the tropics (where they shed diatoms with their skin).

**Food and feeding** Feeds preferentially on Weddell seals (usually in coordinated groups, by 'wave-washing') and will often ignore crabeater and leopard seals. It occasionally takes Antarctic minke whales and elephant seals, and may also take humpback whale calves.

**Group size** Usually fewer than 10.

**ADULT MALE LARGE TYPE B (PACK ICE)**

**ADULT MALE LARGE TYPE B (PACK ICE), WITH DIATOMS (not to scale)**

**ADULT FEMALE AND CALF LARGE TYPE B (PACK ICE)**

## SMALL TYPE B (GERLACHE KILLER WHALE)

- Two-toned gray and white (not black and white)
- Smaller and slimmer body than Large Type B (Pack Ice) killer whale
- Often covered in diatoms (turns white areas yellowish, gray areas brownish)
- Small oval scars from cookiecutter shark bites common
- Eye patch variable but always larger than in any other killer whale (except Pack Ice killer whale)
- Eye patch narrower than Pack Ice killer whale
- Saddle patch usually (but not always) closed
- Eye patch may be oriented parallel to body axis or slightly slanted
- Dark gray dorsal cape present (though can be indistinct), stretching from just in front of eye patch backward to just behind dorsal fin, and continuous with the lower leading edge of the saddle (often demarcated by narrow white border originating in saddle)
- Very similar to Large Type B (Pack Ice) killer whales but half the size

**L:** ♂ unknown, ♀ unknown. **Max:** 7m; estimated to be roughly half the bulk of Pack Ice killer whales.

**Distribution** Known mainly from the Antarctic Peninsula (western side and the western Weddell Sea); the Gerlache Strait and Antarctic Sound are hotspots. Usually in more open water (it tends to avoid pack ice), often near penguin colonies. Spends much of the year in Antarctica, but periodically undertakes rapid (6–7-week) round-trip 'maintenance migrations' to tropical and sub-tropical waters (30–37°S).

**Food and feeding** Has only been observed feeding on penguins, especially gentoo and chinstrap (it eats only the breast muscles and discards the rest), but it probably feeds mainly on fish and squid caught near the ocean floor (it is a deep diver).

**Group size** Often 50-plus.

ADULT MALE SMALL TYPE B (GERLACHE)

ADULT FEMALE AND CALF SMALL TYPE B (GERLACHE)

KILLER WHALE 135

## TYPE C (ROSS SEA KILLER WHALE)

- Two-toned gray and white (not black and white)
- May be covered in diatoms (turns white areas yellow or orange, black and gray areas brown)
- Small oval scars from cookiecutter shark bites common
- Smallest killer whale ecotype known
- Dark gray dorsal cape usually visible (often demarcated by narrow white border originating in saddle)
- Small, narrow, wispy eye patch (slanted forwards at 45° angle to body axis)
- Saddle patch usually closed and very distinct

**L:** ♂ 5.6m, ♀ 5.2m. **Max:** 6.1m; weighs several times less than Type A and Type B (Pack Ice) killer whales (which could conceivably prey on it).

**Distribution** Known mainly from East Antarctica, predominantly in the Ross Sea but also west along the Adélie Land to Wilkes Land coasts, with smaller numbers as far west as Prydz Bay. Commonly reported in McMurdo Sound. It lives in dense pack ice, polynyas and leads in the fast ice (often many kilometers from open water) and is concentrated where most sea ice remains year-round. Spends most of the year in Antarctica (and has been recorded in the sea ice during winter) but the presence of cookiecutter shark bites and sightings off New Zealand and Australia indicate that there are at least some migrations to lower latitude and even tropical and sub-tropical waters.

**Food and feeding** Only known to feed on fish, primarily large 2m-long Antarctic toothfish, but also takes at least two much smaller species of icefish and may take super-abundant (but very small) Antarctic silverfish. Limited evidence of hunting penguins. It routinely dives to 200–400m, with a maximum of at least 700m.

**Group size** 10–120 (up to 200); group size appears to have been decreasing in recent years (current average about 14).

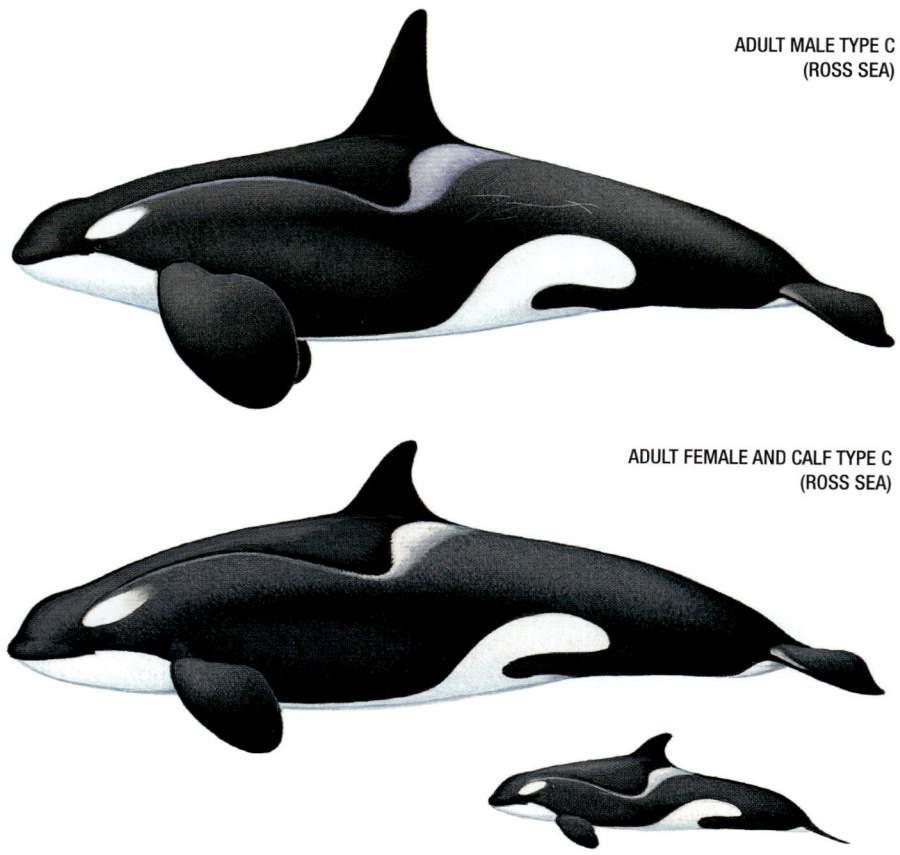

**ADULT MALE TYPE C (ROSS SEA)**

**ADULT FEMALE AND CALF TYPE C (ROSS SEA)**

# TYPE D (SUB-ANTARCTIC KILLER WHALE)

- 'Typical' black-and-white killer whale
- Distinctive minuscule white eye patch (parallel to body axis) – sometimes absent – makes identification easy
- Noticeably bulbous melon, cf. other killer whales (more like pilot whale in some individuals)
- Marked sexual dimorphism in dorsal fin size and shape (as in other ecotypes)
- Male dorsal fin relatively short, narrow, noticeably swept back and with sharply pointed tip (more falcate and pointed than other Antarctic ecotypes)
- Moderately conspicuous saddle patch
- No conspicuous dorsal cape
- No yellowish or brownish coloring (caused in some ecotypes by diatoms)

**L:** ♂ unknown, ♀ unknown. **Max:** 7.3m.

**Distribution** Circumpolar in sub-Antarctic waters, ranging from 40°S to 60°S; predominantly offshore, though sometimes associated with islands. There have been many live sightings in recent years, at the northern edge of the Southern Ocean (including at Crozet, South Georgia and New Zealand sub-Antarctic islands); and it is now being seen almost annually in the Drake Passage and between the Falkland Islands and South Georgia (one particular hotspot is an area along the continental shelf edge, south of Cape Horn, in Chilean waters).

**Food and feeding** Little known, but certainly includes fish (it takes Patagonian toothfish from longline fisheries near the Crozet Archipelago and off Chile).

**Group size** Range 9–35, average 18 (but based on little information).

**Remarks** This is the most distinctive-looking killer whale ecotype, immediately recognizable by its extremely small (or sometimes absent) white eye patch.

# SHORT-FINNED PILOT WHALE
*Globicephala macrorhynchus*     Gray, 1846

The short-finned pilot whale is a distinctive-looking animal but, at sea, is virtually impossible to tell apart from its close relative, the long-finned pilot whale (the main distinguishing feature is a subtle difference in the length and shape of their flippers). Both species are highly sexually dimorphic in both size and appearance.

**IUCN status** Least Concern (2018).
**Population** Unknown. Approximate regional estimates total *c.* 700,000, but large swathes of the range have not been surveyed. Trend unknown.
**Classification** Odontoceti, family Delphinidae.
**Taxonomy** No recognized subspecies. However, there appear to be three morphologically and genetically distinct forms with distinguishable vocal repertoires of undetermined status (possibly subspecies): Atlantic Ocean (largely isolated by the Benguela Barrier off South Africa, separating the Atlantic and Indian Oceans); western/central Pacific and Indian Oceans (naisa type); and eastern Pacific Ocean and northern Japan (shiho type, separated by the East Pacific Barrier – the large, deep open ocean that limits the dispersal of many tropical species between the eastern Pacific Ocean and the central/western Pacific Ocean).
**Other names** Pothead, blackfish (term normally used for non-taxonomic group of six dark-colored members of Delphinidae with 'whale' in their name).
**DISTRIBUTION** Widely distributed in deep tropical, sub-tropical and warm temperate waters worldwide. Does not normally range north of 50°N or south of 40°S. Absent from the Mediterranean Sea (where the long-finned pilot whale is resident). There are long-term residents in some areas, but other populations may move long distances (individual whales have been recorded traveling up to 2,400km). Seasonal inshore–offshore (winter/early spring–summer/fall) movements in some regions are related to the seasonal spawning migrations of squid. Prefers the continental shelf break, continental

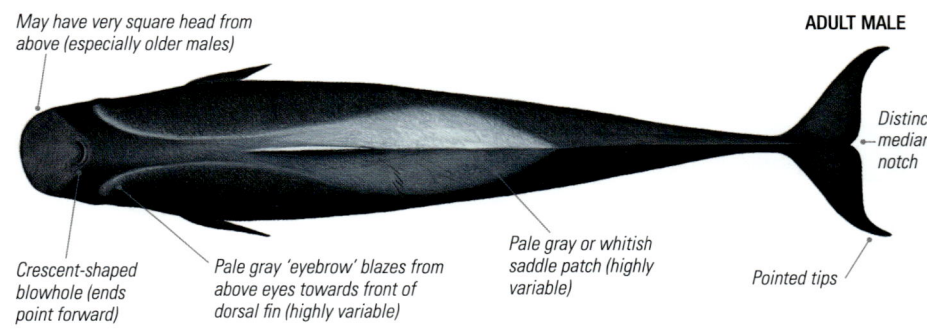

*May have very square head from above (especially older males)*     **ADULT MALE**

*Distinct median notch*

*Crescent-shaped blowhole (ends point forward)*

*Pale gray 'eyebrow' blazes from above eyes towards front of dorsal fin (highly variable)*

*Pale gray or whitish saddle patch (highly variable)*

*Pointed tips*

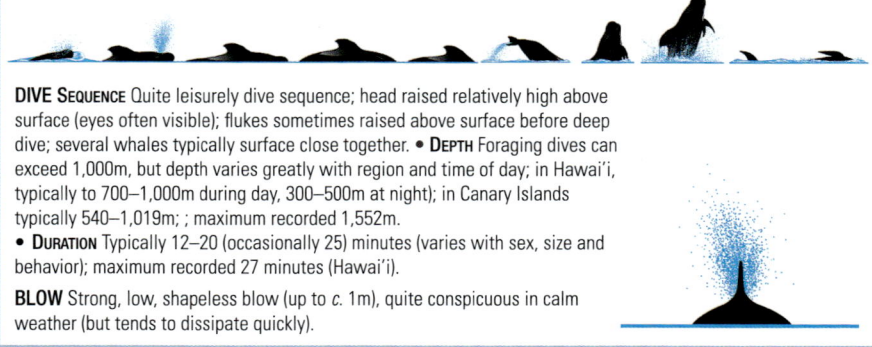

**DIVE SEQUENCE** Quite leisurely dive sequence; head raised relatively high above surface (eyes often visible); flukes sometimes raised above surface before deep dive; several whales typically surface close together. • **DEPTH** Foraging dives can exceed 1,000m, but depth varies greatly with region and time of day; in Hawai'i, typically to 700–1,000m during day, 300–500m at night); in Canary Islands typically 540–1,019m; ; maximum recorded 1,552m.
• **DURATION** Typically 12–20 (occasionally 25) minutes (varies with sex, size and behavior); maximum recorded 27 minutes (Hawai'i).

**BLOW** Strong, low, shapeless blow (up to *c.* 1m), quite conspicuous in calm weather (but tends to dissipate quickly).

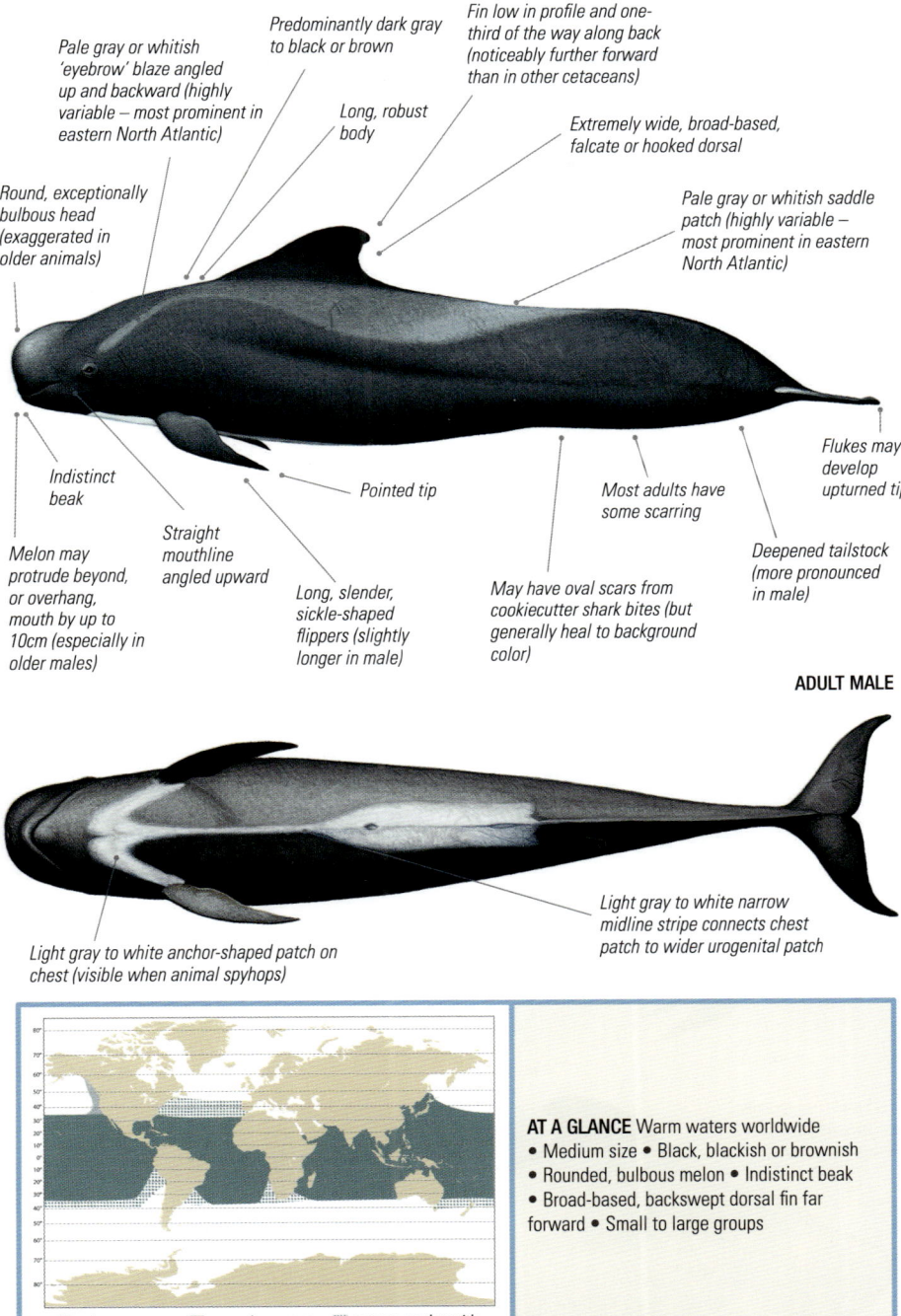

slope and island slope waters, and areas with complex topography such as seamounts and ridges. Abundance is lower in deep oceanic environments. Will approach nearshore areas where the water is sufficiently deep.

**BEHAVIOR** Often observed in aggregations with other cetacean species. There are accounts of pilot whales behaving aggressively toward some cetaceans and they tend to harass larger whales, but the role is sometimes reversed with smaller species (such as melon-headed whales), when the pilot whales are the victims. Oceanic whitetip sharks often follow pilot whales, to scavenge on lost or discarded prey (or they rely on the whales to find prey at depth). More aerially active than the long-finned pilot whale, breaching occasionally (though not as often as many smaller delphinids), spyhopping and lobtailing. Spends much of the day logging (resting) at the surface. More prone to mass strandings than almost any other cetacean (apart from the long-finned pilot whale), probably in part because of its strong social bonds. Will grieve for dead members of the group and will carry a dead calf around for hours or days. Reaction to boats varies according to location.

**FOOD AND FEEDING** Mainly squid; some octopus, and mid- and deepwater fish. While foraging may spread out in 'chorus line' up to 3km long.

**TEETH** Upper jaw 14–18; lower jaw 14–18.

**GROUP SIZE AND STRUCTURE** Highly social, living in matrilineal groups (consisting of a matriarch with her immediate kin) similar in structure to killer whale groups (though not quite as stable). Typical group size is 15–50 (average 18 around the main Hawaiian islands, 15 in Madeira), including all ages and sexes (though there tend to be more adult females). They usually remain in the family group for life. Males will mate during temporary aggregations of separate family groups. Several family groups may join together to form a pod or school, typically with 30–90 members (up to several hundred). Rarely seen alone (very occasionally, lone adult males are reported).

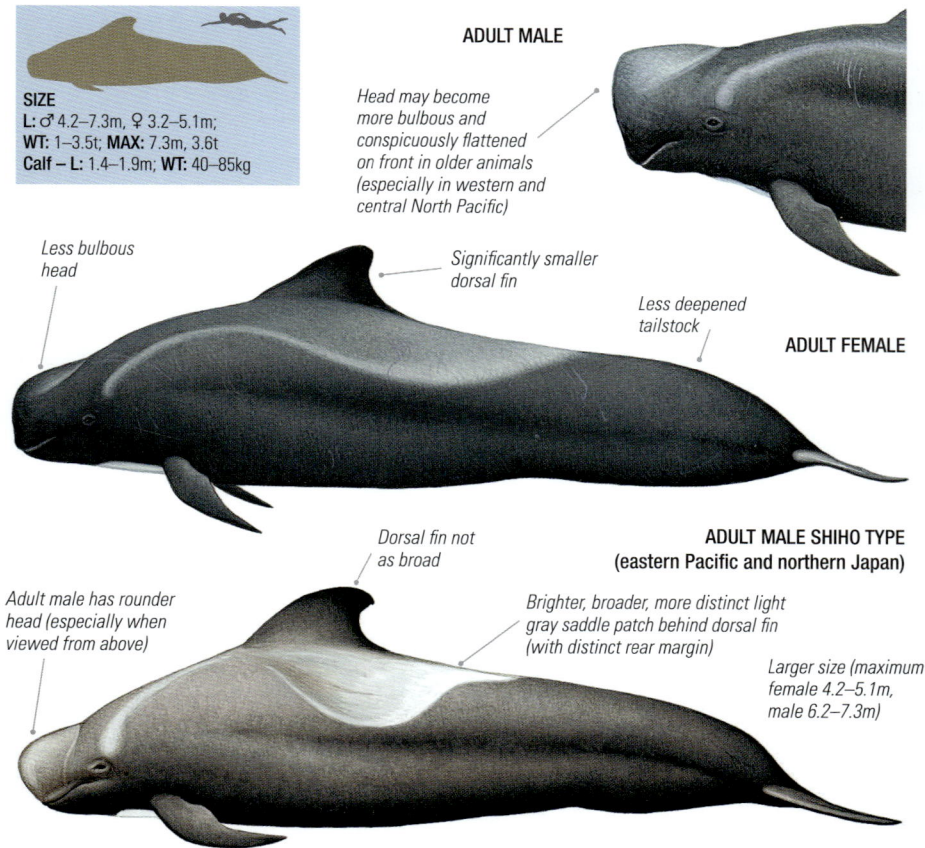

**SIZE**
**L:** ♂ 4.2–7.3m, ♀ 3.2–5.1m;
**WT:** 1–3.5t; **MAX:** 7.3m, 3.6t
**Calf – L:** 1.4–1.9m; **WT:** 40–85kg

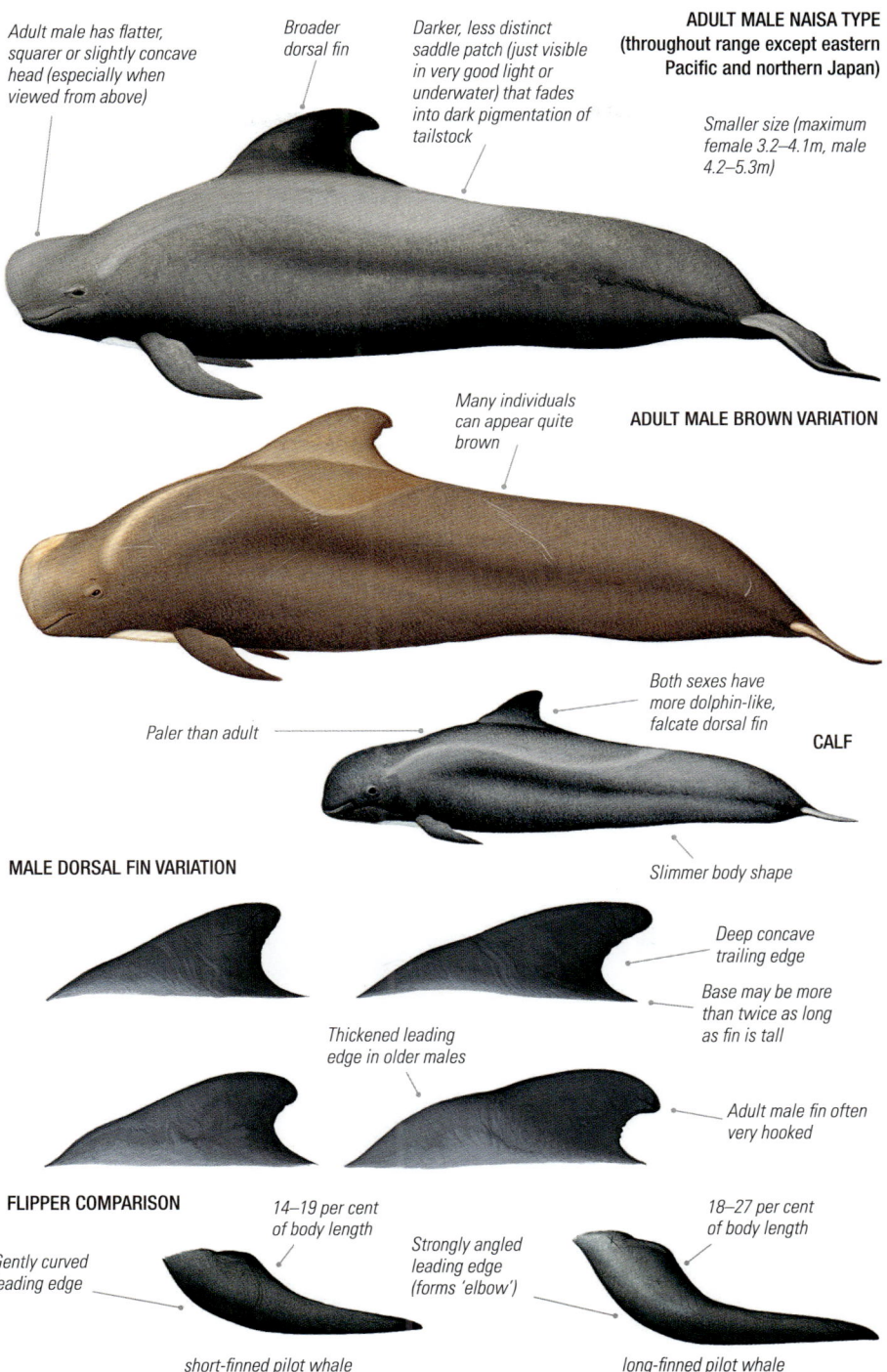

# LONG-FINNED PILOT WHALE
*Globicephala melas* (Traill, 1809)

With experience, it is possible to tell the sex and approximate age of long-finned and short-finned pilot whales by looking at their dorsal fins: they change shape as they grow older and are quite different on females and males. The adult male pilot whale's fin is like no other: low in profile and exceptionally broad-based.

**IUCN status** Least Concern (2018). Mediterranean sub-population Endangered (2021). Strait of Gibraltar sub-population Critically Endangered (2021).
**Population** Unknown, a guesstimate of *c.* 1 million suggested (based on available regional estimates) including *c.* minimum ½ million in the North Atlantic. Trend unknown.
**Classification** Odontoceti, family Delphinidae.
**Taxonomy** Two subspecies recognized: North Atlantic long-finned pilot whale (*G. m. melas*) and southern long-finned pilot whale (*G. m. edwardii*). An undescribed subspecies in Japan and Alaska thousands of years ago (now extinct) was informally known as the North Pacific long-finned pilot whale.
**Other names** Pothead, caaing whale, blackfish (term normally used for non-taxonomic group of six dark-colored members of Delphinidae with 'whale' in their name).
**DISTRIBUTION** Widely distributed in deep cold temperate to sub-polar waters, but separated by a wide tropical belt. Prefers the continental shelf break, continental slope and island slope waters, and areas with complex topography such as seamounts and ridges. Most sightings are in waters deeper than 2,000m.
**BEHAVIOR** Often observed in mixed-species aggregations. Less aerially active than the short-finned pilot whale, often spyhopping and lobtailing but only occasionally breaching. Spends much of the day logging (resting) at the surface. Like short-finned pilot whale, more prone to mass strandings than almost any other cetacean, probably partly because of its strong social bonds. Reaction to boats varies according to location.
**FOOD AND FEEDING** Mainly squid and other cephalopods; some small to medium-sized fish; occasionally shrimps; however, great variation according to location. Deep foraging tends to be at night in most regions.
**TEETH** Upper jaw 16–26; lower jaw 16–26.
**GROUP SIZE AND STRUCTURE** Highly social, living in matrilineal groups similar in structure to killer whale groups (though not quite as stable). Typical family group size is 8–20, with considerable geographical variation, including all ages and sexes. They remain in the group for life. Several family groups may join to form a pod or school, frequently with up to 50 members (sometimes 100+ and occasionally up to 1,200).

**ADULT MALE SOUTHERN HEMISPHERE**

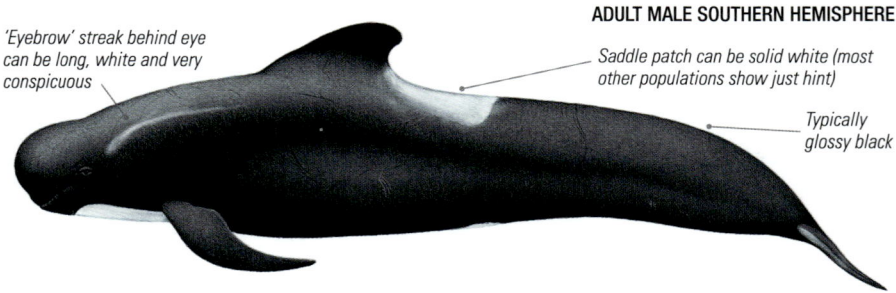

'Eyebrow' streak behind eye can be long, white and very conspicuous

Saddle patch can be solid white (most other populations show just hint)

Typically glossy black

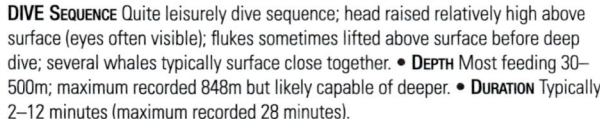

**DIVE SEQUENCE** Quite leisurely dive sequence; head raised relatively high above surface (eyes often visible); flukes sometimes lifted above surface before deep dive; several whales typically surface close together. • **DEPTH** Most feeding 30–500m; maximum recorded 848m but likely capable of deeper. • **DURATION** Typically 2–12 minutes (maximum recorded 28 minutes).

**BLOW** Strong, low, shapeless blow (up to *c.* 1m height); quite conspicuous in calm weather, but tends to dissipate quickly.

## ADULT MALE NORTH ATLANTIC

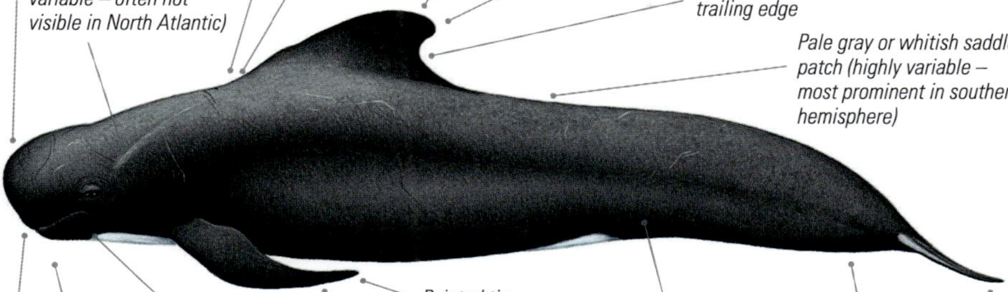

- Round, exceptionally bulbous head (exaggerated and sometimes with flattened front in older animals)
- Predominantly dark gray to jet black or brown
- Extremely wide, broad-based, falcate or hooked dorsal fin
- Fin low in profile and one-third of the way along back (noticeably further forward than in other cetaceans except short-finned pilot whale)
- Pale gray or whitish 'eyebrow' blaze angled up and backward (highly variable – often not visible in North Atlantic)
- Long, robust body
- Deep, concave trailing edge
- Pale gray or whitish saddle patch (highly variable – most prominent in southern hemisphere)
- Indistinct beak
- Straight mouthline angled upward
- Pointed tip
- Long, slender, sickle-shaped flippers (slightly longer in male)
- Most adults have some scarring
- Deepened tailstock (more pronounced in male)
- Flukes may develop upturned tips
- Melon may protrude beyond, or overhang, mouth by up to 10cm (especially in older males)

## ADULT FEMALE NORTH ATLANTIC

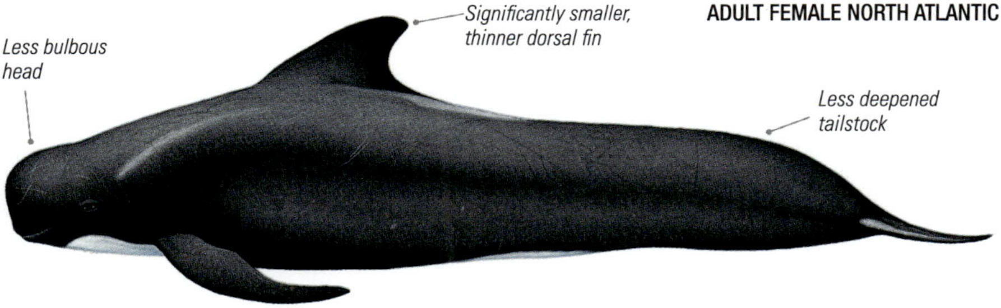

- Less bulbous head
- Significantly smaller, thinner dorsal fin
- Less deepened tailstock

**SIZE**
L: ♂ 4–6.7m, ♀ 3.8–5.7m;
WT: 1.3–2.3t; MAX: 6.7m, 2.3t
Calf – L: 1.7–1.8m; WT: c. 75–80kg

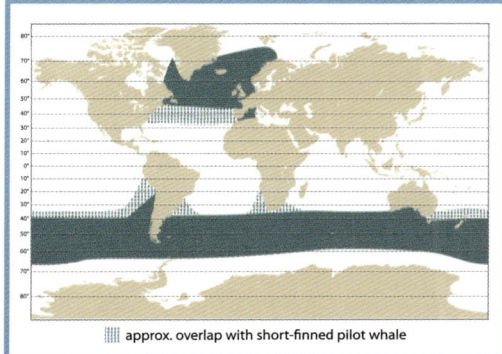

approx. overlap with short-finned pilot whale

**AT A GLANCE** Cold waters of North Atlantic and southern hemisphere • Medium size • Black, blackish or brownish • Rounded, bulbous melon • Indistinct beak • Broad-based, backswept dorsal fin positioned far forward • Small to large groups

LONG-FINNED PILOT WHALE

# FALSE KILLER WHALE
*Pseudorca crassidens* (Owen, 1846)

Despite its name, the false killer whale belongs taxonomically to the dolphin family, Delphinidae, and it often behaves more like one of its energetic and sprightly smaller relatives.

**IUCN status** Near Threatened (2018).
**Population** Unknown. Considered among the least abundant delphinids, even in locations with the highest densities. Trend unknown (but the best studied population, around Hawai'i, has been declining).
**Classification** Odontoceti, family Delphinidae.
**Taxonomy** No recognized forms or subspecies.
**Other names** Pseudorca, blackfish (term normally used for non-taxonomic group of six dark-colored members of Delphinidae with 'whale' in their name).

**DISTRIBUTION** Tropical to warm temperate waters worldwide, mainly between *c.* 50°N and *c.* 50°S. Density is much higher in lower latitudes. Sightings in cooler temperate waters (such as the Baltic Sea, off the UK and Canada's British Columbia) are generally considered extralimital. Primarily favors deep oceanic waters, particularly those deeper than 1,000m; it also occurs where deep water approaches the coast, especially around oceanic islands. However, certain populations appear to be more coastal and occur in shallower water.

**BEHAVIOR** An exuberant, fast-swimming cetacean. Often leaps clear of the water, especially when attacking prey; will often breach with prey in the mouth, and may throw it quite high into the air. Mass strandings are fairly common (likely due to strong social bonds). Regularly associates with other cetaceans and long-term associations with bottlenose dolphins have been documented. Not shy of boats.

**FOOD AND FEEDING** Varies by region, but mainly large fish and squid; will occasionally attack and eat other small cetaceans, but more often associates in a non-aggressive manner. Cooperative feeder and will share prey; hunts day and night.

**TEETH** Upper jaw 14–22; lower jaw 16–24.

**GROUP SIZE AND STRUCTURE** Highly variable depending on location, typically ranging from 10–60 (less commonly 2–100); exceptionally, larger groups have been reported (the largest mass stranding involved at least 835 animals). Where one small group is present, often other small groups are scattered over a wider area. Smaller groups are generally stable and consist primarily of closely related individuals and are of mixed age and sex; females (and possibly males) seem to remain within the social group in which they were born. Larger groups are likely temporary associations of smaller, more stable groups.

**DORSAL FIN VARIATIONS**

*Fin shape highly variable*

*Usually rounded tip*

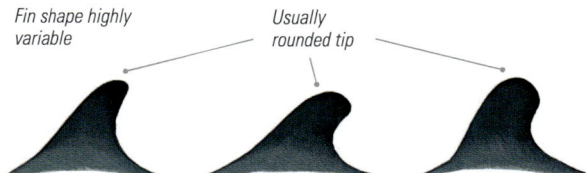

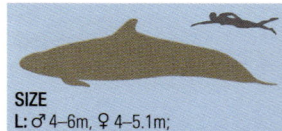

**SIZE**
**L:** ♂ 4–6m, ♀ 4–5.1m;
**WT:** 1.1–2t; **MAX:** 6.1m, 2.2t
**Calf – L:** 1.5–2.1m; **WT:** *c.* 80kg
Males larger than females; regional size differences (e.g. 10–20 per cent larger in Japan than South Africa).

**DIVE SEQUENCE** When swimming slowly, head and melon break surface (eyes may be visible), and may strongly arch tailstock but flukes rarely visible; when swimming quickly, may porpoise just clear of water in low, flat arcs, but often shows little more than splash and dorsal fin. • **DEPTH** Most feeding near surface, but will forage along seafloor; capable of more than 1,000m. • **DURATION** Long dives typically 4–6 minutes; maximum recorded 18 minutes.
**BLOW** Short and bushy but only sometimes visible.

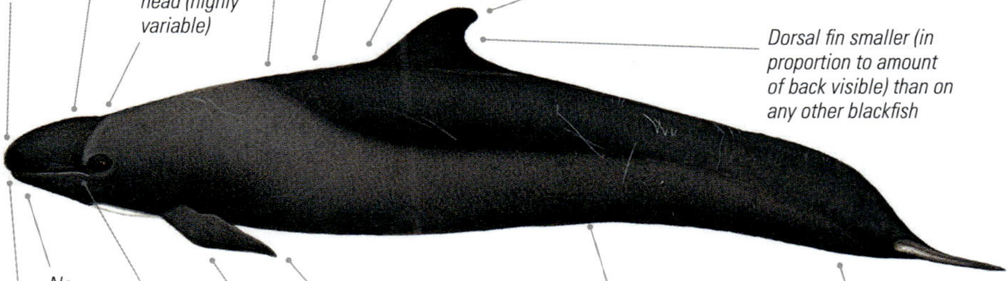

**ADULT**

- Front of head may be flattened in older males
- Small, conical head with non-bulbous melon
- Dark gray to black (may appear slightly paler slate-gray in bright light)
- Slightly darker dorsal cape (visible only in good light)
- Relatively narrow-based, falcate dorsal fin midway along back
- May have light gray areas on sides of head (highly variable)
- Long, slender body
- Dorsal fin taller than it is long (18–41cm high)
- Dorsal fin smaller (in proportion to amount of back visible) than on any other blackfish
- No discernible beak
- Long mouthline
- Melon overhangs lower jaw more in male
- Narrow flippers relatively far forward on body
- Distinct bulge on leading edge (makes S-shape)
- May be light star-shaped scars from cookiecutter shark bites (most heal to same color as background)
- Deep tailstock

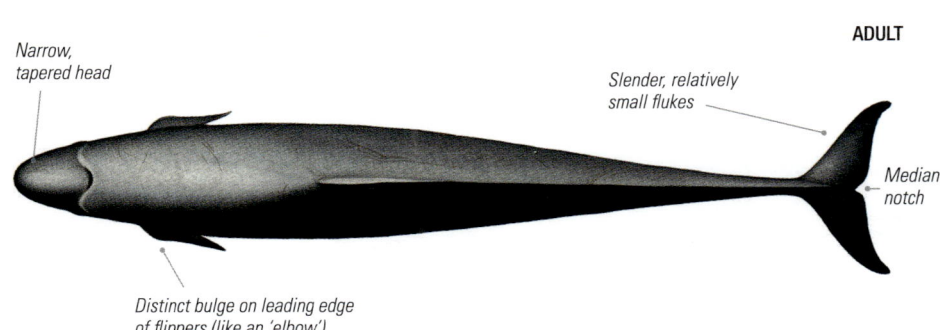

**ADULT**

- Narrow, tapered head
- Slender, relatively small flukes
- Median notch
- Distinct bulge on leading edge of flippers (like an 'elbow')

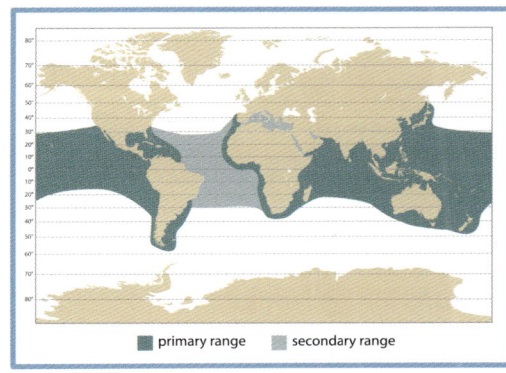

primary range  secondary range

**AT A GLANCE** Warm (mainly offshore) waters worldwide • Medium size • Dark gray to black • Long, slender body • Relatively narrow-based, falcate dorsal fin with rounded tip • Small, conical head with no beak • Distinct bulge on leading edge of flippers • Small, often exuberant groups

FALSE KILLER WHALE

# PYGMY KILLER WHALE
*Feresa attenuata*  Gray, 1874

Despite its name, the pygmy killer whale belongs taxonomically to the dolphin family, Delphinidae. Until 1952 it was known from only two skulls, collected in 1827 and 1874. It is still poorly known, but our knowledge has improved dramatically in recent years.

**IUCN status** Least Concern (2017).
**Population** Unknown. Trend unknown.
**Classification** Odontoceti, family Delphinidae.
**Taxonomy** No recognized forms or subspecies.
**Other names** Blackfish (term normally used for non-taxonomic group of six dark-colored members of Delphinidae with 'whale' in their name).
**DISTRIBUTION** Tropical and sub-tropical waters worldwide, between 40°N and 35°S, overlapping almost exactly with the melon-headed whale. Most sightings are in deep waters offshore and around oceanic islands where deep, clear waters are found near the coast.
**BEHAVIOR** Pygmy killer whales can be quite difficult to spot and, while high breaches have been observed, rarely engage in aerial behavior. Their reaction to boats is extremely variable, but they will occasionally bow-ride slow-moving vessels. They spend most of the day traveling slowly, socializing or resting motionless at the surface.
**FOOD AND FEEDING** Mostly squid and fish; known to attack and possibly eat other dolphins in eastern tropical Pacific. Most feeding seems to occur at night.
**TEETH** Upper jaw 16–22; lower jaw 22–26.
**GROUP SIZE AND STRUCTURE** Usually 12–50, although they have been encountered in pairs and herds of up to several hundred; the average group size in Hawai'i is nine.

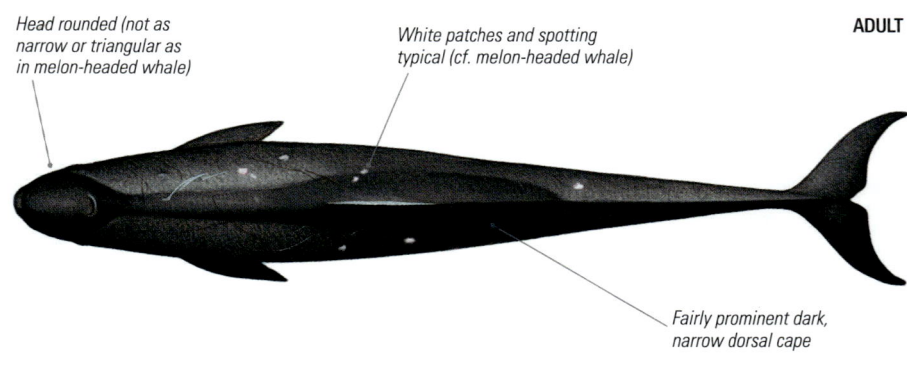

Head rounded (not as narrow or triangular as in melon-headed whale)

White patches and spotting typical (cf. melon-headed whale)

**ADULT**

Fairly prominent dark, narrow dorsal cape

**SIZE**
**L:** ♂ 2–2.6m, ♀ 2–2.6m;
**WT:** 110–170kg; **MAX:** 2.7m, 228kg
**Calf – L:** *c.* 80cm; **WT:** *c.* 15kg

**DIVE SEQUENCE** Surfaces quietly and discreetly (rather a slow, sluggish swimmer), rarely porpoising and keeping low profile; herd often swims shoulder to shoulder in coordinated 'chorus line'. • **DEPTH** Believed to feed at depth (one dive in Gulf of Mexico to 368m); in Hawai'i, most commonly seen in water 500–3,500m deep. • **DURATION** Unknown (more than 9 minutes recorded).
**BLOW** Rarely visible.

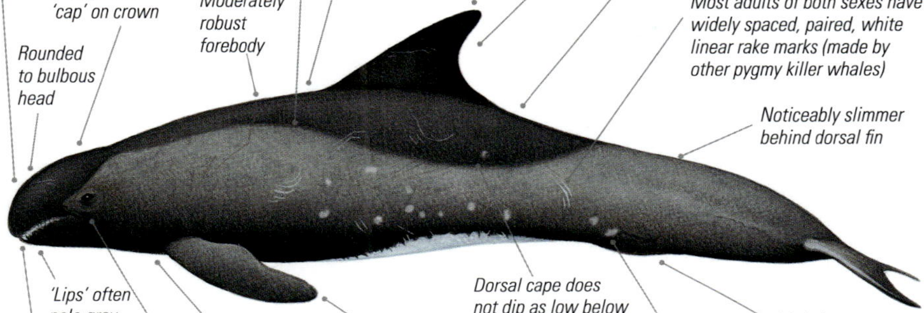

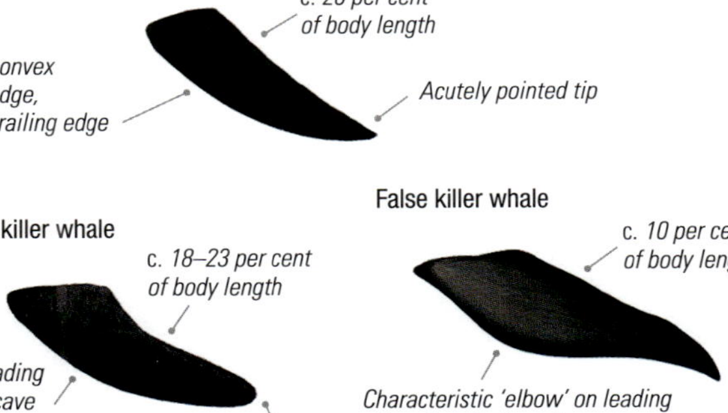

**ADULT**

- Melon may be more extensive (even forward-leaning) in some older individuals
- Cape broadens to large dark 'cap' on crown
- Rounded to bulbous head
- Distinct demarcation between darker cape and lighter gray sides
- Moderately robust forebody
- In poor light appears uniformly dark gray or blackish
- Tall, falcate dorsal fin in middle of back (shape varies with age)
- Trailing edge of dorsal fin may be damaged
- Fairly prominent dark, narrow dorsal cape
- Most adults of both sexes have widely spaced, paired, white linear rake marks (made by other pygmy killer whales)
- Noticeably slimmer behind dorsal fin
- 'Lips' often pale gray to white
- Neck very flexible
- No face 'mask'
- No visible beak (though upper jaw slightly overhangs tip of lower jaw)
- Moderately long, slender flippers with rounded tips
- Dorsal cape does not dip as low below dorsal fin as on melon-headed whale
- Cookiecutter shark scars often present (may be tinged pink from blood flow) especially in older animals
- Male has pronounced ventral keel (not present on female)

**HEAD VARIATIONS**

Melon-headed whale

Pygmy killer whale

**FLIPPER COMPARISON**

c. 20 per cent of body length
- Slightly convex leading edge, straight trailing edge
- Acutely pointed tip

c. 18–23 per cent of body length
- Convex leading edge, concave trailing edge
- Rounded tip

False killer whale

c. 10 per cent of body length
- Characteristic 'elbow' on leading edge gives S-shaped appearance

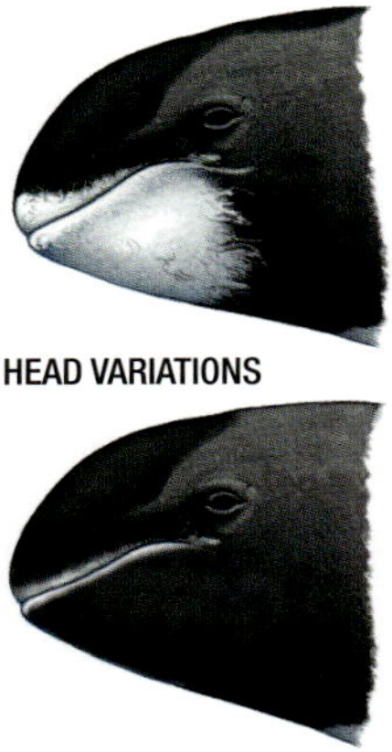

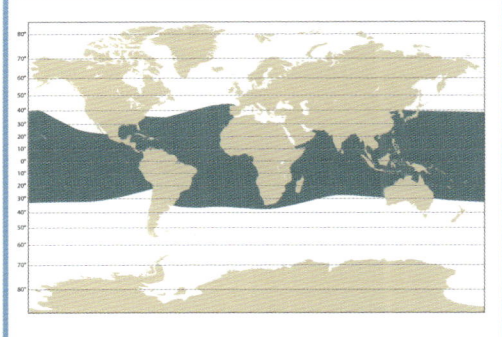

**AT A GLANCE** Tropical and subtropical waters worldwide • Small size • Appears uniformly dark in poor light • Distinct dark dorsal cape and 'cap' and no face 'mask' • Relatively large, broad dorsal fin in middle of back • Frequently paired, white linear rake marks • Rounded to bulbous head • Generally slow and lethargic • Typically in small herds of fewer than 50

# MELON-HEADED WHALE
*Peponocephala electra* (Gray, 1846)

Despite its name, the melon-headed whale belongs taxonomically to the dolphin family, Delphinidae. It was known only from skeletons until the 1960s, but nowadays is seen regularly in several parts of the world.

**IUCN status** Least Concern (2019).
**Population** Unknown. One guesstimate suggests low hundreds of thousands; considered to be relatively common in some parts of its range. Trend unknown.
**Classification** Odontoceti, family Delphinidae.
**Taxonomy** No recognized forms or subspecies.
**Other names** Electra dolphin, little blackfish, Hawaiian blackfish ('blackfish' is normally used for non-taxonomic group of six dark-colored members of Delphinidae with 'whale' in their name); sometimes affectionately called a 'pep', after the scientific name.

**DISTRIBUTION** Tropical and sub-tropical waters worldwide, overlapping almost exactly with the pygmy killer whale. Most records are between 20°N and 20°S, and it is rarely seen north of 40°N or south of 35°S (rare records from higher latitudes are usually associated with incursions of warm-water currents). Most sightings are in deep waters offshore and around oceanic islands (typically in depths of 300–2,000m). There is some evidence of inshore movements during the day (for resting and socializing) and offshore movements to feed at night.

**BEHAVIOR** Usually encountered in large, dense, fast-swimming herds, which are notorious for suddenly changing direction. They are often in mixed aggregations with Fraser's dolphins, and have been observed with other cetaceans. They tend to flee from approaching vessels in the eastern tropical Pacific, but will bow-ride elsewhere. Breaching and spyhopping are fairly common. During the day in calm seas, large groups often rest at the surface.

**FOOD AND FEEDING** Mainly squid, but also small fish and crustaceans; may also prey opportunistically on dolphins in some areas. Most feeding seems to occur at night.

**TEETH** Upper jaw 40–52; lower jaw 40–52.

**GROUP SIZE AND STRUCTURE** Usually in large, tight-knit groups of 100–500, with exceptional sightings of up to 2,000 individuals. Females remain in groups with their mother and sisters, while males move between groups.

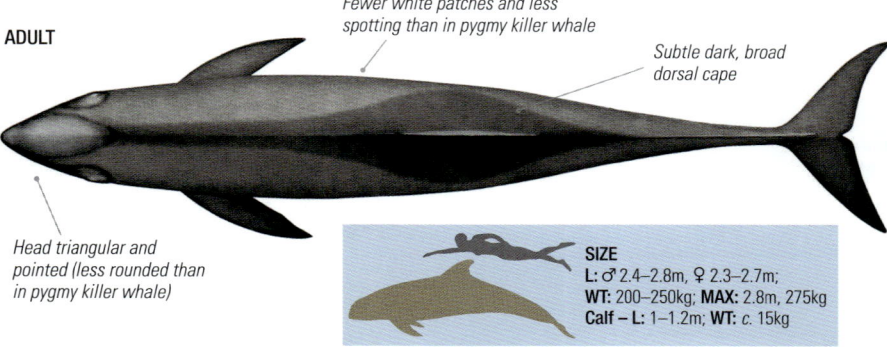

ADULT
Fewer white patches and less spotting than in pygmy killer whale
Subtle dark, broad dorsal cape
Head triangular and pointed (less rounded than in pygmy killer whale)

**SIZE**
**L:** ♂ 2.4–2.8m, ♀ 2.3–2.7m;
**WT:** 200–250kg; **MAX:** 2.8m, 275kg
**Calf – L:** 1–1.2m; **WT:** c. 15kg

**DIVE SEQUENCE** When swimming slowly, head and melon break surface briefly before dorsal fin appears, and flukes rarely visible; when swimming quickly, porpoises clear of water or skims surface, producing much spray.
• **DEPTH** Typically prefers depths greater than 1,000m and feeds deep in water column; maximum recorded dive 472m. • **DURATION** Maximum recorded 12 minutes.
**BLOW** Rarely visible.

### ADULT MALE

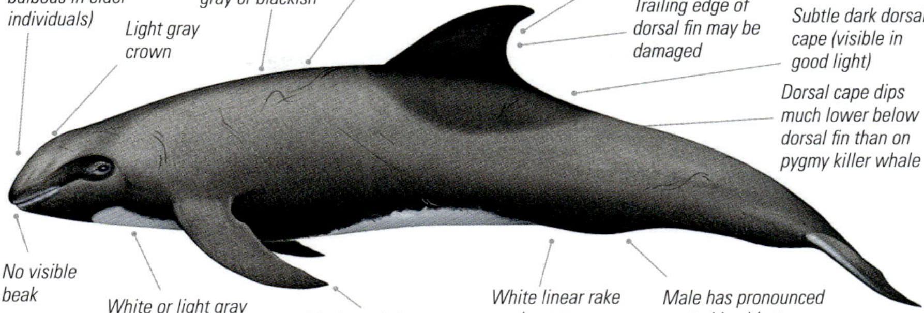

- Triangular head shape (can become more bulbous in older individuals)
- Light gray crown
- In poor light appears uniformly dark gray or blackish
- Moderately robust body
- Tall, broad-based dorsal fin in middle of back
- Dorsal fin often more falcate in older individuals
- Trailing edge of dorsal fin may be damaged
- Subtle dark dorsal cape (visible in good light)
- Dorsal cape dips much lower below dorsal fin than on pygmy killer whale
- No visible beak
- White or light gray areas common around throat (may extend along underside)
- Moderately long, sharply pointed, sickle-shaped flippers
- White linear rake marks rare
- Male has pronounced ventral keel (not present in female)

### ADULT FEMALE

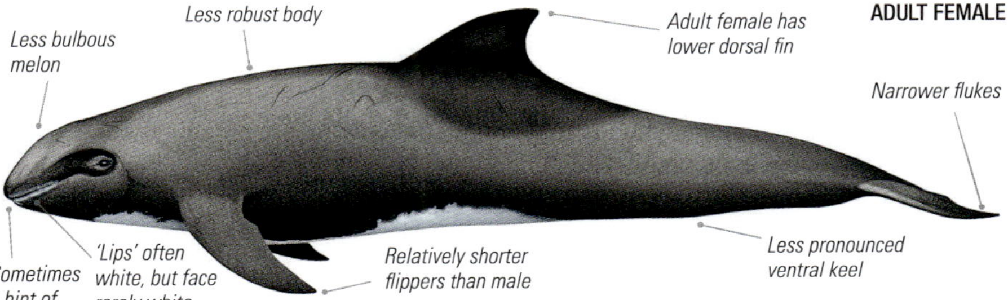

- Less bulbous melon
- Less robust body
- Adult female has lower dorsal fin
- Narrower flukes
- Sometimes a hint of beak
- 'Lips' often white, but face rarely white (both sexes)
- Relatively shorter flippers than male
- Less pronounced ventral keel

### HEAD COMPARISON

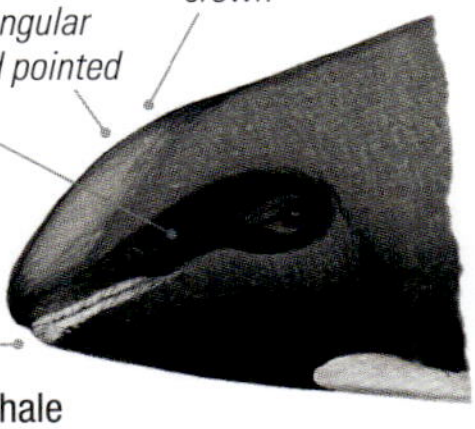

**Melon-headed whale**
- Head more triangular and pointed
- Light gray crown
- Subtle dark face 'mask' (more visible in good light)
- 'Lips' (and, in some older animals, tip of lower jaw) often pale gray to white

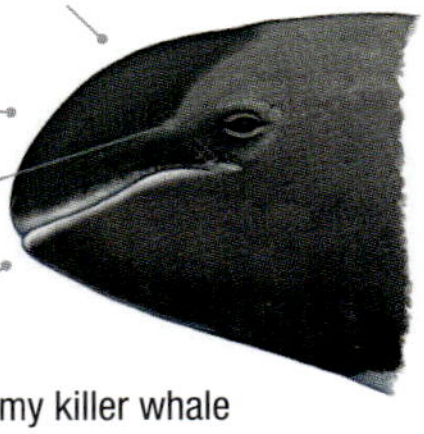

**Pygmy killer whale**
- Dark cap on crown
- Head more bulbous and rounded
- No face 'mask'
- 'Lips' (and sometimes entire chin) pale gray to white

**AT A GLANCE** Tropical and subtropical waters worldwide • Small size • Indistinct dark dorsal cape and face 'mask' visible in good light • Appears uniformly dark in poor light • Tall, broad-based dorsal fin in middle of back • Triangular, pointed head • May swim at high speed • Typically in large herds of 100+

# RISSO'S DOLPHIN
*Grampus griseus* (G. Cuvier, 1812)

Risso's dolphin is the most heavily scarred of all the dolphins and the largest species called 'dolphin'. There is huge variation in color – between individuals, age classes and regions – and this is one of the most distinctive characteristics of the species. It is quite easy to identify at close range – the only smallish, blunt-headed, heavily scarred cetacean that is typically light in color.

**IUCN status** Least Concern (2018). Mediterranean sub-population Endangered (2020).
**Population** Unknown. The sum of existing estimates is 425,000–433,000, but this is likely to be an underestimate. Trend unknown.
**Classification** Odontoceti, family Delphinidae.
**Taxonomy** No recognized forms or subspecies. Most closely related to the blackfish.
**Other names** Grampus; in older literature, the killer whale was also called grampus.

**DISTRIBUTION** Widely distributed in both hemispheres, from the tropics to cool temperate waters, occurring in all habitats from coastal to oceanic and including many semi-enclosed seas. However, it shows a strong range-wide preference for mid-temperate waters between *c.* 30° and 45° latitude. Favors water warmer than 12°C (it is rarely found in water less than 10°C). This results in seasonal movements in some regions. Prefers deep waters of the continental shelf break, upper slopes and submarine canyons, especially in areas with steep seafloor topography (generally 400–1,000m deep). It also occurs in some oceanic areas beyond the continental slope and will enter shallow coastal waters to feed seasonally on cuttlefish. There is evidence that habitat use is coordinated to avoid spatial and temporal overlap with other deep-diving odontocetes, including Cuvier's beaked whales and sperm whales.

**BEHAVIOR** During daytime, it is usually socializing, resting or traveling. When socializing it can be aerially active and will breach, spyhop (often revealing the entire head and body down to the flippers), head-slap, lobtail and flipper-slap. Commonly associates with other cetaceans. Readily bow-rides, wake-rides and associates with boats in some areas, but elsewhere it does not approach boats; it is not particularly shy or nervous, but typically maintains a 'personal space' and slowly turns away.

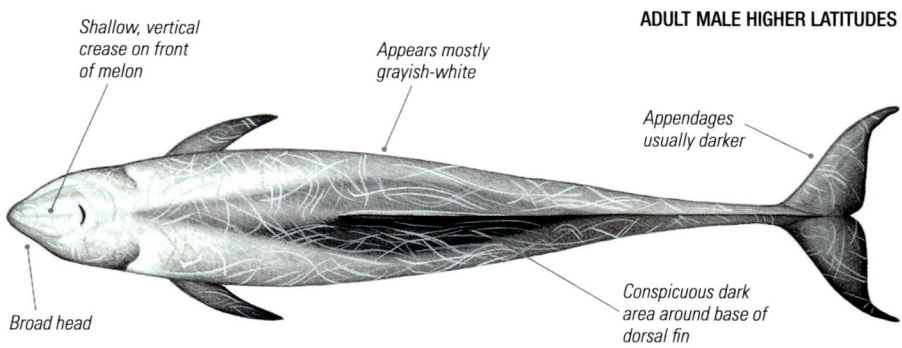

**ADULT MALE HIGHER LATITUDES**

- Shallow, vertical crease on front of melon
- Appears mostly grayish-white
- Appendages usually darker
- Broad head
- Conspicuous dark area around base of dorsal fin

**DIVE SEQUENCE** Usually surfaces slowly at 45° angle; eye usually appears above surface; tall dorsal fin conspicuous as back arches slightly; members of cluster often travel and surface in synchrony. • **DEPTH** Often less than 50m, but up to 400m; maximum recorded 656m. • **DURATION** Typically 1–10 minutes, likely capable of more (one report – unverified – of 30 minutes).

**ADULT MALE HIGHER LATITUDES**

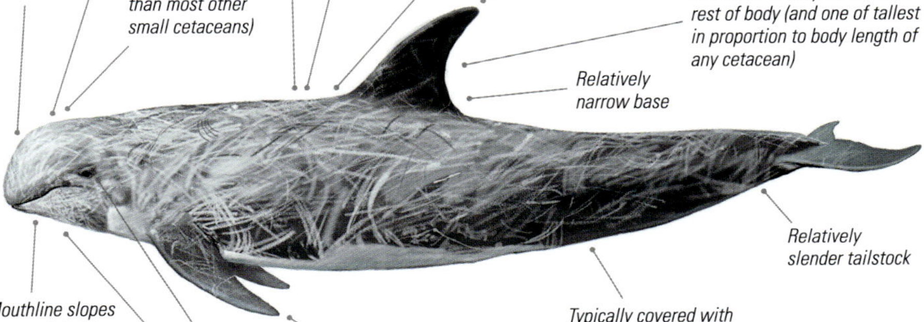

- Some healed wounds on head may be from beaks and suckers of squid prey
- Robust body shape (appears to have most of bulk in front of dorsal fin)
- Tends to be lighter overall in higher latitudes
- Underlying body color dark to pale gray (highly variable between individuals and regions)
- Very tall (up to c. 45cm), erect, moderately falcate dorsal fin midway along back (highly variable shape)
- Dorsal fin usually darker than rest of body (and one of tallest in proportion to body length of any cetacean)
- Relatively narrow base
- Front of melon has distinct vertical cleft or furrow (visible only at close range)
- Bulbous, squarish head with indistinct beak (squarer profile than most other small cetaceans)
- Relatively slender tailstock
- Mouthline slopes upward (slightly downturned at corners)
- Chin often white
- May have dark eye patch
- Long, pointed, sickle-shaped flippers (usually darker than rest of body)
- Typically covered with white scratches and blotches caused by teeth of conspecifics during social interaction

**ADULT MALE**

- Distinct vertical V-shaped groove on front of melon (unique to Risso's dolphin, function unknown)

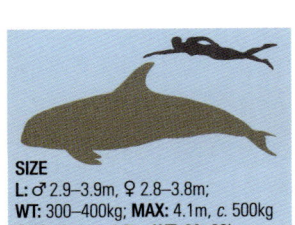

**SIZE**
L: ♂ 2.9–3.9m, ♀ 2.8–3.8m;
WT: 300–400kg; MAX: 4.1m, c. 500kg
Calf – L: 1.1–1.7m; WT: 20–30kg

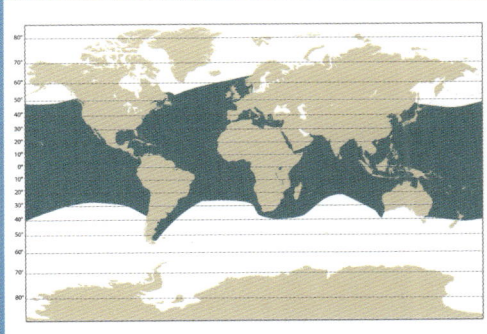

**AT A GLANCE** Worldwide from tropics to cool temperate waters • Small size • Robust body • Squarish head (side view) with indistinct beak • Cleft melon • Extensive linear scarring • Highly variable coloration within single group • Older animals almost white • Appendages usually darker than rest of body • Very tall, erect dorsal fin

**FOOD AND FEEDING** Mainly deepwater squid and octopuses, but also some cuttlefish and, rarely, krill. Most feeding appears to be during late afternoon and at night; evidence of cooperative feeding.

**TEETH** Upper jaw 0–4 (vestigial – usually unerupted); lower jaw 4–14. Teeth present in both sexes near the front of the lower jaw (usually 6–8); may be worn down (or missing) in older adults.

**GROUP SIZE AND STRUCTURE** Typically 5–30, but often up to 100; there are reports of as many as 4,000 together. There appears to be a 'stratified' group structure, with an average of 3–12 individuals in stable clusters grouped by age and sex classes. Young animals appear to remain in the vicinity of their natal group for some years after being weaned, then form pods of sub-adults at the age of 6–8 years.

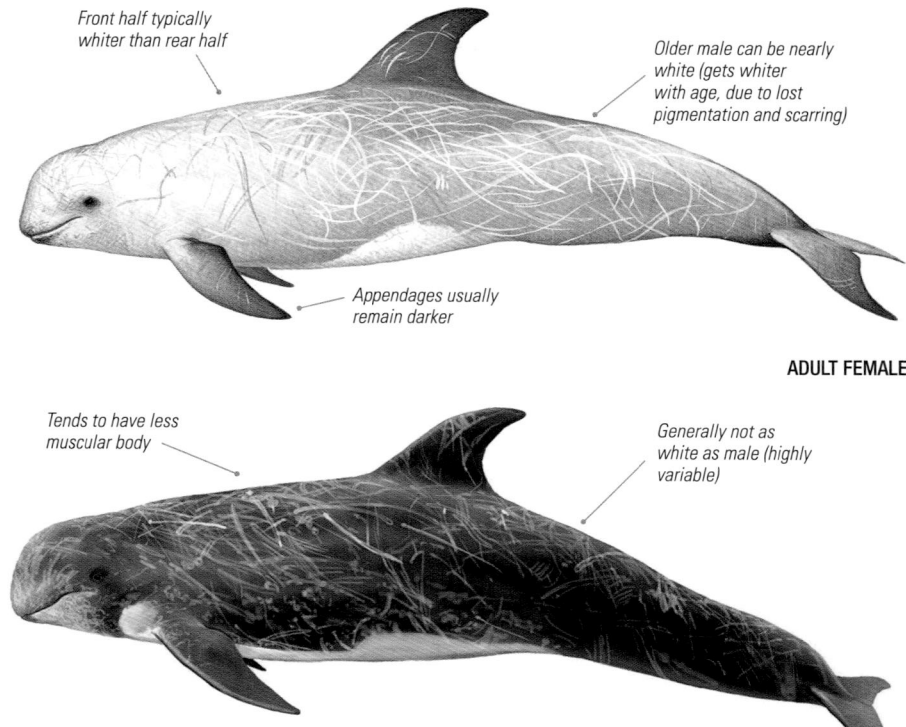

**OLDER MALE**
- Front half typically whiter than rear half
- Older male can be nearly white (gets whiter with age, due to lost pigmentation and scarring)
- Appendages usually remain darker

**ADULT FEMALE**
- Tends to have less muscular body
- Generally not as white as male (highly variable)

**ADULT MALE LOWER LATITUDES**
- Tends to be much darker than in higher latitudes
- Darker gray cape very distinctive

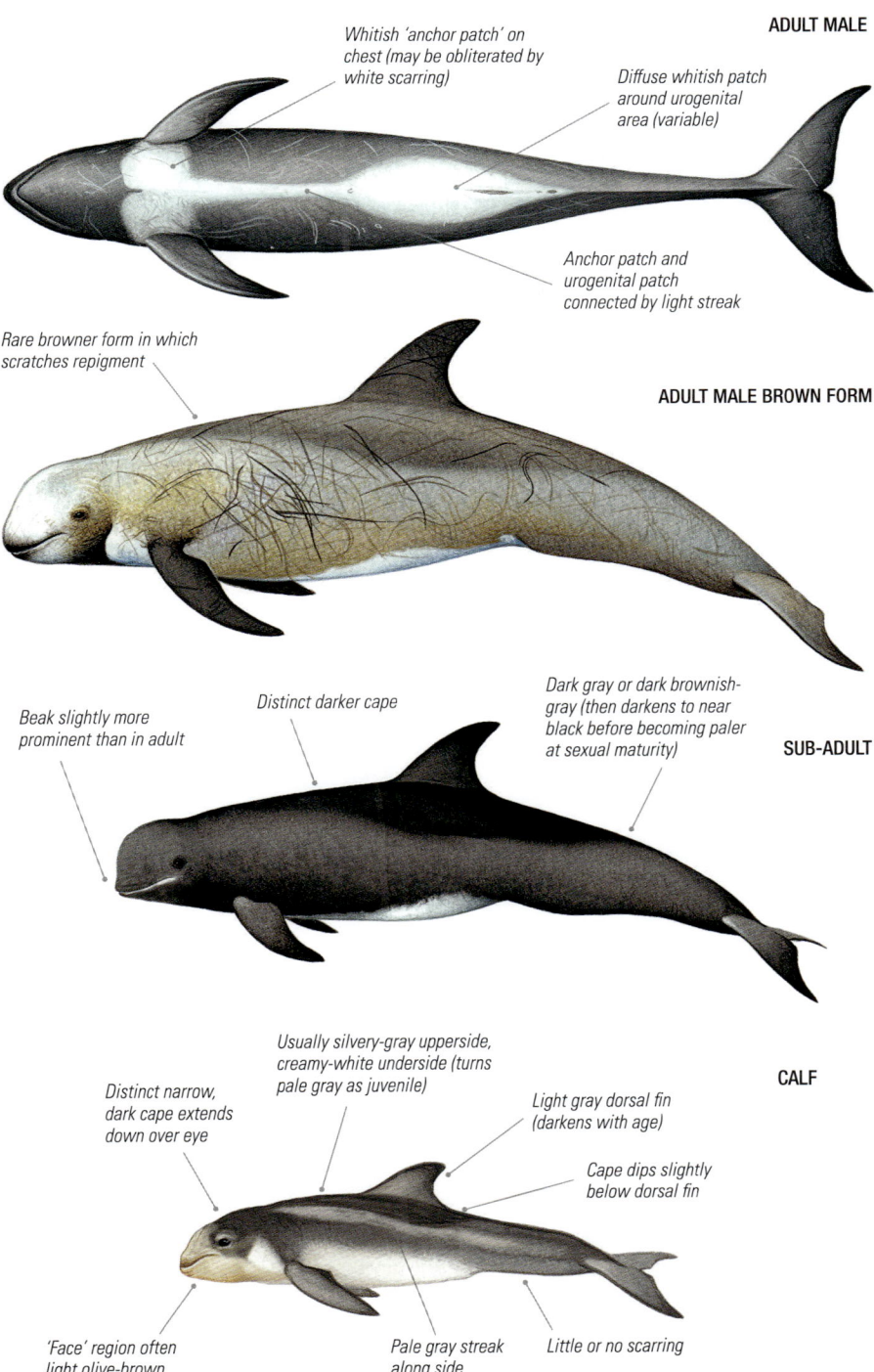

# FRASER'S DOLPHIN
*Lagenodelphis hosei*  Fraser, 1956

For many years, Fraser's dolphin was known only from a partial skeleton found on a beach in Sarawak, Malaysian Borneo, some time before 1895. The first official record in the wild was in 1971, but nowadays it is a fairly familiar sight in several parts of the world.

**IUCN status**  Least Concern (2018).
**Population**  Unknown. Approximate regional estimates total 350,000, but large swathes of the range have not been surveyed. Trend unknown.
**Classification**  Odontoceti, family Delphinidae.
**Taxonomy**  No recognized forms or subspecies.
**Other names**  Sarawak dolphin, Bornean dolphin.

**DISTRIBUTION**  Tropical, sub-tropical and occasionally warm temperate waters. Recent sightings in the Azores (*c.* 38°N) and Madeira (*c.* 33°N) may reveal the species as a potential bio-indicator of climate change (as it expands its range into warming waters further north). Mainly oceanic, in water deeper than 1,000m; sometimes close to shore where deep water approaches the coast.

**BEHAVIOR**  Active and energetic swimmer, usually in tight, fast-moving schools that whip the sea surface into a froth. Frequently found in association with other cetaceans. Occasionally performs low, relatively undemonstrative breaches. Response to boats varies from avoidance to quite approachable (will bow-ride, briefly, in some areas).

**FOOD AND FEEDING**  Mesopelagic fish, cephalopods, crustaceans. Feeding techniques unknown.

**TEETH**  Upper jaw 72–88; lower jaw 68–88.

**GROUP SIZE AND STRUCTURE**  Herds tend to be large and tightly packed, typically with 40–300 animals, but groups as small as 4–15 and as large as 1,000 are occasionally seen.

**DIVE SEQUENCE**  When swimming slowly, only blowhole, part of back and dorsal fin exposed; rolls forward with slight arching of back; when swimming quickly, porpoises in long, low-angled, splashy leaps. • **DEPTH** Near surface to *c.* 600m; physiological studies indicate capable of deep diving. • **DURATION** Unknown.

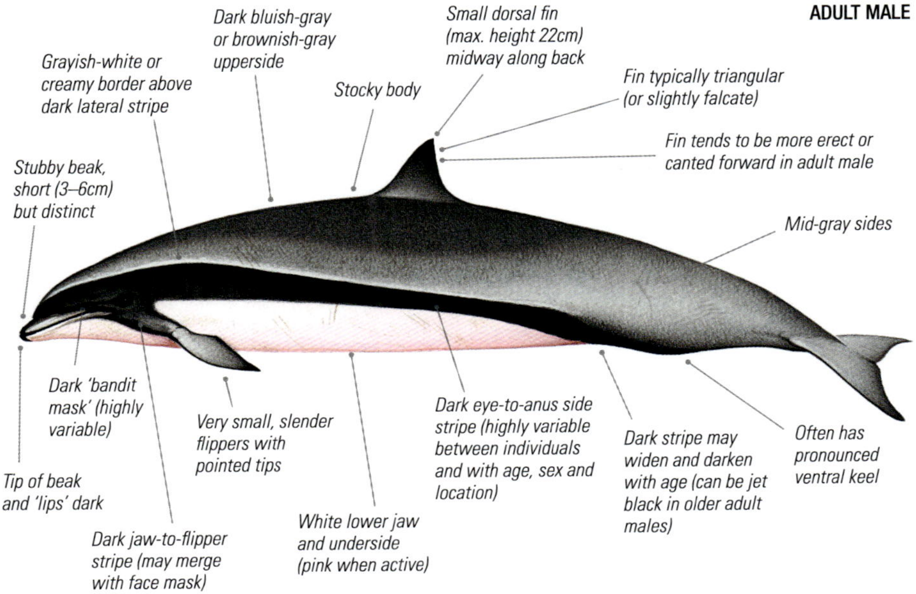

# Fraser's Dolphin

**ADULT MALE**

Labels (adult male illustration):
- Grayish-white or creamy border above dark lateral stripe
- Dark bluish-gray or brownish-gray upperside
- Stocky body
- Small dorsal fin (max. height 22cm) midway along back
- Fin typically triangular (or slightly falcate)
- Fin tends to be more erect or canted forward in adult male
- Mid-gray sides
- Stubby beak, short (3–6cm) but distinct
- Dark 'bandit mask' (highly variable)
- Tip of beak and 'lips' dark
- Dark jaw-to-flipper stripe (may merge with face mask)
- Very small, slender flippers with pointed tips
- White lower jaw and underside (pink when active)
- Dark eye-to-anus side stripe (highly variable between individuals and with age, sex and location)
- Dark stripe may widen and darken with age (can be jet black in older adult males)
- Often has pronounced ventral keel

**SIZE**
L: ♂ 2.2–2.7m, ♀ 2.1–2.6m;
WT: 130–200kg; **MAX:** 2.7m, 209kg
Calf – L: 1–1.1m; WT: 15–20kg

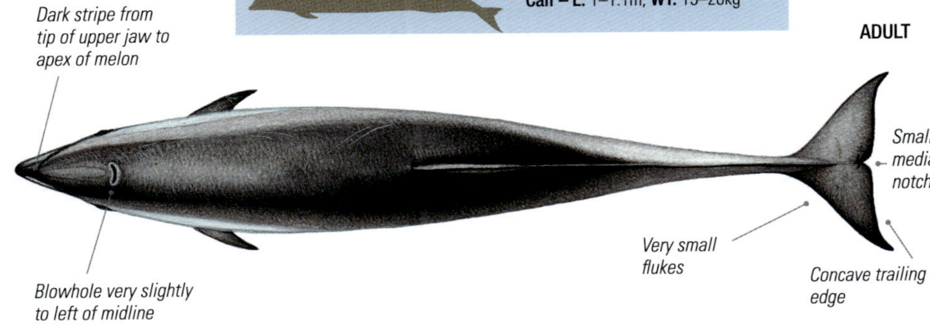

**ADULT**

Labels (adult dorsal view):
- Dark stripe from tip of upper jaw to apex of melon
- Blowhole very slightly to left of midline
- Very small flukes
- Small median notch
- Concave trailing edge

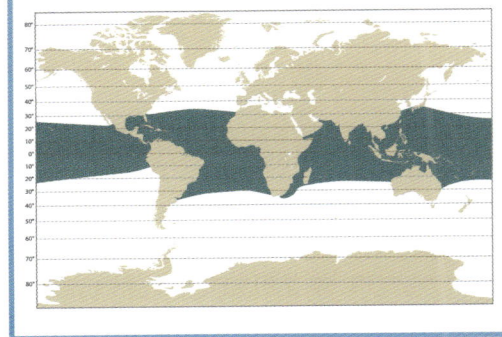

**AT A GLANCE** Deep tropical and sub-tropical waters worldwide • Small size • Stocky body • Short but distinct beak • Male often has dark 'bandit mask' and lateral stripe • Small, triangular dorsal fin, flippers and flukes • Much individual variation within herds • Splashy, tight-knit herds leave distinct white water

# ATLANTIC WHITE-SIDED DOLPHIN
*Leucopleurus acutus* (Gray, 1828)

Calling the Atlantic white-sided dolphin 'white-sided' is a bit of misnomer – its markings are complex, bold and more colorful than those of most other dolphins – though the brilliant white patch on either side is one of the most striking features of this gregarious dolphin.

**IUCN status** Least Concern (2019).
**Population** Minimum 300,000 may be a reasonable guesstimate (including *c.* 93,000 in US waters, *c.* 187,000 in the North Sea (up to the Barents Sea) and *c.* 15,500 west of Britain). Trend unknown.
**Classification** Odontoceti, family Delphinidae.
**Taxonomy** Recently moved from *Lagenorhynchus* to its own genus, *Leucopleurus*. No recognized forms or subspecies.
**Other names** White-side, springer, jumper; affectionately called a 'lag' by researchers (from the generic name).
**DISTRIBUTION** Cold temperate to sub-Arctic waters of the North Atlantic, typically within a temperature range of 1–16°C (preferring 5–11°C). The extreme limits of the range are poorly known. Prefers fairly deep waters (primarily 100–500m) with high seabed relief of the outer continental shelf and slope, but also occurs in oceanic waters and will enter fjords and inlets less than 50m deep. There are large-scale seasonal shifts in abundance in some regions: typically to more northerly latitudes and/or closer to shore during warmer months.
**BEHAVIOR** Lively and acrobatic, especially in larger groups. Will often leap and, less often, tail slap. Its leaps are either simple (no spinning or twisting – clearing and re-entering the water with a smooth arc) or complex (higher, involving twists and turns in the air). Will associate and feed with large baleen whales and sometimes forms mixed groups with other cetaceans. Keen bow-rider and wake-rider, and will ride the bow waves of mysticetes.
**FOOD AND FEEDING** Mainly small schooling fish, squid, shrimps. Known to feed cooperatively on sand lance off New England, by herding prey into tight ball against surface.
**TEETH** Upper jaw 58–80; lower jaw 62–76.
**GROUP SIZE AND STRUCTURE** Small, relatively fluid, sub-groups of 2–10; aggregations of 30–100 typical; large aggregations of up to 500 not uncommon; and exceptional records of up to 4,000. Group sizes tend to be larger during travel and social interaction.

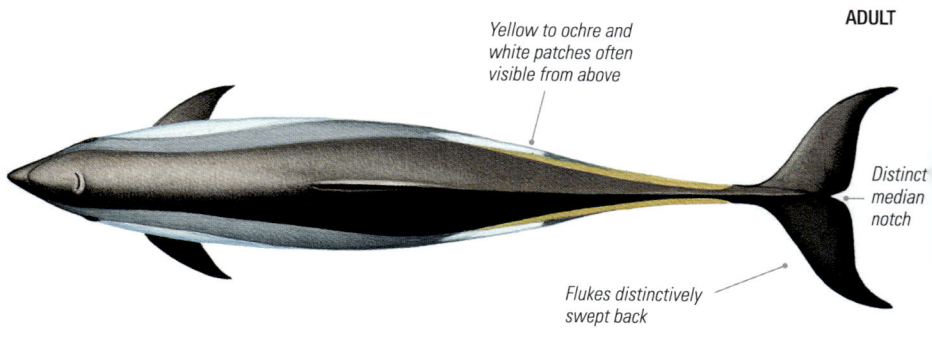

ADULT
Yellow to ochre and white patches often visible from above
Distinct median notch
Flukes distinctively swept back

**DIVE SEQUENCE** Bubbles often appear before head breaks surface; briefly shows much of beak, head and eyes; yellow and white patches often visible simultaneously; strongly arches back. • **DEPTH** Unknown, but likely fairly shallow. • **DURATION** Usually less than one minute; maximum recorded four minutes.

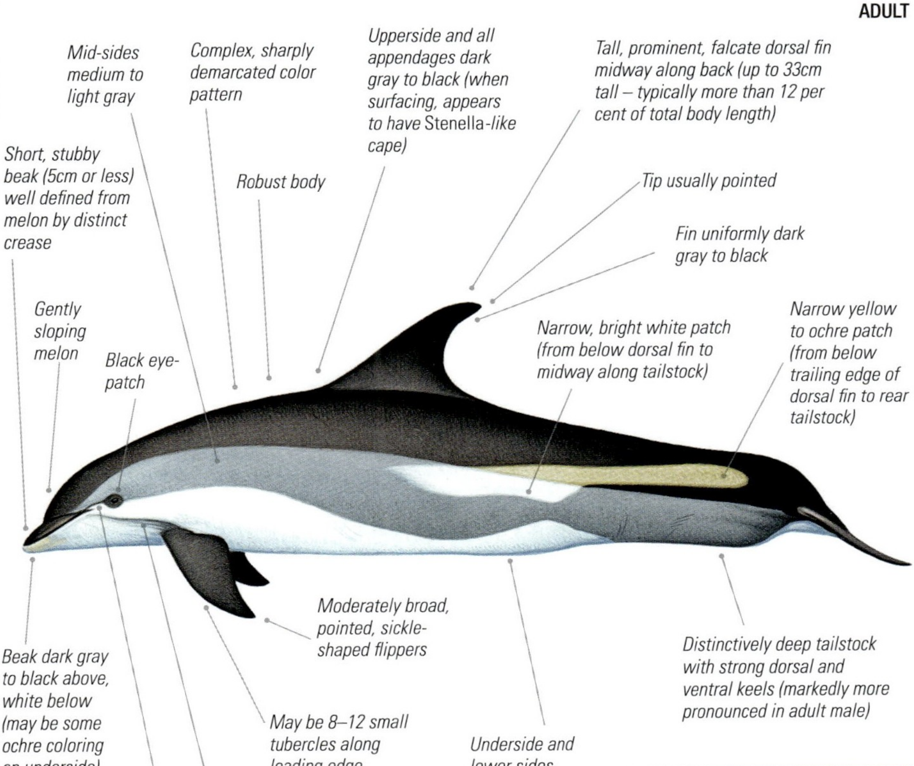

**ADULT**

- Mid-sides medium to light gray
- Complex, sharply demarcated color pattern
- Upperside and all appendages dark gray to black (when surfacing, appears to have Stenella-like cape)
- Tall, prominent, falcate dorsal fin midway along back (up to 33cm tall – typically more than 12 per cent of total body length)
- Short, stubby beak (5cm or less) well defined from melon by distinct crease
- Robust body
- Tip usually pointed
- Fin uniformly dark gray to black
- Gently sloping melon
- Black eye-patch
- Narrow, bright white patch (from below dorsal fin to midway along tailstock)
- Narrow yellow to ochre patch (from below trailing edge of dorsal fin to rear tailstock)
- Beak dark gray to black above, white below (may be some ochre coloring on underside)
- Moderately broad, pointed, sickle-shaped flippers
- May be 8–12 small tubercles along leading edge
- Underside and lower sides predominantly white (to urogenital area)
- Distinctively deep tailstock with strong dorsal and ventral keels (markedly more pronounced in adult male)
- Thin black line from beak to eye patch (variable)
- Oblique medium to light gray stripe from rear margin of lower jaw to leading edge of flipper

**SIZE**
L: ♂ 2.2–2.7m, ♀ 2–2.5m;
WT: 170–230kg; MAX: 2.8m, 235kg
Calf – L: 1–1.2m; WT: c. 24–30kg

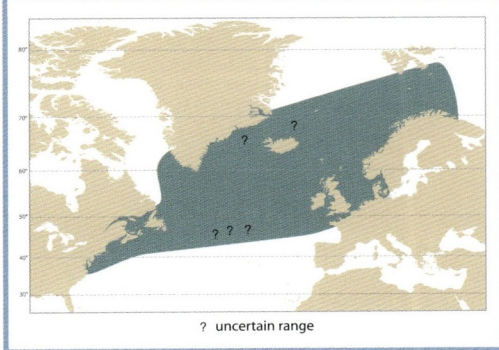

? uncertain range

**AT A GLANCE** Cold temperate to sub-Arctic waters of the North Atlantic • Small size • Complex, sharply demarcated color pattern • Yellow to ochre patch on tailstock • Bold, bright white patch on side • Very tall, pointed, falcate dorsal fin midway along back • Short, stubby beak • Distinctive dorsal and ventral keels on tailstock • Often lively and acrobatic

ATLANTIC WHITE-SIDED DOLPHIN

# PACIFIC WHITE-SIDED DOLPHIN
*Aethalodelphis obliquidens*     Gill, 1865

The Pacific white-sided dolphin is remarkably lively and energetic, repeatedly leaping high out of the water and doing a variety of somersaults, backflips, spins and cartwheels. A large school of these gregarious dolphins often throws up so much spray that their splashes can be seen long before the dolphins themselves.

**IUCN status** Least Concern (2018).
**Population** Possibly in excess of 1 million (including *c.* 25,000 in British Columbia). Trend unknown.
**Classification** Odontoceti, family Delphinidae.
**Taxonomy** Recently moved from *Lagenorhynchus* to a new genus, *Aethalodelphis*. No recognized subspecies, but there may be as many as six geographical forms (indistinguishable in the field). There are also a number of uncommon anomalous color patterns or 'morphs', including all-black and largely all-white (though not albinistic) individuals; the commonest is the 'Brownell type'.
**Other names** Affectionately called a 'lag' by researchers (from the generic name).
**DISTRIBUTION** Found in a continuous band across cool temperate waters of the North Pacific and some adjacent seas. Open oceans and coastal waters; in nearshore waters where deeper water approaches closer to shore.
**BEHAVIOR** Highly acrobatic, especially while traveling, with single leaps more common while feeding or socializing. Breaches may include side-slaps and belly-flops, and it will also flipper-slap and tail-slap. Often seen in association with other marine mammals (especially northern right whale dolphins). Can be extremely inquisitive and an avid bow-rider and wake-rider.
**FOOD AND FEEDING** Small schooling fish and cephalopods; occasionally shrimps. Large herds cooperatively corral schools of fish into bait balls close to the surface.
**TEETH** Upper jaw 46–72; lower jaw 46–72.
**GROUP SIZE AND STRUCTURE** Highly gregarious, typically in herds of up to 100 of both sexes and all ages, but sometimes up to several thousand. Large herds often segregate into sub-groups according to age and sex.

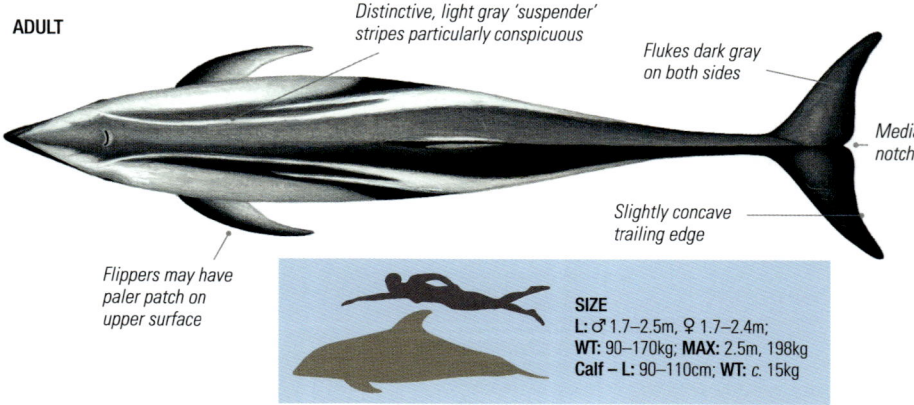

ADULT
Distinctive, light gray 'suspender' stripes particularly conspicuous
Flukes dark gray on both sides
Median notch
Slightly concave trailing edge
Flippers may have paler patch on upper surface

**SIZE**
**L:** ♂ 1.7–2.5m, ♀ 1.7–2.4m; **WT:** 90–170kg; **MAX:** 2.5m, 198kg
**Calf – L:** 90–110cm; **WT:** *c.* 15kg

**DIVE SEQUENCE** Typically surfaces quite fast; may produce Dall's porpoise-like rooster tail of spray; may cut through water with just dorsal fin showing (shark-like). • **DEPTH** Offshore populations pursue fish found at depths of 500–1,000m; coastal populations mostly eat surface-schooling prey. • **DURATION** Average 24 seconds, with longer dives rarely more than 3 minutes; maximum 6.2 minutes.

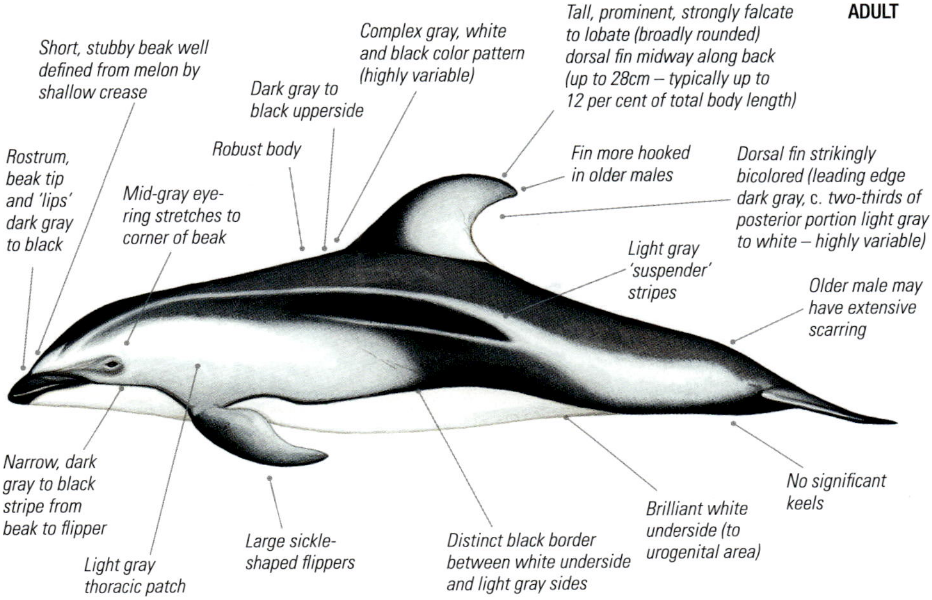

**ADULT**

- Short, stubby beak well defined from melon by shallow crease
- Complex gray, white and black color pattern (highly variable)
- Tall, prominent, strongly falcate to lobate (broadly rounded) dorsal fin midway along back (up to 28cm – typically up to 12 per cent of total body length)
- Dark gray to black upperside
- Robust body
- Fin more hooked in older males
- Dorsal fin strikingly bicolored (leading edge dark gray, c. two-thirds of posterior portion light gray to white – highly variable)
- Rostrum, beak tip and 'lips' dark gray to black
- Mid-gray eye-ring stretches to corner of beak
- Light gray 'suspender' stripes
- Older male may have extensive scarring
- Narrow, dark gray to black stripe from beak to flipper
- Large sickle-shaped flippers
- Distinct black border between white underside and light gray sides
- Brilliant white underside (to urogenital area)
- No significant keels
- Light gray thoracic patch

**ADULT BROWNELL MORPH (highly variable)**

- Wide, pure white stripe above eye runs back towards urogenital patch
- Very dark thoracic patch
- Less distinct (or absent) black border between white underside and light gray sides

**AT A GLANCE** Cool temperate waters of the North Pacific • Small size • Tall, prominent, strikingly bicolored dorsal fin • Complex gray, white and black coloration • Brilliant white underside • Pale gray thoracic patch • Light gray 'suspender' stripes along back • Short, stubby beak • Acrobatic and demonstrative • Tends to approach boats

PACIFIC WHITE-SIDED DOLPHIN

# WHITE-BEAKED DOLPHIN
*Lagenorhynchus albirostris* (Gray, 1846)

Despite their name, not all white-beaked dolphins have white beaks – many are actually quite dark or flecked. The species' Greenlandic name (*aarluarsuk*) means 'killer whale look-alike'.

**IUCN status** Least Concern (2018).
**Population** Minimum 100,000 (the sum of available abundance estimates). Trend unknown.
**Classification** Odontoceti, family Delphinidae.
**Taxonomy** The genus *Lagenorhynchus* has recently been revised: the white-beaked dolphin is the only remaining member; no recognized forms or subspecies.
**Other names** Squidhound, jumper, springer; as with all *Lagenorhynchus* dolphins, affectionately called a 'lag' by researchers (from the generic name).
**DISTRIBUTION** Cold temperate to ice-free polar waters of the North Atlantic. More common in European waters than North American. Four areas of high density have been identified: the Labrador Shelf (including south-west Greenland); Iceland; Scotland (including the northern Irish Sea and northern North Sea); and the northern coast of Norway (extending north into the White Sea). Sometimes occurs up to the edge of the pack ice. Mainly coastal, in water less than 200m deep, but also occurs in deeper, offshore waters (up to 1,000m in the Barents Sea and off West Greenland). There has been a northward shift in distribution in recent years.
**BEHAVIOR** Acrobatic, frequently leaping out of the water and performing a range of aerial behaviors. Can be quite elusive in some areas, but in others frequently approaches boats from a distance to bow-ride and jump in the wake.
**FOOD AND FEEDING** Pelagic schooling and benthic fish; may also take squid, octopus and benthic crustaceans. Feeds alone deep underwater and cooperatively to herd fish against the sea surface.
**TEETH** Upper jaw 46–56; lower jaw 44–56. First three teeth in each row often concealed within the gum.
**GROUP SIZE AND STRUCTURE** Usually 5–30; average 9 in Iceland, 6 in Svalbard, 4–6 in Denmark. Rarely seen alone. Groups of several hundred known, especially offshore.

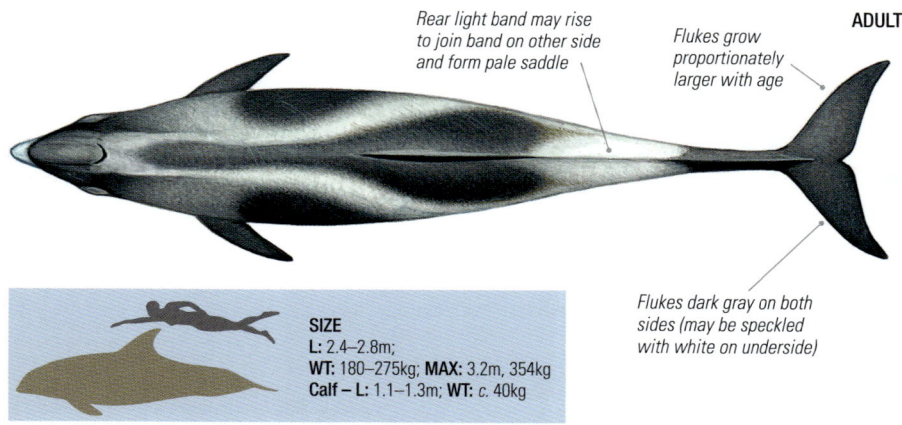

*Rear light band may rise to join band on other side and form pale saddle*

*Flukes grow proportionately larger with age*

**ADULT**

*Flukes dark gray on both sides (may be speckled with white on underside)*

**SIZE**
**L:** 2.4–2.8m;
**WT:** 180–275kg; **MAX:** 3.2m, 354kg
**Calf – L:** 1.1–1.3m; **WT:** *c.* 40kg

**DIVE SEQUENCE** When traveling fast, tends not to porpoise cleanly out of water – skims over surface, producing distinctive 'rooster-tail' spray; when swimming slowly: head, back and top of beak appear above surface, then dorsal fin appears before it rolls gently to dive. • **DEPTH** Unknown; one Icelandic individual reached 45m.
• **DURATION** Very little information; average in Iceland 24–28 seconds; maximum recorded 78 seconds.

## ADULT

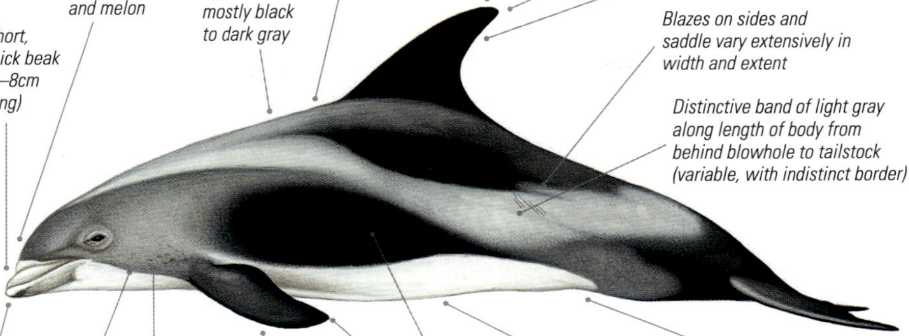

- Short, thick beak (5–8cm long)
- Shallow crease between beak and melon
- Upperside mostly black to dark gray
- Very robust body (especially male)
- Tall, dark, strongly falcate dorsal fin in middle of back (more prominent in male and with age)
- Tip pointed and often hooked
- May be off-white patch on dorsal fin
- Blazes on sides and saddle vary extensively in width and extent
- Distinctive band of light gray along length of body from behind blowhole to tailstock (variable, with indistinct border)
- Beak variable color
- May have dark or light 'speckles' between eye and flipper
- Long, pointed, dark gray flippers (up to 19 per cent of body length)
- May have dark gray stripe between corner of mouth and flipper
- Falcate trailing edge
- Large dark patch above and behind flipper
- Many individuals have medium gray patch on chest and abdomen (with variable longitudinal white line down middle)
- Underside mostly white to pale gray

### BEAK VARIATIONS

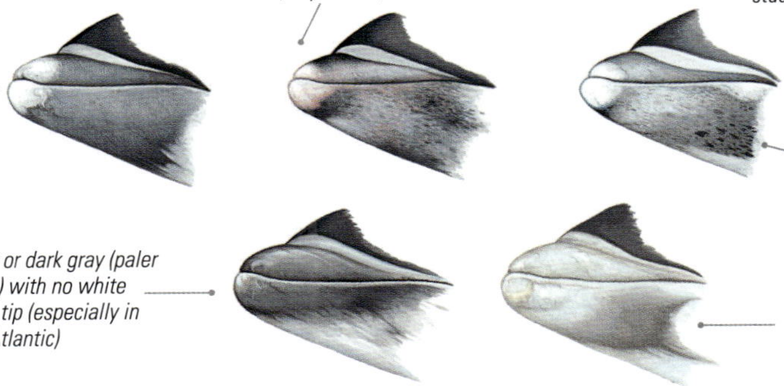

Percentages based on studies in Iceland (i.e. eastern Atlantic)

- Dark gray with white tip (52 per cent)
- White mottled with dark gray and with pinkish tip (32 per cent)
- White mottled with dark gray and with white tip (9 per cent)
- Ashy-gray or dark gray (paler than head) with no white or pinkish tip (especially in western Atlantic)
- Pure white (7 per cent)

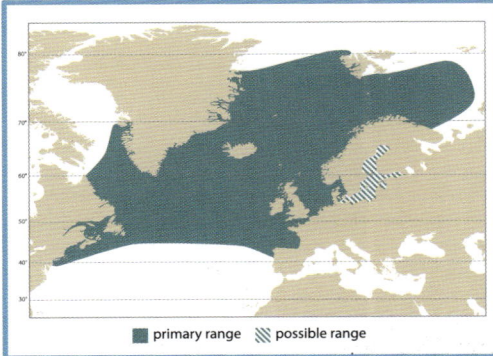

primary range    possible range

**AT A GLANCE** Cool waters of the North Atlantic • Small size • Complex, diffuse (and variable) gray, black and white coloration • Distinctive band of light gray along side • Very robust body • Grayish-white 'saddle' behind dorsal fin • Short, thick beak (often white) • Tall, dark, falcate dorsal fin

# NORTHERN RIGHT WHALE DOLPHIN
*Lissodelphis borealis* (Peale, 1848)

With its striking black-and-white markings, slender body and no dorsal fin, the northern right whale dolphin is easily identifiable within its range. The two species of right whale dolphins may look superficially similar, but their markings are strikingly different and they are widely separated geographically.

**IUCN status** Least Concern (2018).
**Population** c. 68,000 (including c. 26,000 in US waters). Trend unknown.
**Classification** Odontoceti, family Delphinidae.
**Taxonomy** No recognized forms or subspecies; some populations are characterized by a 'swirled' color morph (which may resemble the southern right whale dolphin).
**Other names** Affectionately called a 'lisso' by researchers (from the generic name).
**DISTRIBUTION** Cool to warm temperate waters in the North Pacific, mainly from 31–50°N in the east and 35–51°N in the west. Favors deep oceanic waters from the outer continental shelf and beyond, but also occurs where deep waters approach the coast. Most abundant where sea surface temperatures are 8–19°C. Some seasonal movements north (summer) and south (winter).
**BEHAVIOR** Fast swimmer, capable of bursts of speed up to 34km/h. Frequently associates with at least 14 other cetacean species. May erupt into bouts of high excitement, with much aerial activity such as breaching, spyhopping, belly-flopping, and side- and fluke-slapping, and bursts of energetic swimming. Response to boats varies enormously: it will bow-ride (especially in the presence of other bow-riding dolphins) but can be skittish and easily startled.
**FOOD AND FEEDING** Mainly fish, but some squid. Feeding techniques unknown.
**TEETH** Upper jaw 74–104; lower jaw 84–108.
**GROUP SIZE AND STRUCTURE** Highly gregarious: 100–200 common, but sometimes up to 2,000–3,000. Rarely seen alone.

**ADULT**

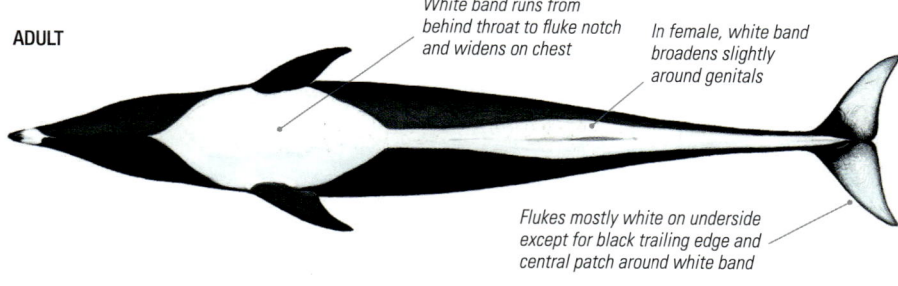

White band runs from behind throat to fluke notch and widens on chest

In female, white band broadens slightly around genitals

Flukes mostly white on underside except for black trailing edge and central patch around white band

**SIZE**
**L:** ♂ 2.3–3m, ♀ 2.3–2.6m;
**WT:** 60–100kg; **MAX:** 3.1m, 113kg
**Calf – L:** c. 0.8–1m; **WT:** unknown

**DIVE SEQUENCE** When swimming quickly, makes graceful, 'bouncing', low-angled leaps (up to 7m) or splashy belly flops, creating much surface disturbance; can appear almost eel-like; when swimming slowly, makes low-profile roll, barely breaking surface to breathe (easy to miss). • **DEPTH** Probably capable of 200+m. • **DURATION** 10–75 seconds; maximum recorded 6.2 minutes.

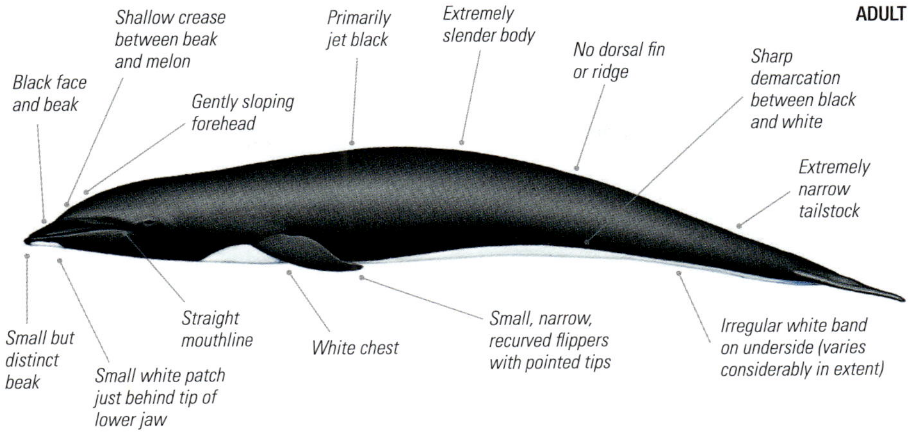

**ADULT**

- Black face and beak
- Shallow crease between beak and melon
- Gently sloping forehead
- Primarily jet black
- Extremely slender body
- No dorsal fin or ridge
- Sharp demarcation between black and white
- Extremely narrow tailstock
- Small but distinct beak
- Straight mouthline
- Small white patch just behind tip of lower jaw
- White chest
- Small, narrow, recurved flippers with pointed tips
- Irregular white band on underside (varies considerably in extent)

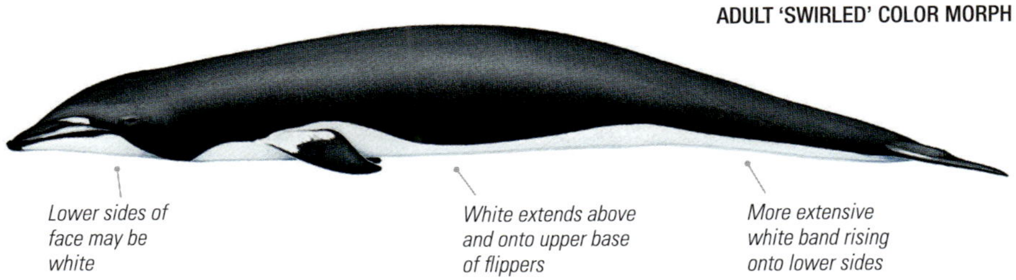

**ADULT 'SWIRLED' COLOR MORPH**

- Lower sides of face may be white
- White extends above and onto upper base of flippers
- More extensive white band rising onto lower sides

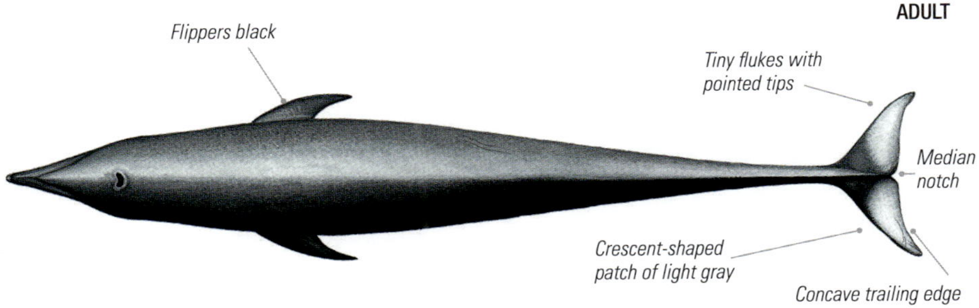

**ADULT**

- Flippers black
- Tiny flukes with pointed tips
- Median notch
- Crescent-shaped patch of light gray
- Concave trailing edge

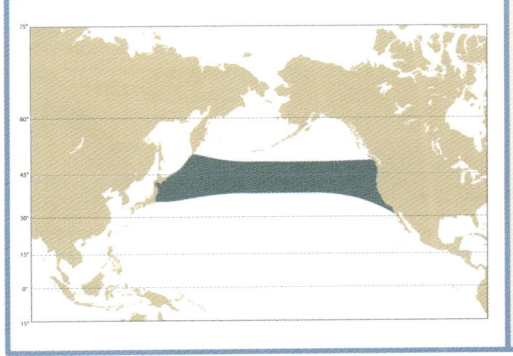

**AT A GLANCE** Deep temperate waters of the North Pacific • Small size (appears even smaller at sea than measurements suggest) • Only dolphin in North Pacific with no dorsal fin • Mainly black with white band on underside • Extremely slender body • Small but distinct beak • Low-angled leaps • Usually in sizeable groups

NORTHERN RIGHT WHALE DOLPHIN

# ROUGH-TOOTHED DOLPHIN
*Steno bredanensis* (Lesson, 1828)

Unmistakable at close range, with its smoothly sloping forehead, the rough-toothed dolphin has been described as looking more like an extinct ichthyosaur (a marine reptile from the age of the dinosaurs) than a cetacean. It is named for the unique ridges on its teeth.

**IUCN status** Least Concern (2018). Mediterranean sub-population Near Threatened (2020).
**Population** Possible minimum *c.* 250,000; but very little information. Trend unknown.
**Classification** Odontoceti, family Delphinidae.
**Taxonomy** No recognized forms or subspecies. There is an argument for Atlantic and Pacific/Indian populations to be recognized as two subspecies.
**Other names** Slopehead.

**DISTRIBUTION** Tropical to sub-tropical (and some warm temperate) waters in the Atlantic, Pacific and Indian Oceans, mainly from *c.* 40°N to *c.* 35°S. Also in many semi-enclosed seas; formerly considered a vagrant in the Mediterranean, but a small, relict population is present in the eastern basin. Prefers deep offshore waters beyond the continental shelf – though often close to land around islands with steep drop-offs (and found in shallow coastal waters in some areas). Shows fidelity to at least some oceanic islands.

**BEHAVIOR** Can appear quite lethargic and inactive, and it is not wildly acrobatic, but does breach fairly regularly and often multiple times in succession (although not particularly high). Spyhopping, surface-slapping and low-angled arced leaping are quite common. Well known for swimming shoulder to shoulder in a synchronized 'chorus line'. Reaction to boats varies. In most places, quite easy to approach as long as it is not actively foraging. Often bow-rides and wake-rides.

**FOOD AND FEEDING** Fish, squid and octopus. Sometimes hunts cooperatively; thought to feed primarily on near-surface species, some evidence of food sharing.

**TEETH** Upper jaw 38–52; lower jaw 38–56.

**GROUP SIZE AND STRUCTURE** Most common group size is 5–20, though occasionally smaller groups or alone; up to 50 not uncommon in the eastern tropical Atlantic and the Azores; some reports of more than 300 (probably aggregations of sub-groups).

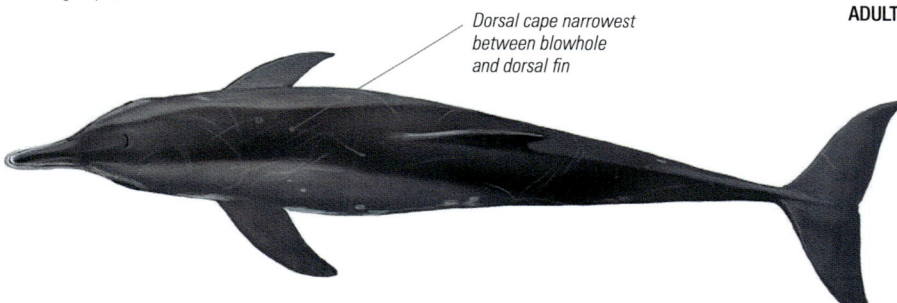

*Dorsal cape narrowest between blowhole and dorsal fin*

**ADULT**

**DIVE SEQUENCE** When swimming slowly, surfaces unobtrusively (although dorsal fin striking), with slight arching of back as it dives; at moderate speed, skims along with head and chin just above surface, forming distinctive walls of spray (looks rather like surfing). • **DEPTH** Primarily uses near-surface waters (in top 30m water column); deepest recorded 400m (though morphologically capable of deeper); dives deeper at night. • **DURATION** Varies with location; average 4–7 minutes in Hawai'i; maximum recorded 15 minutes.

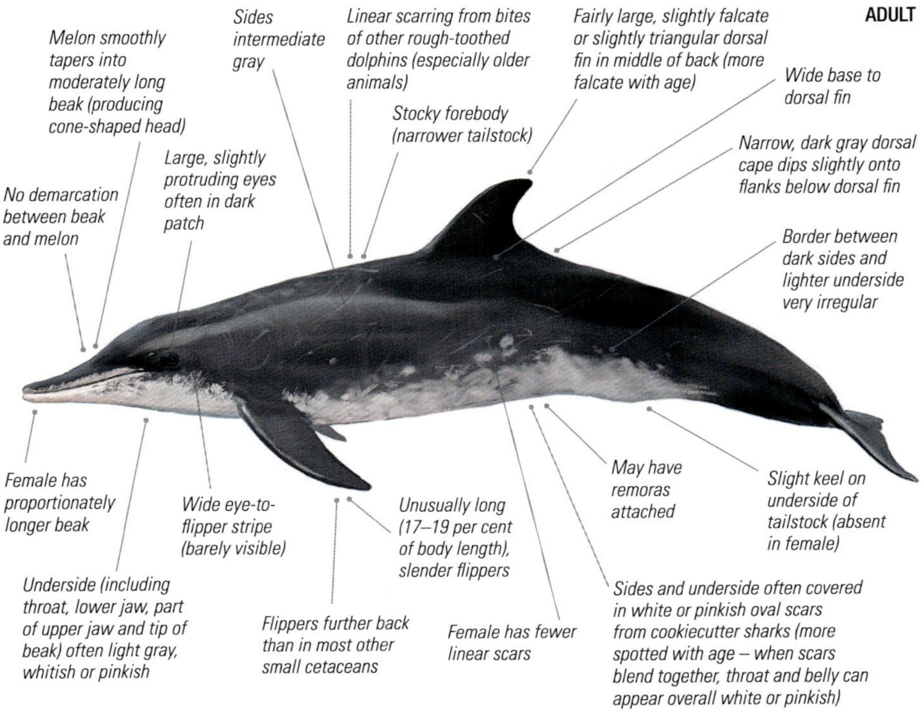

**ADULT**

- Melon smoothly tapers into moderately long beak (producing cone-shaped head)
- Sides intermediate gray
- Linear scarring from bites of other rough-toothed dolphins (especially older animals)
- Fairly large, slightly falcate or slightly triangular dorsal fin in middle of back (more falcate with age)
- Wide base to dorsal fin
- Narrow, dark gray dorsal cape dips slightly onto flanks below dorsal fin
- Large, slightly protruding eyes often in dark patch
- Stocky forebody (narrower tailstock)
- No demarcation between beak and melon
- Border between dark sides and lighter underside very irregular
- Female has proportionately longer beak
- Wide eye-to-flipper stripe (barely visible)
- Unusually long (17–19 per cent of body length), slender flippers
- May have remoras attached
- Slight keel on underside of tailstock (absent in female)
- Underside (including throat, lower jaw, part of upper jaw and tip of beak) often light gray, whitish or pinkish
- Flippers further back than in most other small cetaceans
- Female has fewer linear scars
- Sides and underside often covered in white or pinkish oval scars from cookiecutter sharks (more spotted with age – when scars blend together, throat and belly can appear overall white or pinkish)

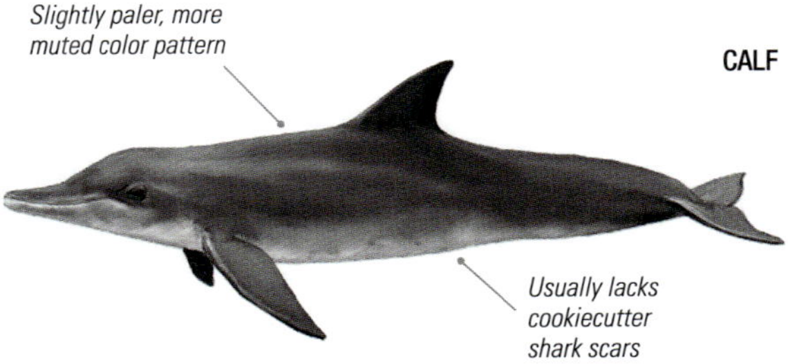

**CALF**

- Slightly paler, more muted color pattern
- Usually lacks cookiecutter shark scars

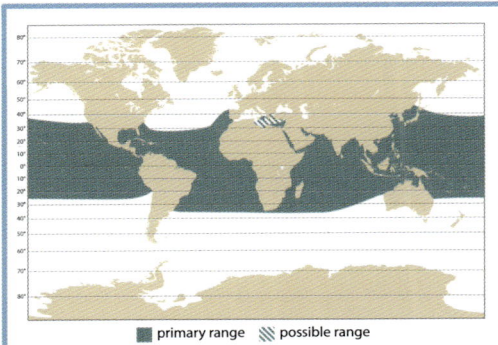

**SIZE**
L: ♂ 2.2–2.7m, ♀ 2.1–2.6m;
WT: 90–155kg; MAX: 2.8m
Calf – L: c. 1–1.2m; WT: c. 15kg

■ primary range  ▨ possible range

**AT A GLANCE** Offshore waters in tropics and sub-tropics • Small size (but chunky) • Complex three-toned coloration • Prominent, slightly falcate dorsal fin • Cone-shaped head • Melon slopes smoothly into moderately long beak • Unusually large flippers • May be covered in pink or white blotches • Almost reptilian in appearance • Often 'skims' along surface

# COMMON BOTTLENOSE DOLPHIN
*Tursiops truncatus* (Montagu, 1821)

The common bottlenose dolphin is the quintessential dolphin and, thanks to its coastal habits, prevalence in captivity and frequent appearances on television, one of the best-known cetaceans. But its taxonomy is still in dispute – due to huge geographical variation in size, shape, skull morphology and coloration. Over the years, more than 20 nominal species have been proposed, but only three are currently recognized (common bottlenose, Indo-Pacific bottlenose and Tamanend's bottlenose). Others could be accepted in the future.

**IUCN status** Least Concern (2018). Lahille's subspecies (*T. t. gephyreus*) Vulnerable (2019). Black Sea subspecies (*T. t. ponticus*) Endangered (2008). Fiordland (New Zealand) sub-population Critically Endangered (2010). Mediterranean sub-population Least Concern (2021). Gulf of Ambracia sub-population Critically Endangered (2020).
**Population** Minimum 600,000 (based on available abundance estimates – but much of the range has not been surveyed). Trend unknown.
**Classification** Odontoceti, family Delphinidae.
**Taxonomy** Four subspecies are recognized: common bottlenose dolphin (*T. t. truncatus*), found in tropical to temperate waters worldwide; Lahille's bottlenose dolphin (*T. t. gephyreus*), a larger, coastal form found in the western South Atlantic (which has been proposed as a separate species); Black Sea bottlenose dolphin (*T. t. ponticus*), known only from the Black Sea, Kerch Strait (and connecting part of the Azov Sea) and the Turkish Straits system; and the offshore Eastern tropical Pacific bottlenose dolphin (*T. t. nuuanu*) in Mexico and central and northern South America. In the eastern Pacific, two further distinct ecotypes are recognized: one in coastal waters of southern California and Baja California, Mexico, the other in northern temperate offshore waters. The 'Burrunan bottlenose dolphin (*Tursiops australis*)', found off southern and southeastern Australia, was proposed as a new species in 2011, but has been rejected. Tamanend's bottlenose dolphin (*Tursiops erebennus*) was separated from the common bottlenose dolphin in 2023.
**Other names** Bottlenose dolphin, bottle-nosed dolphin.

**DISTRIBUTION** Widespread in tropical to temperate offshore and coastal waters worldwide. Most abundant between 45°N and 45°S, except in northern Europe (with significant numbers around the United Kingdom and as far north as the Faroe Islands at 62°N). Most often seen in shallow coastal waters and around oceanic islands, but also out to the continental shelf edge and most abundant in deep offshore waters. Frequently in bays, lagoons, channels and around harbors, and ventures into rivers for brief periods. Frequently near population centers.

**ADULT**

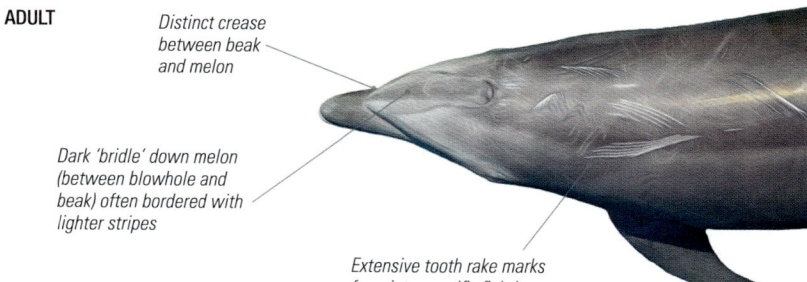

Distinct crease between beak and melon

Dark 'bridle' down melon (between blowhole and beak) often bordered with lighter stripes

Extensive tooth rake marks from intraspecific fighting

**DIVE SEQUENCE** When swimming slowly, tip of beak often breaks surface first; when swimming quickly, porpoises with neat re-entry. • **DEPTH** Highly variable depending on location and prey; typically up to 70m, but offshore often several hundred meters; maximum *c.* 1,000m. • **DURATION** Offshore average *c.* 1–5 minutes (maximum recorded 13.8 minutes); inshore typically 30–120 seconds (maximum recorded 8 minutes).

**ADULT**

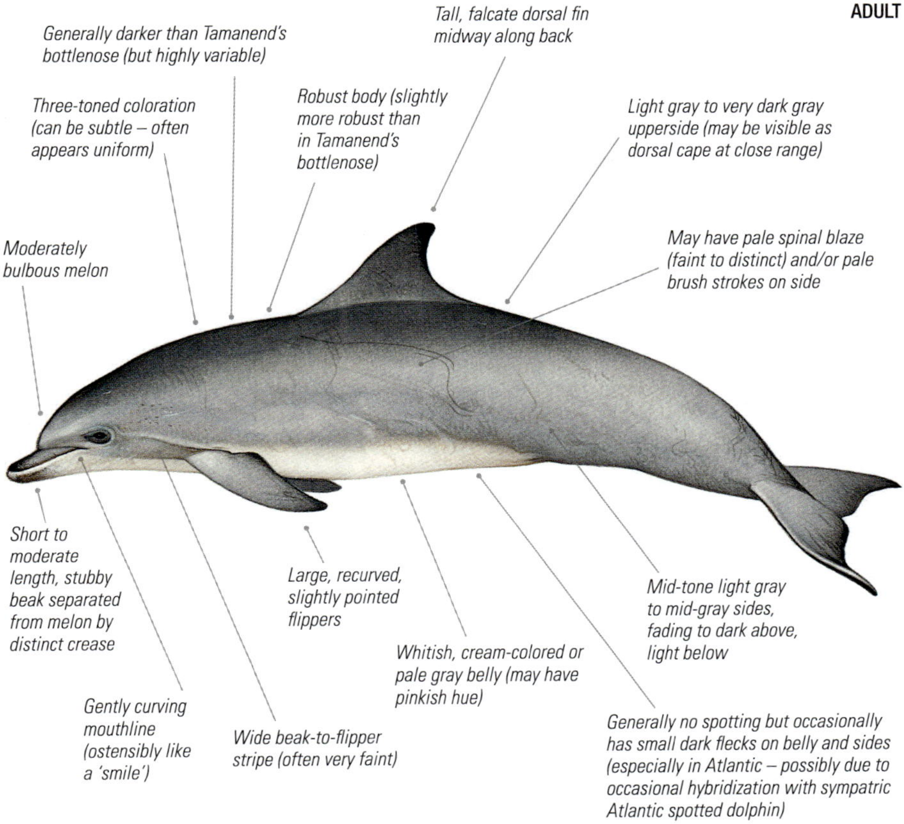

- Generally darker than Tamanend's bottlenose (but highly variable)
- Three-toned coloration (can be subtle – often appears uniform)
- Tall, falcate dorsal fin midway along back
- Robust body (slightly more robust than in Tamanend's bottlenose)
- Light gray to very dark gray upperside (may be visible as dorsal cape at close range)
- Moderately bulbous melon
- May have pale spinal blaze (faint to distinct) and/or pale brush strokes on side
- Short to moderate length, stubby beak separated from melon by distinct crease
- Large, recurved, slightly pointed flippers
- Whitish, cream-colored or pale gray belly (may have pinkish hue)
- Mid-tone light gray to mid-gray sides, fading to dark above, light below
- Gently curving mouthline (ostensibly like a 'smile')
- Wide beak-to-flipper stripe (often very faint)
- Generally no spotting but occasionally has small dark flecks on belly and sides (especially in Atlantic – possibly due to occasional hybridization with sympatric Atlantic spotted dolphin)

**SIZE**
**L:** ♂ 1.9–3.8m, ♀ 1.8–3.5m;
**WT:** 136–600kg; **MAX:** 3.9m, 635kg
**Calf – L:** 1–1.5m; **WT:** *c.* 15–25kg
Wide variation in size between populations.

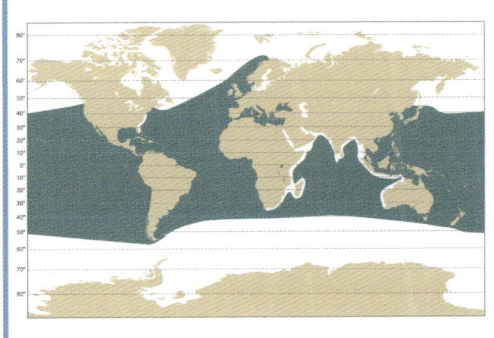

**AT A GLANCE** Tropical to temperate waters worldwide • Small size • Robust body • Short, stubby beak • Archetypal dolphin • Ostensibly 'smiling' mouthline • Three-toned coloration (subtle to distinct) • Rarely has small dark flecks on underside • In coastal waters usually in small groups • Often bow-rides

**BEHAVIOR** Tends to be active much of the time and will leap, lobtail, porpoise, body surf and perform other aerial behaviors. Keen bow-rider and wake-rider. It will bow-ride any vessel, from a small motor boat to a large oceanic cargo vessel or cruise ship, and will even ride the bow-waves in front of large whales ('snout-riding'). Known to attack and kill harbor porpoises in Scotland and Wales (UK) and California (US); the reasons for this behavior are unclear, and it is more widespread than simply the aberrant behavior of one or two individuals. Some wild common bottlenose dolphins have become 'friendly' – solitary individuals, not part of a social group, which hang around harbors and befriend people, apparently preferring to associate with boats, divers and swimmers rather than with other dolphins. Some stay for weeks or months, others for years.

**FOOD AND FEEDING** Generalist feeder overall, but specialization within populations and among individuals; wide variety of fish, cephalopods and crustaceans; will attempt to swallow absurdly large prey. Wide variety of techniques, depending on prey and location, including high-speed chasing, bubble-blowing to herd prey towards surface, 'fish-whacking' (knocking fish out of water with flukes – sometimes catching them in mid-air), 'strand-feeding' (sending wave of water that pushes fish onto mudbanks, then temporarily beaching themselves to grab the fish), 'kerplunking' (scaring fish out from seagrass beds and other vegetative cover with bubble-forming tail slaps), and 'mud-ringing' (one dolphin creates ring-shaped mud plume, then others catch fish in mid-air as they leap out of ring). Will feed behind shrimp trawlers (to eat discarded fish and steal fish from fishing gear). In Mauritania and Brazil, regularly drive mullet towards fishermen holding nets in shallow water.

**TEETH** Upper jaw 36–54; lower jaw 36–54. Teeth often worn down or missing in older animals.

**GROUP SIZE AND STRUCTURE** Typically 1–15, but offshore sometimes in large herds of several hundred; groups of more than 1,000 have been reported. Generally, groups are smaller close to shore. Group structure varies greatly, but tends to be relatively fluid.

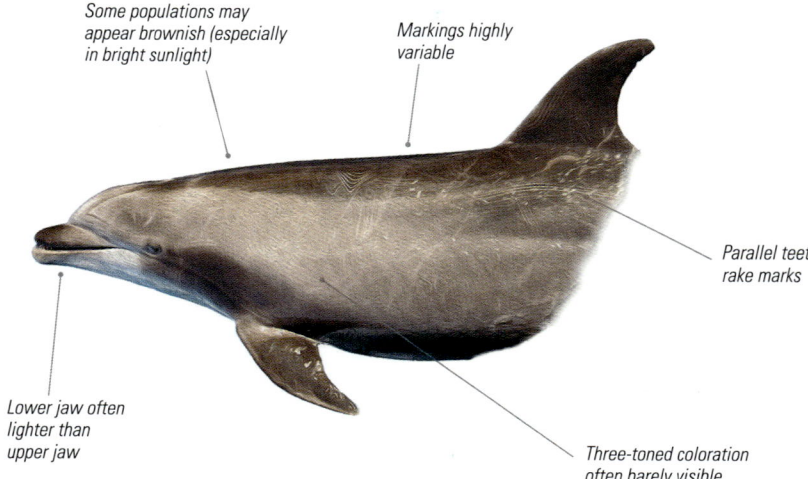

Some populations may appear brownish (especially in bright sunlight)

Markings highly variable

Parallel teeth rake marks

Lower jaw often lighter than upper jaw

Three-toned coloration often barely visible

**CALF**

Similar coloration to adult

Slightly slimmer body

Relatively shorter, stubbier beak

**ADULT VARIATIONS** (offshore populations are generally larger, darker and have proportionally shorter fins and beaks)

COMMON BOTTLENOSE DOLPHIN

# TAMANEND'S BOTTLENOSE DOLPHIN
*Tursiops erebennus* (Cope, 1865)

Tamanend's bottlenose dolphin was formally accepted as a new species in 2023. Known only from coastal waters in the eastern United States, and named after Chief Tamanend (1628–1701) of the Turtle Clan of the Nanticoke Lenni-Lenape Tribal Nation, it is smaller and less robust than the common bottlenose dolphin.

**IUCN status** Not yet evaluated.
**Population** 2021 summer aerial survey estimated *c.* 32,500. Trend unknown (declined 2010–2016, following an unusual mortality event, then increased 2016–2021).
**Classification** Odontoceti, family Delphinidae.
**Taxonomy** Two bottlenose dolphin ecotypes (coastal and offshore) had long been recognized in the western North Atlantic (in particular, exhibiting differences in skull size and shape). Recent DNA research reveals sufficient differences to identify them as distinct species. No recognized forms or subspecies.
**Other names** None.
**DISTRIBUTION** Continuous distribution in shallow waters along the western North Atlantic coast, from New York to the south-east coast of Florida. Found in nearshore coastal waters, including bays and estuaries, and frequently seen from shore. Evidence of long-term site fidelity in some stocks (remaining in the same area for multiple decades and many generations) and seasonal migrations in others. It is unclear how far offshore the distribution extends (in some areas, particularly in the north, there appears to be a significant longitudinal gap between Tamanend's and common bottlenose dolphins, which are further offshore). Additional work is needed to determine if the coastal bottlenose dolphins in the Gulf of Mexico, the Bahamas and the Caribbean are also Tamanend's.
**BEHAVIOR** Tends to be less active than common bottlenose dolphins, but will leap, lobtail, porpoise, body surf and perform other aerial behaviours. Keen bow-rider and wake-rider.
**FOOD AND FEEDING** Wide variety of small benthic fish, some squid and octopus. Wide variety of techniques, including 'strand-feeding' (sending a wave of water that pushes fish onto mudbanks, then temporarily beaching themselves to grab the fish). Sometimes hunts cooperatively (such as 'mud-ringing' – one dolphin creates ring-shaped mud plume, then others catch fish in mid-air as they leap out of ring).
**TEETH** Upper jaw median 48; lower jaw median 46.
**GROUP SIZE AND STRUCTURE** Ranges from 1–25, but typically alone, in pairs or in small groups (larger when socializing or feeding). Averages smaller groups (3–7) than common bottlenose dolphin. Group size may also vary according to the degree of 'openness' of the body of water (larger in more open water).

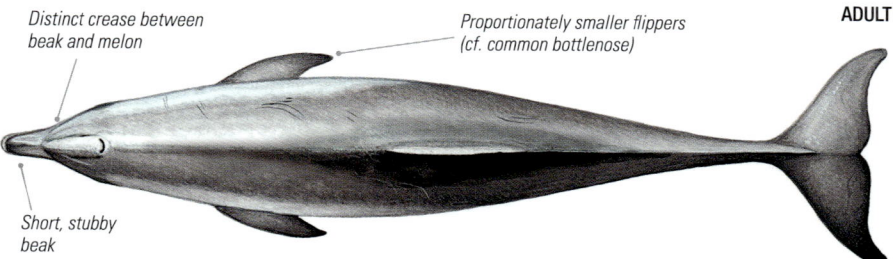

Distinct crease between beak and melon
Proportionately smaller flippers (cf. common bottlenose)
**ADULT**
Short, stubby beak

**DIVE SEQUENCE** When swimming slowly, tip of beak often breaks surface first; when swimming quickly, porpoises with neat re-entry. • **DEPTH** Likely shallow (usually in water shallower than 25m). • **DURATION** Probably no more than a few minutes.

**ADULT**

- May have pale spinal blaze (faint to distinct) and/or pale brush strokes on side
- May appear brownish in bright sunlight
- Tall, falcate dorsal fin midway along back
- Robust body (less robust than common bottlenose)
- Light gray to medium gray upperside (may be visible as dorsal cape at close range)
- Moderately bulbous melon
- Generally no spotting but occasionally has small dark flecks on belly and sides
- Relatively short, narrow, stubby beak separated from melon by distinct crease
- Flippers may be proportionately larger than offshore common bottlenose
- Three-toned coloration (can be subtle – often appears uniform)
- May be lighter than common bottlenose offshore (but variable)
- Mid-tone light gray to medium gray sides, fading to darker above, lighter below

**SIZE**
L: ♂ 2.4–2.7m, ♀ 2.3–2.5m;
WT: *c.* 260–280kg; **MAX:** 2.9m, 300kg
Calf – L: *c.* 1m; WT: *c.* 15kg

**AT A GLANCE** Coastal and estuarine waters along east coast of the US • small size • Robust body • Short, stubby beak • Archetypal dolphin • Ostensibly 'smiling' mouthline • Three-toned coloration (subtle to distinct) • Rarely has small dark flecks on underside • Small groups

TAMANEND'S BOTTLENOSE DOLPHIN

# PANTROPICAL SPOTTED DOLPHIN
*Stenella attenuata* (Gray, 1846)

The pantropical spotted dolphin is highly variable in appearance – between ages, individuals and regions – from virtually unspotted to very heavily spotted. Even though it has been severely depleted by tuna purse-seine fishing in the eastern tropical Pacific, it is still one of the most abundant cetaceans on the planet.

**IUCN status** Least Concern (2018).
**Population** Minimum 2.3 million (not including populations yet to be assessed). At least 4 million killed by commercial tuna fleets. Trend unknown.
**Classification** Odontoceti, family Delphinidae.
**Taxonomy** Two subspecies are currently recognized: offshore (*S. a. attenuata*), which is slightly smaller, more slender and lightly spotted, and coastal (*S. a. graffmani*), which is slightly larger, stockier and more heavily spotted.
**Other names** Spotter, bridled dolphin.

**DISTRIBUTION** Tropical and some sub-tropical waters in the Pacific, Atlantic and Indian Oceans (roughly 30–40°N to 20–40°S). Within this range, most abundant in lower latitudes. The offshore subspecies is found mainly in oceanic waters, beyond the continental shelf edge, and around some oceanic islands, but it does occur nearshore where sufficiently deep water approaches the coast. Primarily inhabits waters with surface water temperatures above 25°C. The coastal subspecies is usually within 130km of shore, along the west coasts of Latin America from southern Mexico to northern Peru, often in water shallower than 50m.

**BEHAVIOR** Fast swimmer. Can be highly acrobatic (though it does not spin) and frequently performs breaches and side-slaps. Readily approaches boats and bow-rides (except on tuna fishing grounds in the eastern tropical Pacific); females and juveniles are more likely to bow-ride than males. In the eastern tropical Pacific and western Indian Oceans, it frequently associates with yellowfin tuna and skipjack tuna (perhaps for foraging efficiency or protection from predators) and with other dolphin species.

**FOOD AND FEEDING** Offshore subspecies – mainly small epipelagic and mesopelagic fish, squid and crustaceans; coastal subspecies – possibly mainly larger, bottom-living fish. Forages mainly at night; offshore subspecies exploits deep scattering layer.

**TEETH** Upper jaw 68–96; lower jaw 68–94.

**GROUP SIZE AND STRUCTURE** Coastal subspecies generally in groups of 10–20 (ranging from one to *c.* 100). Offshore herds can number in the hundreds or thousands, sometimes spread out over several kilometers.

ADULT

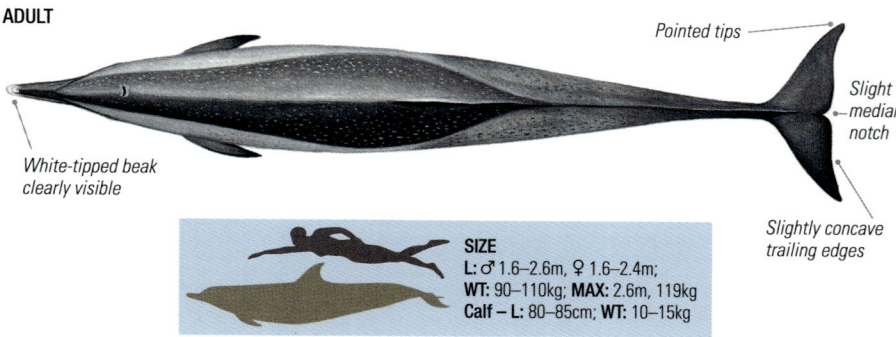

*Pointed tips*
*Slight median notch*
*White-tipped beak clearly visible*
*Slightly concave trailing edges*

**SIZE**
**L:** ♂ 1.6–2.6m, ♀ 1.6–2.4m; **WT:** 90–110kg; **MAX:** 2.6m, 119kg
**Calf – L:** 80–85cm; **WT:** 10–15kg

**DIVE SEQUENCE** When swimming slowly, tip of beak often breaks surface first; when swimming fast, porpoises with neat re-entry. • **DEPTH** Typically shallower (5–50m) during day, deeper (25–250m) at night (varies according to region); maximum recorded 342m. • **DURATION** Typically 30–120 seconds; maximum 5.4 minutes.

# PANTROPICAL SPOTTED DOLPHIN

## ADULT OFFSHORE

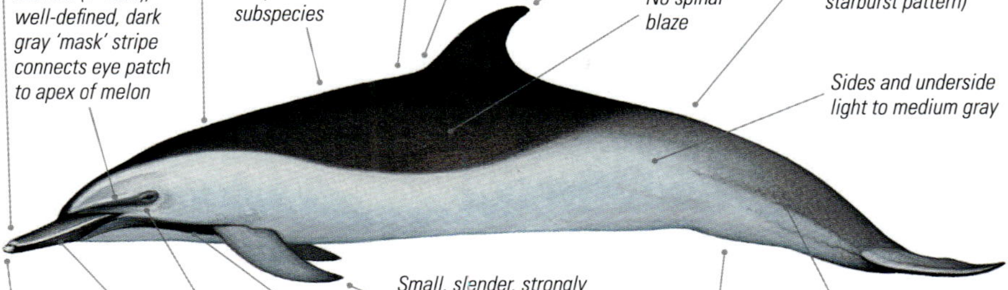

- Long, slender beak (up to 13cm) separated from melon by distinct crease (tends to be slightly more slender than coastal subspecies)
- Typically very little or no white spotting on cape (less than in coastal)
- Dark dorsal cape sweeps very low on side just ahead of dorsal fin
- Two-toned underlying color pattern
- Very narrow, falcate dorsal fin (up to 20cm tall) midway along back (variable shape)
- Typically pointed tip
- Some individuals have small number of cookiecutter shark bites (often heal into starburst pattern)
- More slender body than coastal subspecies
- Narrow (1–2cm), well-defined, dark gray 'mask' stripe connects eye patch to apex of melon
- No spinal blaze
- Sides and underside light to medium gray
- Many larger individuals have brilliant white-tipped beak (up to first 4cm) that gets progressively whiter with age (may flush pink)
- Dark gray eye-patch
- Typically pale 'lips' (may flush pink)
- Small, slender, strongly recurved flippers
- Dark gray beak-to-flipper stripe (variable)
- May be moderate ventral keel in adult male
- Two-toned tailstock (dark upperside, light underside)

## ADULT COASTAL

- Stockier body than offshore subspecies
- Spotting may extend to dorsal fin and flippers
- White spots generally smaller than large ones typical of Atlantic spotted
- Two-toned tailstock
- Much more extensive white spotting on dark dorsal cape (highly variable – may be dense enough to obliterate cape)
- Beak tends to be slightly thicker than offshore subspecies
- Dark spots on underside often fuse and lighten into slightly mottled or (seen from distance) uniform pale gray color in older animals (cf. Atlantic spotted dolphin)
- Typically more spotting than offshore
- Male tends to be more heavily spotted than female

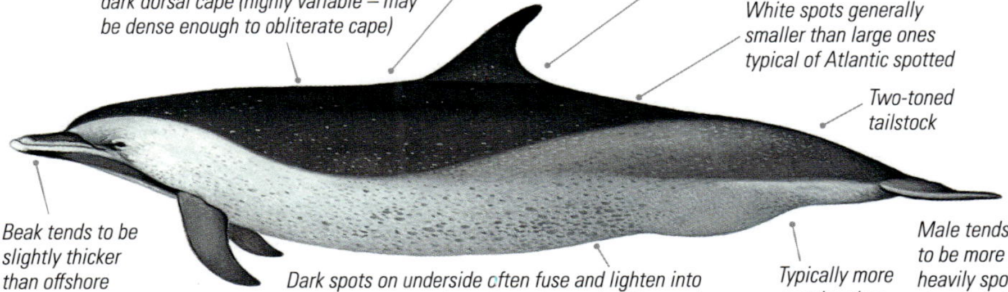

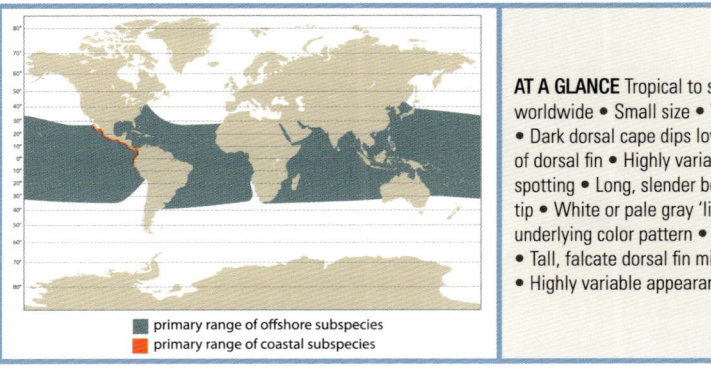

■ primary range of offshore subspecies
■ primary range of coastal subspecies

**AT A GLANCE** Tropical to sub-tropical waters worldwide • Small size • Two-toned tailstock • Dark dorsal cape dips lowest just ahead of dorsal fin • Highly variable dark and light spotting • Long, slender beak with white tip • White or pale gray 'lips' • Two-toned underlying color pattern • No spinal blaze • Tall, falcate dorsal fin midway along back • Highly variable appearance within group

# ATLANTIC SPOTTED DOLPHIN
*Stenella frontalis* (G. Cuvier, 1829)

The Atlantic spotted dolphin is highly variable in appearance – between ages, individuals and regions – from virtually unspotted to very heavily spotted. Young animals are unspotted (rather like slender bottlenose dolphins) and spots develop as they age. Pantropical spotted dolphins generally have a more slender body, a broadly two-toned underlying color pattern, a tailstock divided into dark upper and light lower portions, and lack the light-colored spinal blaze.

**IUCN status** Least Concern (2018).
**Population** Probably low hundreds of thousands (minimum 77,000 in US waters). Trend unknown.
**Classification** Odontoceti, family Delphinidae.
**Taxonomy** No recognized subspecies. However, there appear to be two forms: a larger, heavy-bodied, heavily spotted form occurring mainly over the continental shelf in warmer waters of the western North Atlantic; and a smaller, slimmer, lightly spotted or unspotted form in more oceanic areas over the continental slope in the Gulf Stream and the central North Atlantic (and around some offshore islands, such as the Azores).
**Other names** Spotted dolphin, spotter.

**DISTRIBUTION** Tropical to warm temperate waters in the Atlantic Ocean. Occurs around some oceanic islands, such as the Azores and the Bahamas. The heavily spotted form prefers shallow continental shelf waters (typically at least 8–20km offshore). The lightly spotted form occurs over the outer continental shelf, the upper continental slope and in deep oceanic waters. May prefer shallower waters around oceanic islands, such as over shallow (6–12m) sand banks in the Bahamas.

**BEHAVIOR** Highly acrobatic and capable of some exceptionally high leaps. An avid bow-rider in most of its range.

**FOOD AND FEEDING** Small to large fish and squid. Offshore form cooperates in herding balls of fish against the surface.

**TEETH** Upper jaw 64–84; lower jaw 60–80.

**GROUP SIZE AND STRUCTURE** Generally in small to medium-sized groups of up to 50 animals (occasionally up to 200). Group sizes tend to be smaller (5–15) nearer shore and larger offshore.

**ADULT HEAVILY SPOTTED FORM**

Flukes usually darker with little or no spotting

Median notch

Flippers usually darker with little or no spotting

**SIZE**
**L:** ♂ 1.7–2.3m, ♀ 1.7–2.3m;
**WT:** 110–140kg; **MAX:** 2.3m, 143kg
**Calf – L:** 0.8–1.2m; **WT:** *c.* 10–15kg

**DIVE SEQUENCE** When swimming slowly, tip of beak breaks surface first, upperside of melon briefly visible as it blows, and back and dorsal fin appear briefly (but tailstock often barely visible); when swimming fast, porpoises with neat re-entry. • **DEPTH** Most dives are less than 10m; maximum recorded 60m. • **DURATION** Mostly 2–4 minutes; maximum recorded 6 minutes.

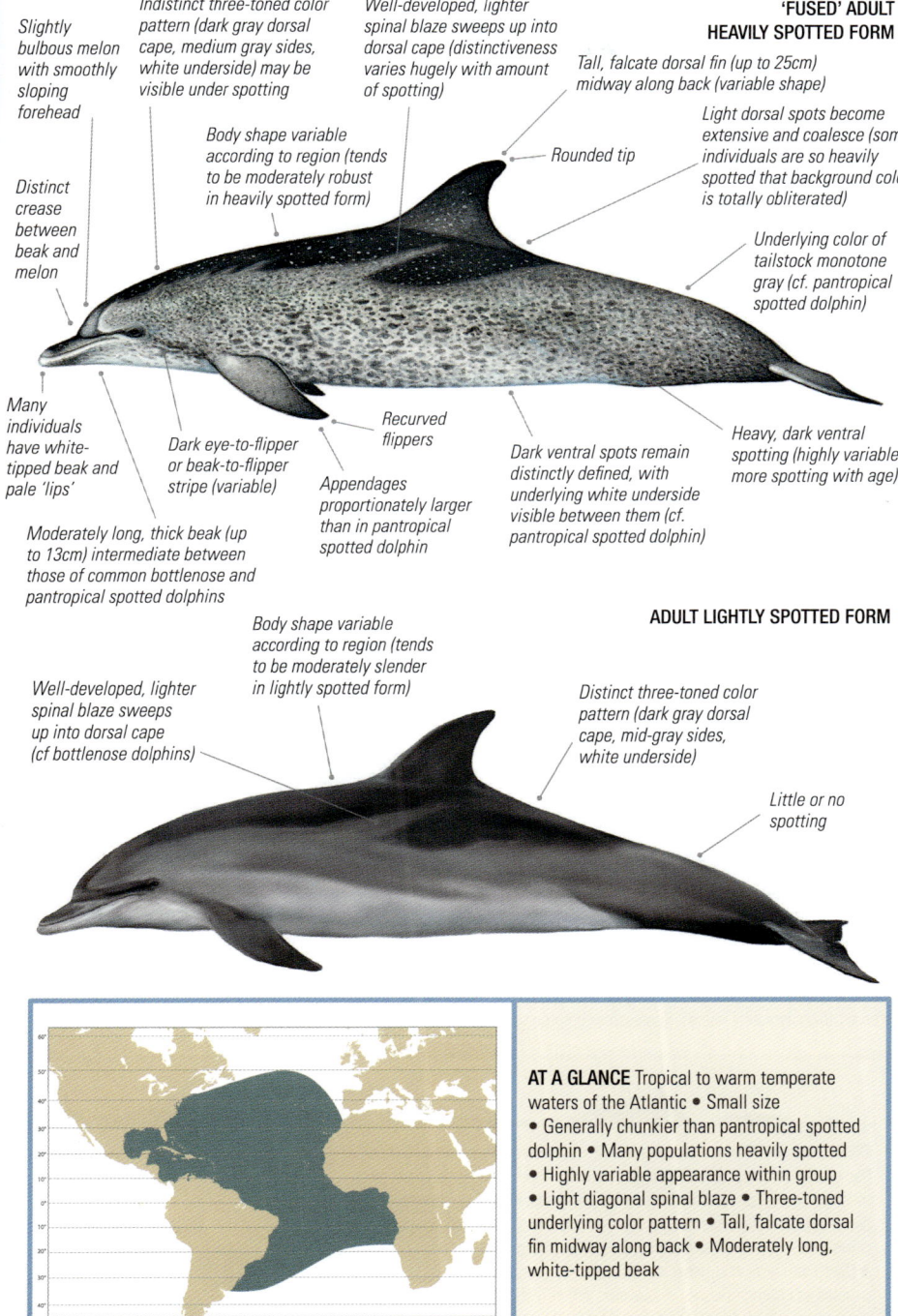

# SPINNER DOLPHIN
*Stenella longirostris* (Gray, 1828)

Named for its spectacular habit of leaping high out of the water and spinning up to seven times longitudinally, before falling back with a great splash, the spinner dolphin is a familiar sight in many parts of the tropics. There is more geographical variation in form and color pattern in this species than in almost any other cetacean.

**IUCN status** Least Concern (2018). Eastern spinner subspecies (*orientalis*) Vulnerable (2008).
**Population** One million+ (based on limited surveys, but most of range unsurveyed). At least 2 million killed by commercial tuna fleets. Trend unknown.
**Classification** Odontoceti, family Delphinidae.
**Taxonomy** Four subspecies currently recognized: Gray's (sometimes 'Hawaiian') spinner dolphin (*S. l. longirostris*), the 'typical' spinner dolphin; Central American (previously 'Costa Rican') spinner dolphin (*S. l. centroamericana*); eastern spinner dolphin (*S. l. orientalis*); and dwarf spinner dolphin (*S. l. roseiventris*). There is also a hybrid – called the whitebelly or white-bellied spinner – that is intermediate between Gray's and eastern spinner dolphins, found in the eastern tropical Pacific where these two 'parent' subspecies meet.
**Other names** Long-snouted spinner dolphin, longsnout, spinner, rollover; see taxonomy for subspecies common names.
**DISTRIBUTION** Found in all tropical and most sub-tropical waters in the Pacific, Atlantic and Indian Oceans. Best known in coastal waters, around oceanic islands and over shallow banks, but also occurs in very large numbers on the high seas and ranges over vast distances of open water. Coastal spinners – especially around oceanic islands – frequently move into shallow sandy bays in the morning and rest until late afternoon or early evening; they venture out to deeper water at night to feed. In some places, they rest actually inside the lagoons of coral atolls.

**SPINNING** Best known for leaping up to 3m into the air, spinning on its longitudinal axis up to seven times and then falling back into the water – often up to 14 times in a descending series (each less vigorous than the previous one). It starts spinning underwater, just before emerging from the surface. Individuals of all ages spin and, once one dolphin starts, others typically join in. Some other dolphin species spin, but not as many times or with the same frequency.

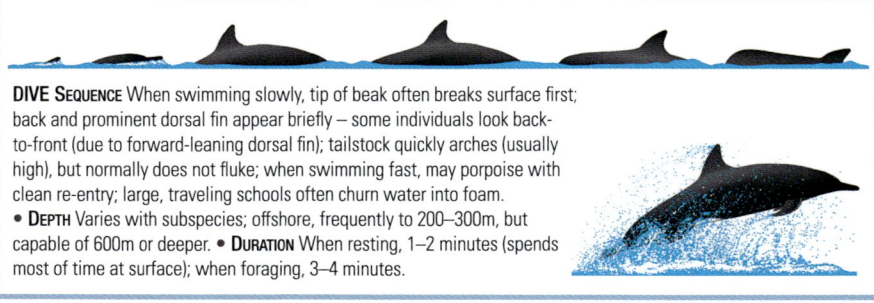

**DIVE Sequence** When swimming slowly, tip of beak often breaks surface first; back and prominent dorsal fin appear briefly – some individuals look back-to-front (due to forward-leaning dorsal fin); tailstock quickly arches (usually high), but normally does not fluke; when swimming fast, may porpoise with clean re-entry; large, traveling schools often churn water into foam.
• **Depth** Varies with subspecies; offshore, frequently to 200–300m, but capable of 600m or deeper. • **Duration** When resting, 1–2 minutes (spends most of time at surface); when foraging, 3–4 minutes.

**ADULT GRAY'S**

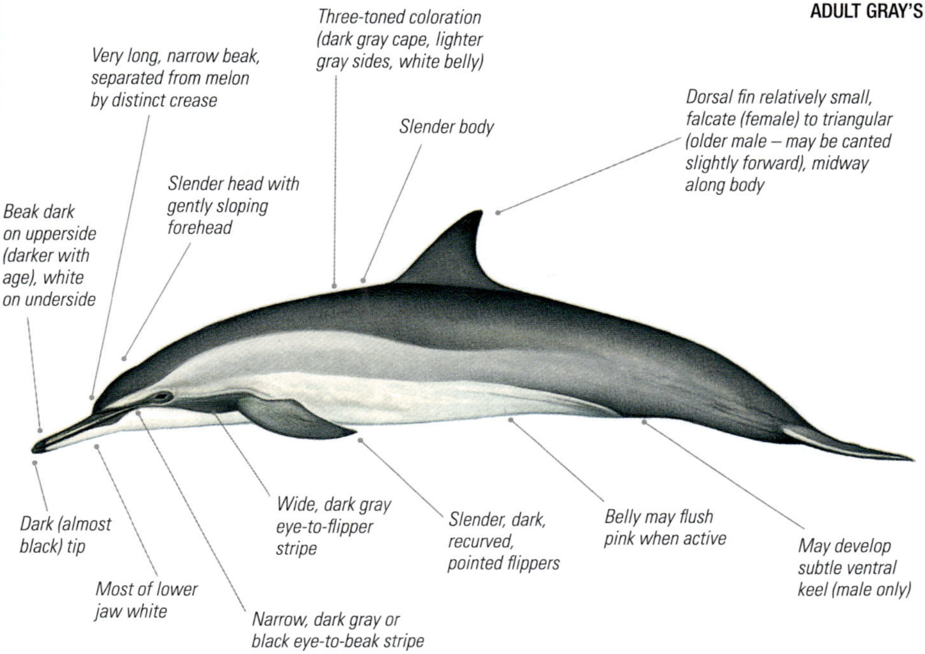

- Three-toned coloration (dark gray cape, lighter gray sides, white belly)
- Very long, narrow beak, separated from melon by distinct crease
- Slender body
- Dorsal fin relatively small, falcate (female) to triangular (older male – may be canted slightly forward), midway along body
- Slender head with gently sloping forehead
- Beak dark on upperside (darker with age), white on underside
- Dark (almost black) tip
- Most of lower jaw white
- Wide, dark gray eye-to-flipper stripe
- Narrow, dark gray or black eye-to-beak stripe
- Slender, dark, recurved, pointed flippers
- Belly may flush pink when active
- May develop subtle ventral keel (male only)

**ADULT GRAY'S**

- Noticeably slender body
- Distinctive black tip

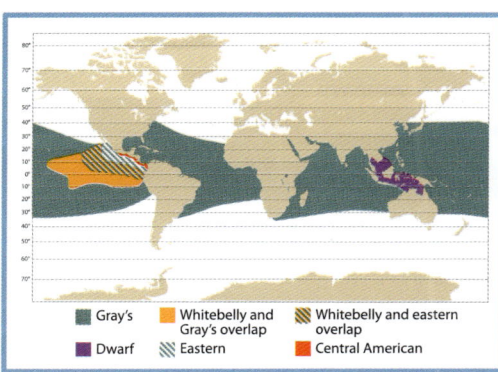

**AT A GLANCE** Tropical and sub-tropical waters worldwide • Small size • Usually slender body • Erect dorsal fin (sometimes canted forward) midway along back • Huge variation in appearance according to region • Long, slender beak • Gently sloping melon • Performs high spinning leaps • Usually quite gregarious

- Gray's
- Dwarf
- Whitebelly and Gray's overlap
- Eastern
- Whitebelly and eastern overlap
- Central American

SPINNER DOLPHIN

**BEHAVIOR** One of the most aerial of all dolphins. As well as spinning, it will perform more traditional breaches, as well as arc-shaped leaps, tail-over-head leaps, side-slaps, fluke-slaps and flipper-slaps. Particularly acrobatic during the change from rest to foraging. In many parts of the range it readily approaches boats and bow-rides (except on tuna fishing grounds in the eastern tropical Pacific, where it generally avoids boats).

**FOOD AND FEEDING** Wide variety of small midwater fish, squid, sergestid shrimps; dwarf spinner feeds on benthic reef fish and invertebrates. Most feeding at night, resting during day (except dwarf spinners – feed during day).

**TEETH** Upper jaw 80–124; lower jaw 80–124. There are small differences in tooth count between subspecies.

**GROUP SIZE AND STRUCTURE** Highly variable according to activity and location, ranging from 10–50 up to several thousand. The largest schools tend to occur offshore, the smallest in coastal waters.

**ADULT MALE CENTRAL AMERICAN**

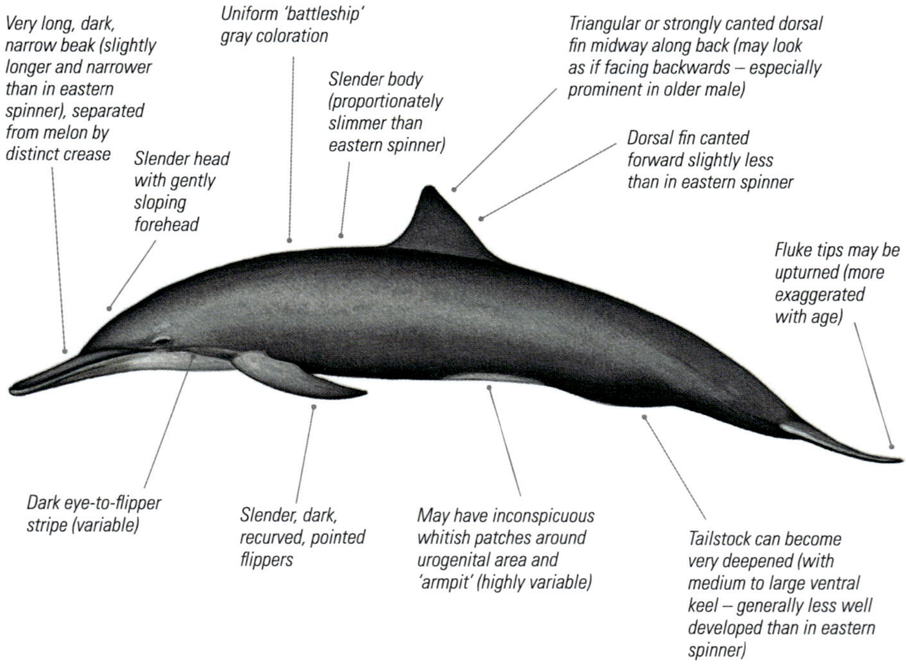

Very long, dark, narrow beak (slightly longer and narrower than in eastern spinner), separated from melon by distinct crease

Uniform 'battleship' gray coloration

Slender body (proportionately slimmer than eastern spinner)

Triangular or strongly canted dorsal fin midway along back (may look as if facing backwards – especially prominent in older male)

Slender head with gently sloping forehead

Dorsal fin canted forward slightly less than in eastern spinner

Fluke tips may be upturned (more exaggerated with age)

Dark eye-to-flipper stripe (variable)

Slender, dark, recurved, pointed flippers

May have inconspicuous whitish patches around urogenital area and 'armpit' (highly variable)

Tailstock can become very deepened (with medium to large ventral keel – generally less well developed than in eastern spinner)

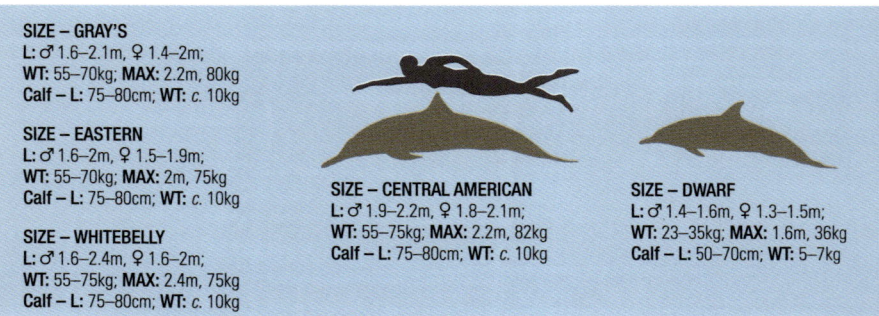

**SIZE – GRAY'S**
L: ♂ 1.6–2.1m, ♀ 1.4–2m;
WT: 55–70kg; MAX: 2.2m, 80kg
Calf – L: 75–80cm; WT: *c.* 10kg

**SIZE – EASTERN**
L: ♂ 1.6–2m, ♀ 1.5–1.9m;
WT: 55–70kg; MAX: 2m, 75kg
Calf – L: 75–80cm; WT: *c.* 10kg

**SIZE – WHITEBELLY**
L: ♂ 1.6–2.4m, ♀ 1.6–2m;
WT: 55–75kg; MAX: 2.4m, 75kg
Calf – L: 75–80cm; WT: *c.* 10kg

**SIZE – CENTRAL AMERICAN**
L: ♂ 1.9–2.2m, ♀ 1.8–2.1m;
WT: 55–75kg; MAX: 2.2m, 82kg
Calf – L: 75–80cm; WT: *c.* 10kg

**SIZE – DWARF**
L: ♂ 1.4–1.6m, ♀ 1.3–1.5m;
WT: 23–35kg; MAX: 1.6m, 36kg
Calf – L: 50–70cm; WT: 5–7kg

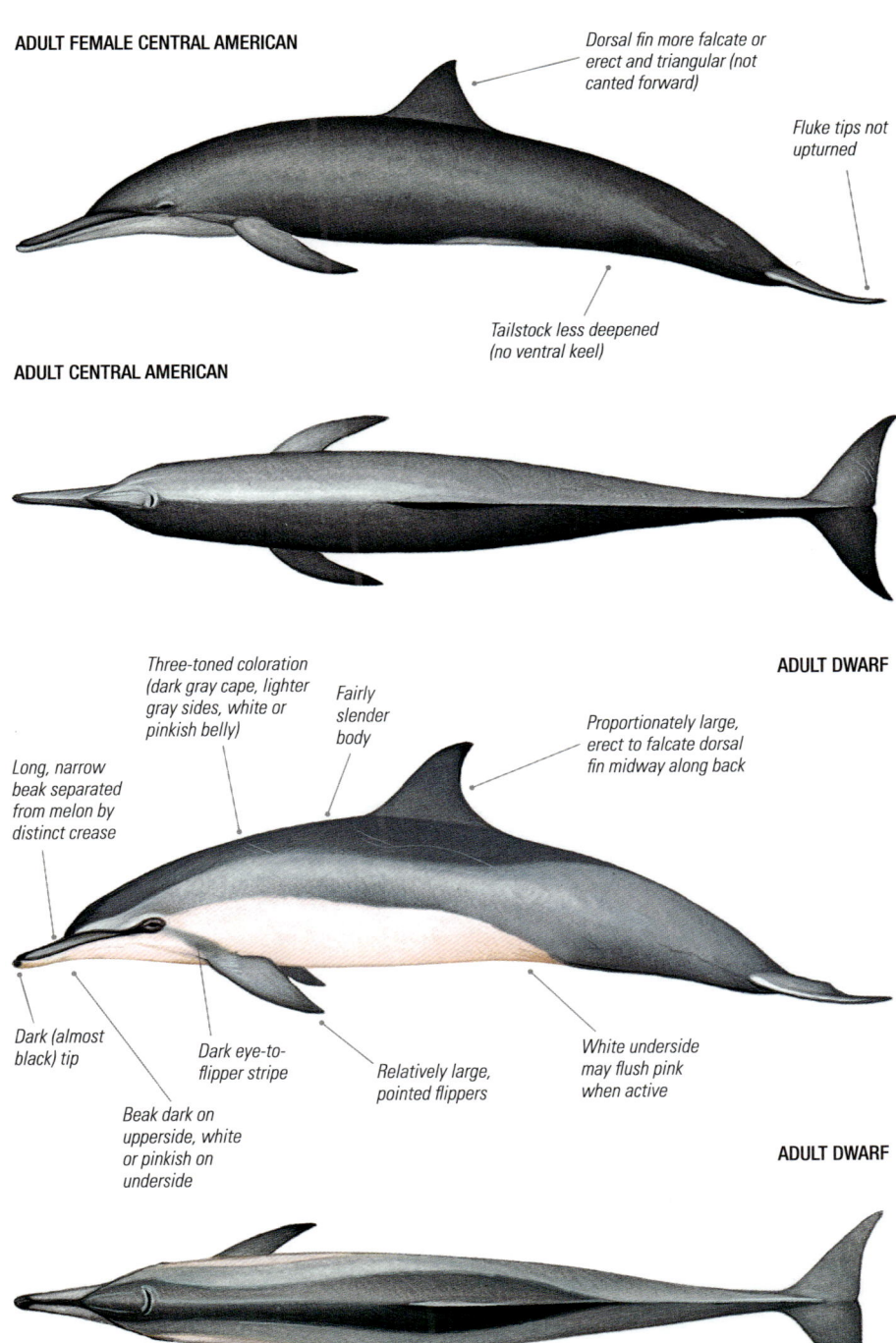

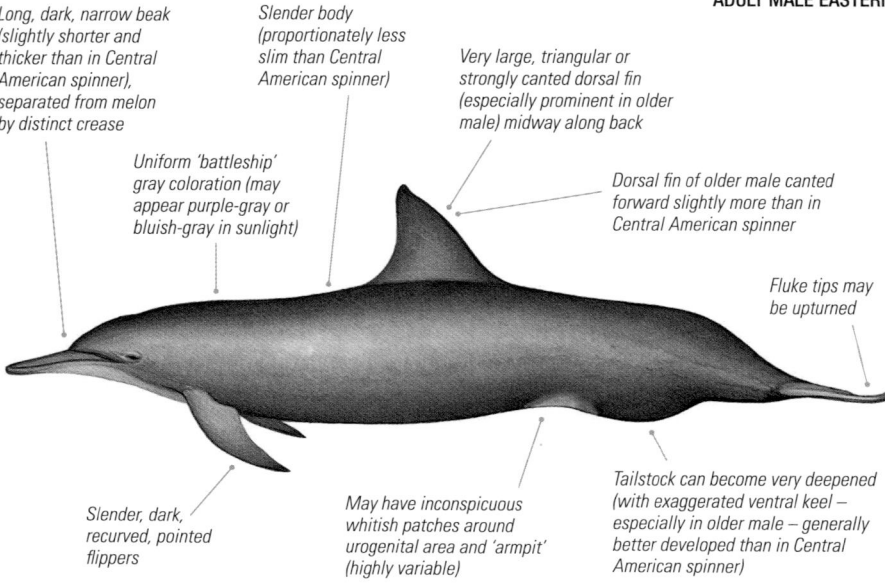
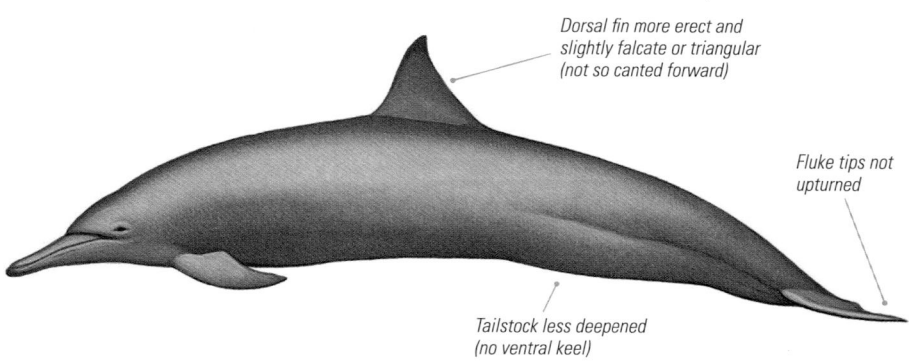
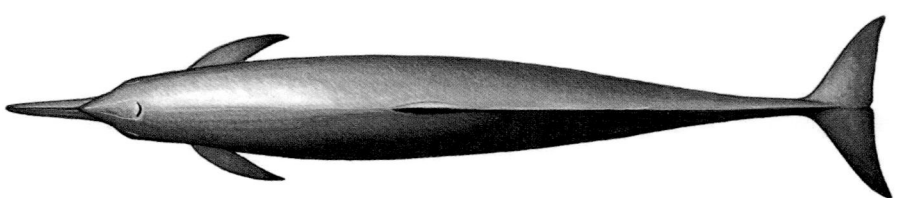

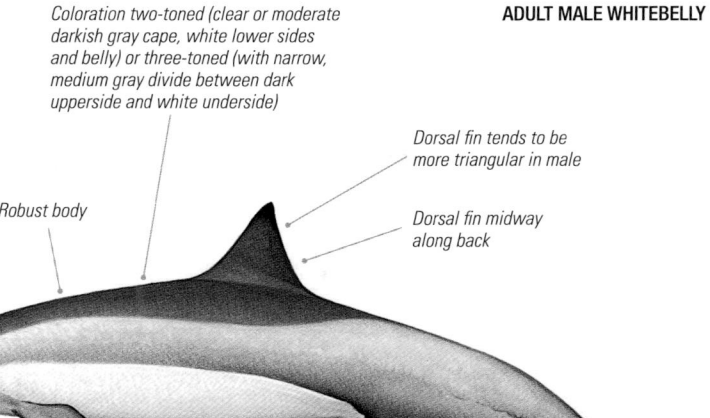

**ADULT MALE WHITEBELLY**

Coloration two-toned (clear or moderate darkish gray cape, white lower sides and belly) or three-toned (with narrow, medium gray divide between dark upperside and white underside)

Dorsal fin tends to be more triangular in male

Dorsal fin midway along back

Robust body

Dark eye-to-flipper stripe

White belly may flush pink when active

May develop small to moderate ventral keel (male only)

**ADULT WHITEBELLY**

**ADULT FEMALE WHITEBELLY**

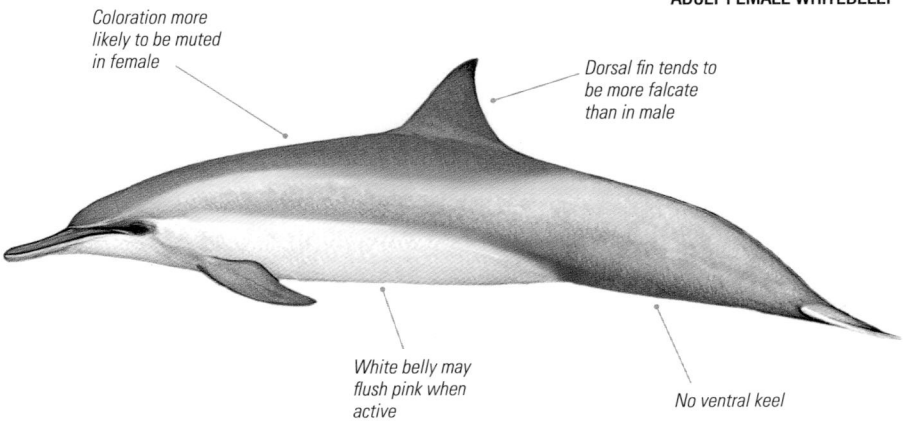

Coloration more likely to be muted in female

Dorsal fin tends to be more falcate than in male

White belly may flush pink when active

No ventral keel

SPINNER DOLPHIN **181**

# CLYMENE DOLPHIN
*Stenella clymene* (Gray, 1850)

Molecular studies reveal that the Clymene dolphin (normally pronounced 'Cly-me-nee') probably evolved through extensive hybridization between spinner and striped dolphins and, in many ways, it appears almost intermediate between the two species. If so, it would be the first marine mammal known to have arisen in this way. Initially believed to be a variant of the spinner dolphin, it was not fully accepted as a distinct species until 1981.

**IUCN status** Least Concern (2018).
**Population** Unknown, but appears to be relatively common in at least parts of its range. Trend unknown.
**Classification** Odontoceti, family Delphinidae.
**Taxonomy** No recognized forms or subspecies.
**Other names** Short-snouted spinner dolphin, Atlantic spinner dolphin, Senegal dolphin, helmet dolphin.
**DISTRIBUTION** Deep tropical, sub-tropical and occasionally warm temperate waters in the Atlantic Ocean, including the Caribbean Sea and Gulf of Mexico. There are only a few mid-Atlantic records, but it is assumed to have a continuous range. An oceanic species, occurring mainly seaward of the continental shelf (preferring the slope and beyond) and rarely seen near shore (except where deep water approaches the coast).
**BEHAVIOR** Quick, agile and often aerially active. Breaches and spins longitudinally; may spin up to four times (though the leaps are lower and less frequent, and the spins less elaborate and acrobatic than those of the spinner dolphin – the only other species that routinely exhibits longitudinal rotations). Response to boats varies from avoidance to quite inquisitive. Avid bow-rider in some areas and will often approach vessels from a distance.
**FOOD AND FEEDING** Small mesopelagic fish and squid. Feeds mostly at night; cooperative feeding has been observed.
**TEETH** Upper 78–104 jaw; lower jaw 76–96.
**GROUP SIZE AND STRUCTURE** Typically 4–150 but ranges from 1 to 200, with an overall average of *c.* 70–80.

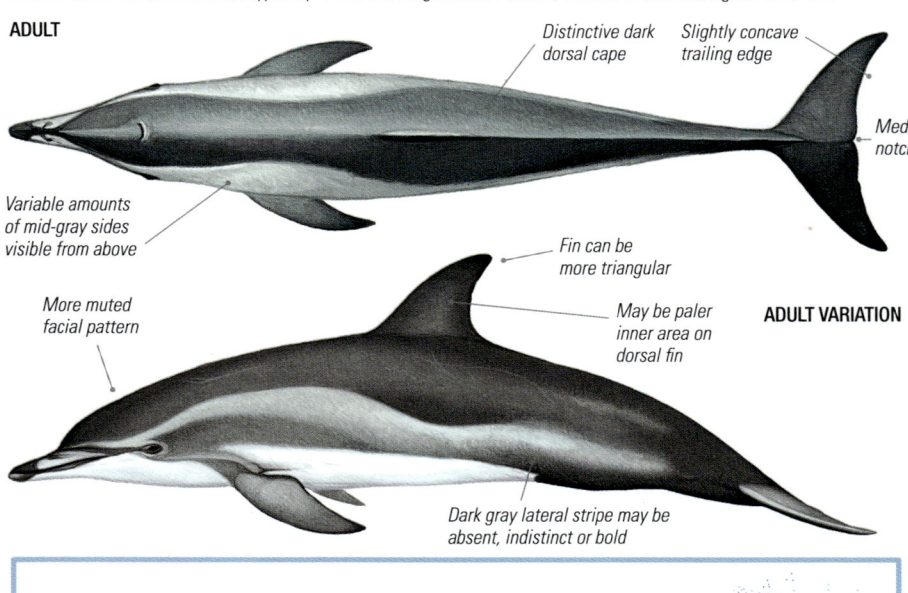

ADULT
- Distinctive dark dorsal cape
- Slightly concave trailing edge
- Median notch
- Variable amounts of mid-gray sides visible from above
- Fin can be more triangular
- More muted facial pattern
- May be paler inner area on dorsal fin
- ADULT VARIATION
- Dark gray lateral stripe may be absent, indistinct or bold

**DIVE SEQUENCE** Porpoises when swimming fast. • **DEPTH** Unknown. • **DURATION** Unknown.

# CLYMENE DOLPHIN

**ADULT**

- Small and triangular to tall and falcate dorsal fin (highly variable)
- May be paler inner area on dorsal fin
- Fin midway along back
- Gray cape dips above eye
- Dark gray dorsal cape
- Gray cape dips below dorsal fin (or in some individuals further forward, below base of leading edge of dorsal fin)
- Dark gray to black 'moustache' (dark streaks) on rostrum (variable)
- Stocky body
- Dark eye-ring
- May be oval scars from cookiecutter shark bites (mainly on sides and underside)
- Distinct crease between beak and melon
- Mid-gray sides (varies in extent and distinctiveness)
- Distinct dark gray lateral stripe between white underside and mid-gray side may be absent, indistinct or bold (sometimes prompting references to color pattern as 'four-toned')
- White or creamy underside, throat and lower jaw
- Beak tip and 'lips' dark gray or black
- Dark gray stripe between eye and beak
- Variable dark gray eye-to-flipper stripe (usually broadens toward flipper)
- Slender, curved flippers with pointed tips
- Moderately long (less than 12cm) beak (shorter and thicker than on spinner dolphin)
- Narrow white eye-to-flipper stripe (above dark stripe)

**ADULT**

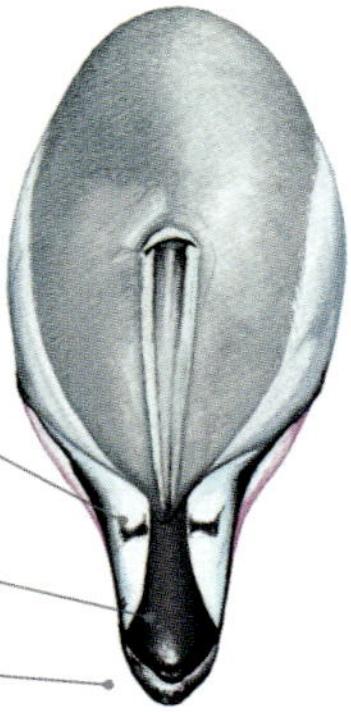

- Dark gray to black 'moustache' (dark streaks) on rostrum visible on bow-riders (varies in extent and intensity)
- Dark stripe from tip of beak to base of melon (continues to blowhole as lighter gray band)
- Blackish-tipped beak

**SIZE**
L: ♂ 1.8–2m, ♀ 1.7–1.9m;
**WT:** 50–80kg; **MAX:** 2m, 80kg
Calf – L: 0.9–1.2m; **WT:** c. 10kg

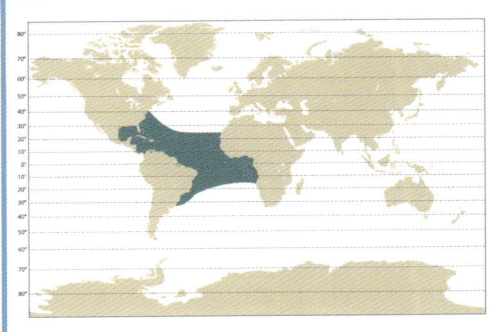

**AT A GLANCE** Warm waters of Atlantic Ocean • Small size • Stocky body • Three-toned color pattern (highly variable) • Two distinctive dips in the dark cape (giving wavy appearance) • Nearly triangular or slightly falcate dorsal fin • Medium-length, robust beak • Dark 'moustache' on surface of beak

# STRIPED DOLPHIN
*Stenella coeruleoalba* (Meyen, 1833)

Ancient Greeks marvelled at the beautiful 'brushstrokes' and colors of striped dolphins, and depicted them in their frescoes several thousand years ago. Widely distributed in warm waters in both hemispheres, the species is a familiar sight in many parts of the world.

**IUCN status** Least Concern (2018). Mediterranean sub-population Least Concern (2020). Gulf of Corinth sub-population Endangered (2022).
**Population** Minimum 2.4 million. Trend unknown.
**Classification** Odontoceti, family Delphinidae.
**Taxonomy** No recognized forms or subspecies.
**Other names** Streaker, euphrosyne dolphin, whitebelly.

**DISTRIBUTION** Widely distributed mainly in deep tropical to warm temperate waters (roughly between 50°N and 40°S), though extends into higher latitudes than other *Stenella* dolphins (it is the only member of the genus that routinely reaches northern Europe). Typically in water deeper than 1,000m (in many areas sighting rates increase dramatically with greater depth). Generally occurs outside the continental shelf but also close to shore where waters sufficiently deep. In the Mediterranean, where it is the most abundant dolphin, it is sometimes in shallower water relatively close to shore.

**BEHAVIOR** Very acrobatic, frequently breaching up to 5–7m high, porpoising upside down and chin-slapping. Performs a unique behavior called 'roto-tailing', in which it makes a high arcing leap while vigorously whipping its tail in a circle. Especially nervous of vessels in the eastern tropical Pacific; elsewhere, it will bow-ride and wake-ride, but can be more easily 'spooked' than other tropical dolphins and often dashes away (with low, splashy leaps) for no apparent reason.

**FOOD AND FEEDING** Variety of small fish, squid and some crustaceans. Probably mostly nocturnal.
**TEETH** Upper jaw 78–110; lower jaw 78–110.
**GROUP SIZE AND STRUCTURE** Typically 20–50, but ranges from 10–100 in a dense school, sometimes up to 500. Great variation between regions. Group size tends to increase the further offshore and deeper the water.

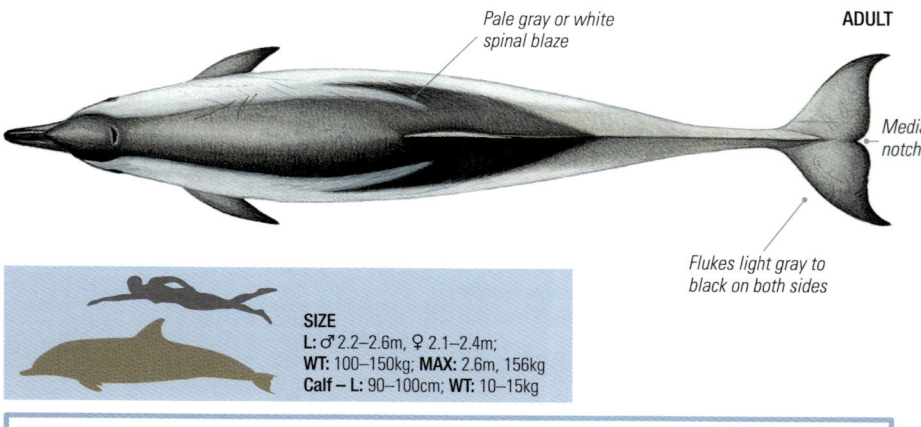

ADULT
Pale gray or white spinal blaze
Median notch
Flukes light gray to black on both sides

**SIZE**
L: ♂ 2.2–2.6m, ♀ 2.1–2.4m;
**WT:** 100–150kg; **MAX:** 2.6m, 156kg
Calf – L: 90–100cm; **WT:** 10–15kg

**DIVE SEQUENCE** Usually seen swimming fast, in long, low arcing leaps; surfaces at shallow angle, sometimes throwing tail high in air mid-leap; when not clearing surface, typically throws up Dall's porpoise-like rooster tail of spray. • **DEPTH** May be capable of 700m, but limited information. • **DURATION** Unknown.

**ADULT**

- Slightly bulbous melon with smoothly sloping forehead
- Generally more robust than other *Stenella* species (but ranging to quite slender)
- Dark gray, bluish-gray or brownish-gray dorsal cape
- Moderately tall, falcate dorsal fin midway along back (up to 27cm)
- Moderately long, fairly stubby beak (up to 11cm or 4.5–5.8 per cent of body length), with distinct crease between beak and melon
- Pale gray thorax
- Pale gray or white spinal blaze extends from thoracic area towards base of dorsal fin (highly variable)
- Dark gray to black eye-to-flipper stripe
- Slender, recurved, light gray to black flippers
- Pointed tip
- Short accessory stripe (not always present)
- White or pinkish belly, throat and lower jaw
- Dark gray to bluish-black lateral stripe from beak, through eye, to anus (widening and slightly fading at rear)
- May have indistinct wounds from cookiecutter shark bites (which heal to background color)

**IMMATURE**

- Dorsal color variable (may be browner)
- Lateral stripe and eye-to-flipper stripes may be more muted

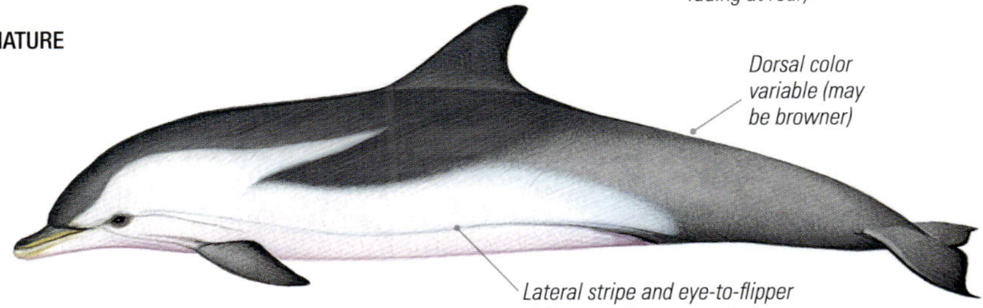

**AT A GLANCE** Deep tropical to warm temperate waters worldwide • Small size • Complex three-toned color pattern • Long, dark lateral stripe • Underside bright white or pinkish • Light gray spinal blaze sweeps back and up towards dorsal fin • Moderately tall, falcate dorsal fin midway along back • Active, energetic and fast swimmer

STRIPED DOLPHIN

# COMMON DOLPHIN
*Delphinus delphis*                                                               Linnaeus, 1758

Aristotle and Pliny the Elder described the common dolphin in great detail – and it was the first dolphin species to be scientifically described – yet there has been ongoing debate ever since about whether it should be classified as one, two or more species. The unique criss-cross or hourglass color pattern on the sides of all common dolphins (except a rare dark morph) should distinguish them from other dolphins.

**IUCN status** Least Concern (2020). Mediterranean sub-population Endangered (2003). Inner Mediterranean sub-population Endangered (2020). Black Sea sub-population Vulnerable (2008). Gulf of Corinth sub-population Critically Endangered (2019).
**Population** At least 4–5 million (based on relatively dated regional estimates). Trend unknown, but there have been dramatic declines in some areas (such as the Mediterranean).
**Classification** Odontoceti, family Delphinidae.
**Taxonomy** Much confusion, with more than 20 species described since 1758. Most experts considered it to be a single species (*Delphinus delphis*) until 1994, when it was split into two – the short-beaked common dolphin (*D. delphis*) and the long-beaked common dolphin (*D. capensis*). Recent research questioned this split and, since 2016, it has once again been considered a single species. There are currently four recognized subspecies: common dolphin (*D. d. delphis*), Indo-Pacific common dolphin (*D. d. tropicalis*), Black Sea common dolphin (*D. d. ponticus*), and Eastern Pacific long-beaked common dolphin (*D. d. bairdii*). It is possible that the Eastern Pacific long-beaked may yet constitute a separate species.
**Other names** Crisscross dolphin, common porpoise, saddleback dolphin.

**DISTRIBUTION** Patchily distributed in tropical to temperate waters worldwide, roughly from 45°N (North Pacific) and 60°N (North Atlantic) to 50°S. May occasionally follow warm-water currents outside the normal latitudinal distribution. From nearshore waters to thousands of kilometers offshore, although absent from Gulf of Mexico and the Caribbean and from much of Atlantic and Indian Oceans. Separate populations occur in some semi-enclosed seas, such as the Mediterranean and Black Seas. Generally where sea surface water temperature above 14°C.

ADULT COMMON

Beak light brownish-gray (often with black tip or black band near tip)

Dark line along crease between beak and melon (extends posteriorly to encircle eyes)

Flukes usually dark brownish-gray above

Deep media notch

Dark line from tip of beak to apex of melon

Concave trailing edge

**DIVE SEQUENCE** When swimming slowly, surfaces at shallow angle, and often begins to blow underwater; when faster, underside of beak skims along surface; may be faint walls of water on either side of head. • **DEPTH** Most foraging shallower than 50m, but dives to 280m recorded (deeper dives at night). • **DURATION** Typically *c.* 10 seconds to 3 minutes; maximum *c.* 8 minutes.

**ADULT COMMON** (Atlantic – including Mediterranean – and Pacific)

- Rounded, bulbous, quite steeply rising melon
- Criss-cross color pattern on side (where dark brownish-gray cape, tan to pale yellow or ochre thoracic patch, light to medium gray flank patch and white underside meet at point below dorsal fin)
- Relatively robust body
- Tall dorsal fin midway along back (variable shape)
- Pointed tip
- Beak sharply demarcated from melon by deep crease
- Black 'lips'
- Relatively shorter and stubbier beak (but still moderately long)
- Black eye-patch continues as black stripe to crease between beak and melon (clearly separated from flipper-to-beak and flipper-to-anus stripes)
- Slender, dark flipper-to-beak stripe joins 'lip' patch at varying locations on underside of beak
- Small, slender, recurved flippers (pointed at tips)
- Flippers can be dark brownish-gray, pale yellow to ochre, pale gray (or any combination in between)
- Flipper-to-anus stripe usually weakly developed or absent
- May have prominent ventral keel (male only)

**ADULT COMMON DORSAL FIN VARIATIONS**

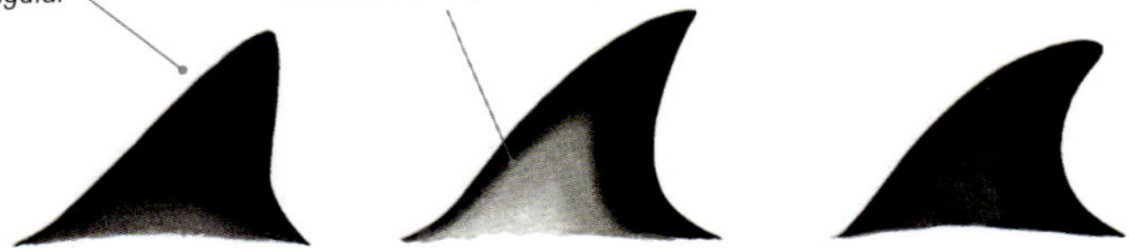

- Dorsal fin highly variable, from falcate to erect or triangular
- May have pale gray patch in center (varies from absent to covering nearly entire fin)

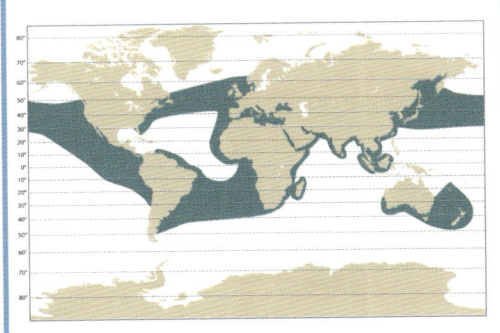

**AT A GLANCE** Tropical to temperate waters worldwide • Small size • Criss-cross or 'hourglass' color pattern on sides • Dark brownish-gray cape dips to 'V' under dorsal fin • Tan to pale yellow or ochre thoracic patch • Light to medium gray flank patch • White underside • Details of color pattern highly variable • Tall, moderately falcate dorsal fin midway along back • Often in fast-moving, splashy groups

COMMON DOLPHIN

**BEHAVIOR** Aerially active – frequently performing a variety of acrobatic leaps and somersaults (sometimes up to 6–7m high), as well as flipper-slapping and lobtailing. Will also 'pitch-pole', a distinctive aerial display in which the dolphin leaps straight out of the water and slams back lengthwise to create maximum splash. Can also make clean entries, with no splash. Typically porpoises out of the water when traveling at high speed. An enthusiastic and energetic bow-rider and wake-rider – even riding the 'snout waves' of large mysticete whales; however, not all individuals in a herd bow-ride. Large, boisterous groups of common dolphins, whipping the ocean's surface into a froth, are a common sight in many parts of the world.

**FOOD AND FEEDING** Wide variety of small schooling fish and squid; some crustaceans. Cooperative feeding techniques often used to herd fish schools.

**TEETH** Upper jaw 82–134; lower jaw 82–128.

**GROUP SIZE AND STRUCTURE** Highly gregarious, in groups of 10 to several thousand. Group composition is poorly known, but large herds are believed to be composed of smaller social units of 20–30 individuals that are not necessarily genetically related (but may be segregated by age and sex). Individuals may swim with preferred associates, but group membership appears to be fluid.

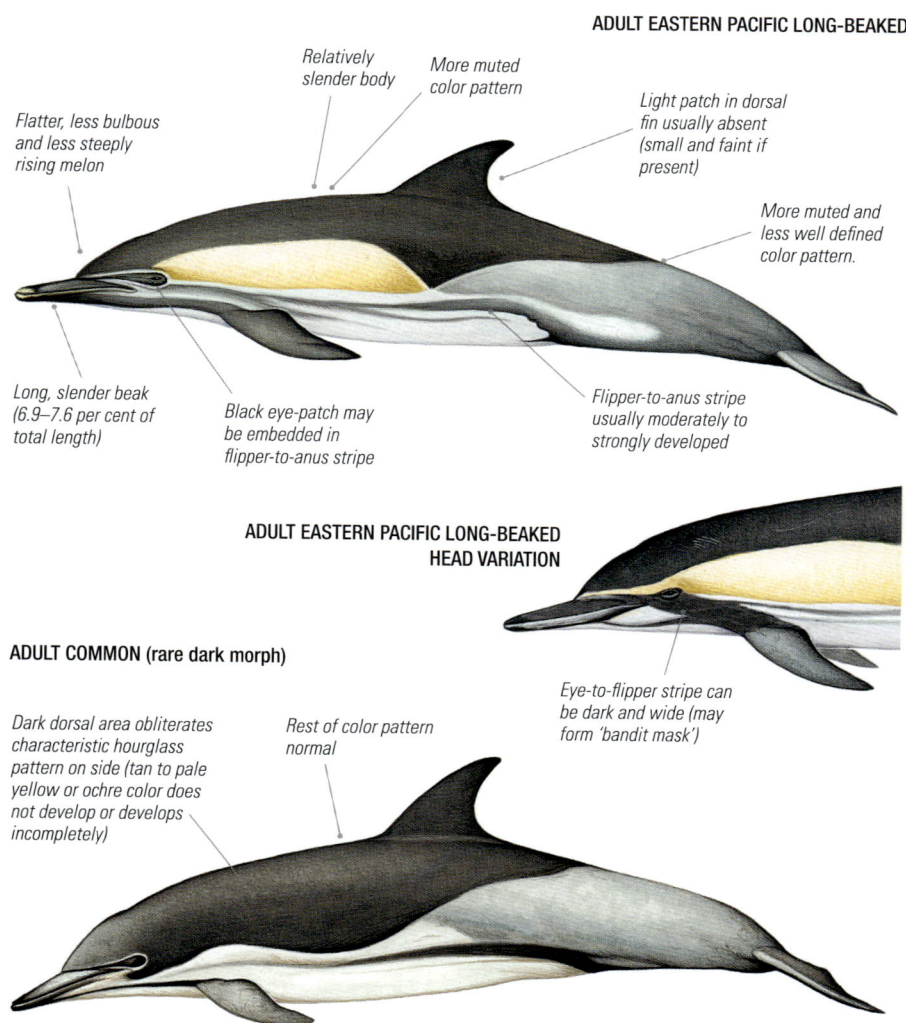

188　LONGER-BEAKED OCEANIC DOLPHINS

## ADULT INDO-PACIFIC (Indian Ocean – including Red Sea, Persian Gulf, Gulf of Thailand – and far western Pacific Ocean)

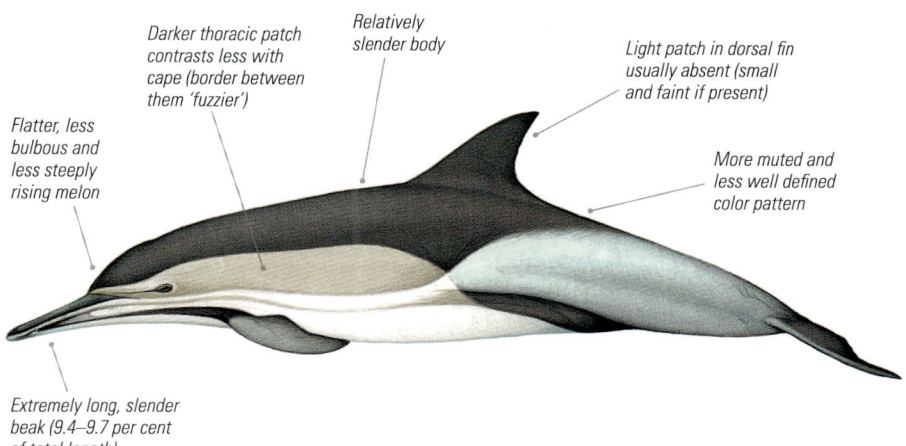

- Darker thoracic patch contrasts less with cape (border between them 'fuzzier')
- Relatively slender body
- Light patch in dorsal fin usually absent (small and faint if present)
- Flatter, less bulbous and less steeply rising melon
- More muted and less well defined color pattern
- Extremely long, slender beak (9.4–9.7 per cent of total length)

## ADULT BLACK SEA

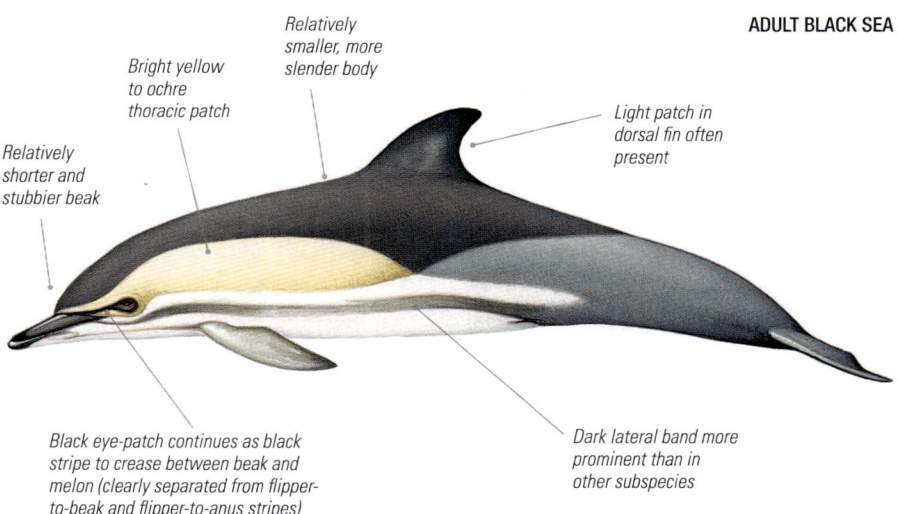

- Relatively smaller, more slender body
- Bright yellow to ochre thoracic patch
- Light patch in dorsal fin often present
- Relatively shorter and stubbier beak
- Black eye-patch continues as black stripe to crease between beak and melon (clearly separated from flipper-to-beak and flipper-to-anus stripes)
- Dark lateral band more prominent than in other subspecies

**SIZE – Common**
L: ♂ 1.7–2.5m, ♀ 1.6–2.4m;
WT: 150–200kg; MAX: 2.7m, 235kg
Calf – L: 76–115cm
Larger in the North Atlantic than the North Pacific.

**SIZE – Black Sea**
L: ♂ 1.5–1.8m, ♀ 1.5–1.7m;
WT: c. 150kg; MAX: 2.2m

**SIZE – Eastern Pacific long-beaked**
L: ♂ 2–2.6m, ♀ 1.9–2.2m;
WT: 150–235kg; MAX: 2.6m

**SIZE – Indo-Pacific**
L: ♂ 2–2.6m, ♀ 1.9–2.2m;
WT: 150–235kg; MAX: 2.6m
Calf – L: 80–93cm; WT: c. 7–10kg

# DALL'S PORPOISE
*Phocoenoides dalli* (True, 1885)

Dall's porpoise is probably the fastest small cetacean, typically seen as a splashy blur when it breaks the surface at high speed. Unlike other porpoises, it often approaches boats and readily bow-rides and wake-rides.

**IUCN status** Least Concern (2017).
**Population** Minimum 1.2 million (including *c.* 83,000 in Alaska). Trend unknown.
**Classification** Odontoceti, family Phocoenidae.
**Taxonomy** Two subspecies are recognized (based on body color pattern): *P. d. dalli* ('*dalli*-type' – the nominate form) and *P. d. truei* ('*truei*-type'). The *dalli*-type also has two minor color morphs, distinguished by the size of their white flank patch (larger in North Pacific–Bering Sea populations and smaller in Sea of Japan–Sea of Okhotsk populations).
**Other names** True's porpoise, True porpoise, spray porpoise, white-flank porpoise.
**DISTRIBUTION** Deep cool temperate to sub-Arctic waters of the northern North Pacific and adjacent seas. Prefers water colder than 17°C, with peak abundance below 13°C. Mainly offshore, but also in coastal areas where the water is deeper than 100m. *Dalli*-type rarer where it overlaps with the *truei*-type (accounting for 4–20 per cent of individuals, according to location).
**BEHAVIOR** An energetic porpoise that can be almost hyperactive, darting jerkily and zigzagging around at high speed (typically 20–30km/h but possibly up to 55km/h). Over short bursts, it may be the fastest small cetacean. A keen bow-rider – indeed, the only porpoise that often bow-rides – it prefers fast-moving vessels and will lose interest in slower ones. It will also ride the stern waves of fast boats. Aerial behavior such as breaching, tail-slapping or porpoising is extremely rare.
**FOOD AND FEEDING** Range of surface to mid-water fish and squid; may rarely take krill, shrimps and other crustaceans. Mostly at night; recent research suggests daytime feeding in some areas.
**TEETH** *Dalli*-type: upper jaw 46–56; lower jaw 48–56. *Truei*-type: upper jaw 38–46; lower jaw 40–48.
**GROUP SIZE AND STRUCTURE** Usually in fluid groups of 2–10 (usually fewer than 5). Larger temporary aggregations around prey concentrations (but these lack the cohesion of dolphin schools); largest groups occur in oceanic populations.

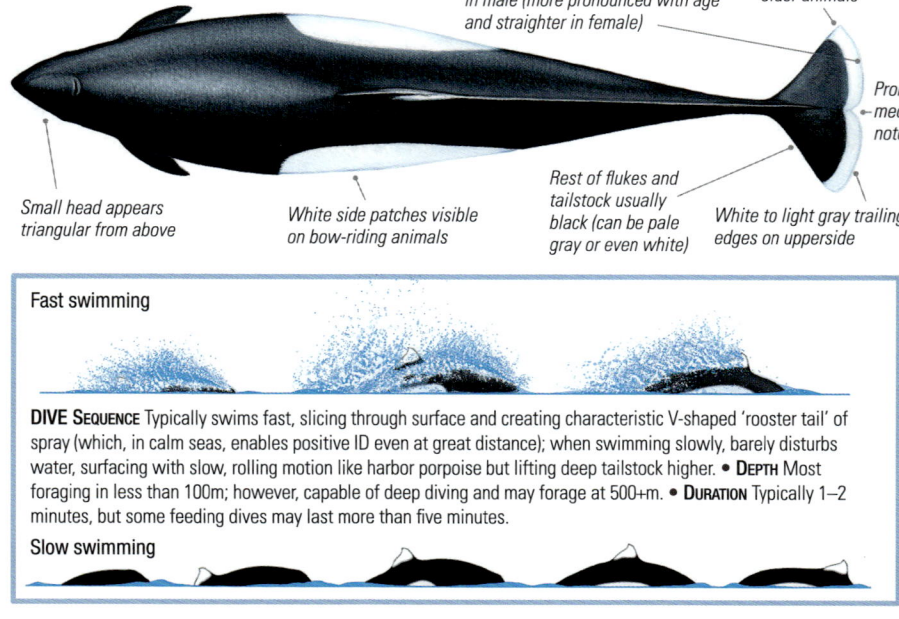

**ADULT MALE *DALLI*-TYPE**

Small flukes have convex trailing edges (giving backward appearance) in male (more pronounced with age and straighter in female)

Tips may become very rounded in older animals

Prominent median notch

Small head appears triangular from above

White side patches visible on bow-riding animals

Rest of flukes and tailstock usually black (can be pale gray or even white)

White to light gray trailing edges on upperside

**Fast swimming**

**DIVE SEQUENCE** Typically swims fast, slicing through surface and creating characteristic V-shaped 'rooster tail' of spray (which, in calm seas, enables positive ID even at great distance); when swimming slowly, barely disturbs water, surfacing with slow, rolling motion like harbor porpoise but lifting deep tailstock higher. • **DEPTH** Most foraging in less than 100m; however, capable of deep diving and may forage at 500+m. • **DURATION** Typically 1–2 minutes, but some feeding dives may last more than five minutes.

**Slow swimming**

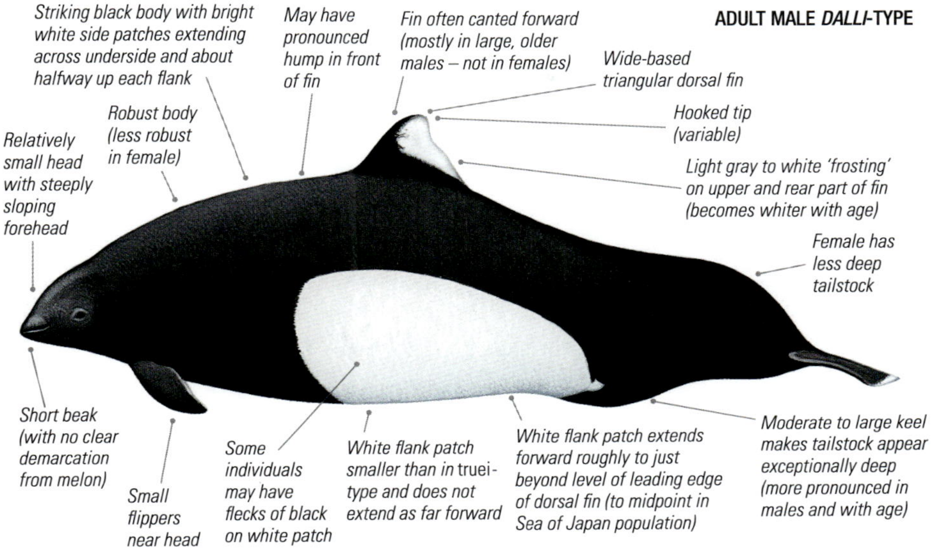

## ADULT MALE *DALLI*-TYPE

- Striking black body with bright white side patches extending across underside and about halfway up each flank
- May have pronounced hump in front of fin
- Fin often canted forward (mostly in large, older males – not in females)
- Wide-based triangular dorsal fin
- Hooked tip (variable)
- Light gray to white 'frosting' on upper and rear part of fin (becomes whiter with age)
- Robust body (less robust in female)
- Relatively small head with steeply sloping forehead
- Female has less deep tailstock
- Short beak (with no clear demarcation from melon)
- Small flippers near head
- Some individuals may have flecks of black on white patch
- White flank patch smaller than in truei-type and does not extend as far forward
- White flank patch extends forward roughly to just beyond level of leading edge of dorsal fin (to midpoint in Sea of Japan population)
- Moderate to large keel makes tailstock appear exceptionally deep (more pronounced in males and with age)

## ADULT MALE *TRUEI*-TYPE

- Slightly slimmer but longer body than dalli-type (at least in Japanese coastal waters)
- Larger flank patch extends further forward (at least to level of flippers)

**SIZE**
**L:** ♂ 1.8–2.3m, ♀ 1.7–2.2m;
**WT:** 135–200kg; **MAX:** 2.4m, 218kg
**Calf – L:** 0.9–1.2m; **WT:** c. 11kg

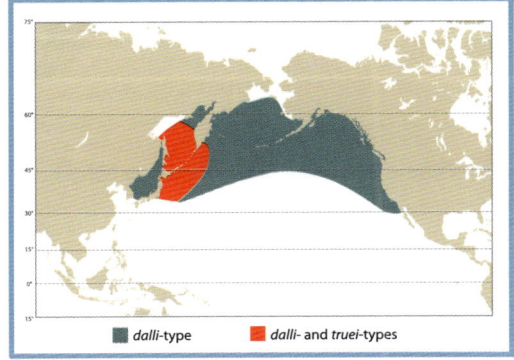

■ dalli-type    ■ dalli- and truei-types

**AT A GLANCE** Cool, deep waters of the North Pacific and adjacent seas • Striking black body with prominent white side patch • Highly distinctive 'rooster tail' of spray common on surfacing • Can be energetic and almost hyperactive (behavior more dolphin-like) • Wide-based, triangular two-toned dorsal fin • Small size • Exceptionally deep tailstock

DALL'S PORPOISE   191

# HARBOR PORPOISE
*Phocoena phocoena* (Linnaeus, 1758)

The harbor porpoise may be the most widespread and commonly seen of all the porpoises, but it can be surprisingly difficult to observe properly. It normally surfaces briefly, shows little of itself and rarely approaches boats, so a typical sighting is not much more than a fleeting glimpse.

**IUCN status** Least Concern (2020). Black Sea subspecies Endangered (2008). Baltic Sea sub-population Critically Endangered (2023).
**Population** Minimum *c.* 1 million. Trend unknown globally, but many populations decreasing.
**Classification** Odontoceti, family Phocoenidae.
**Taxonomy** Three subspecies are recognized: Atlantic harbor porpoise (*P. p. phocoena*), Black Sea harbor porpoise (*P. p. relicta*) and Pacific harbor porpoise (*P. p. vomerina*). There may be two more (as yet unnamed): Western Pacific harbor porpoise, and Afro-Iberian harbor porpoise (from the southern Iberian Peninsula and Mauritania).
**Other names** Harbour porpoise (English spelling), common porpoise; rarely – herring hog (especially Maine, USA), puffing pig (especially Atlantic Canada) or puffer, after its sneeze-like blow.
**DISTRIBUTION** Discontinuous range in cool temperate and sub-Arctic waters of the northern hemisphere. Favors coastal waters and frequents relatively shallow bays, estuaries, fjords, tidal channels and even harbors (and will also swim a considerable distance upriver in some areas). Rarely in seas deeper than 200m – although it is known in deep waters in some inshore regions and the West Greenland population was recently found in deep North Atlantic waters (to 1,000m) during winter. Favors areas of strong tidal currents, usually near islands or headlands.
**BEHAVIOR** Usually avoids boats, or is indifferent, so it can be difficult to approach and follow (although it is more approachable in some areas, such as the San Francisco Bay area and in the Bay of Fundy, eastern Canada). Most approachable during extended periods of inactivity, especially on calm days. Very rarely bow-rides or wake-rides. Acrobatics are uncommon, although it sometimes makes arc-shaped leaps when chasing prey and very occasionally tail-slaps when socializing. Will breach occasionally (such as in the Bay of Fundy).
**FOOD AND FEEDING** Mainly small schooling fish; some squid and octopuses; calves will eat small crustaceans during early phase of weaning. Opportunistic, taking prey mainly from near seabed, but will also forage in water column and close to surface.
**TEETH** Upper jaw 42–58; lower jaw 40–58. Teeth are spatulate, as in all porpoises.
**GROUP SIZE AND STRUCTURE** Usually in mother–calf pairs or loose, fluid groups of 1–3 (larger groups of 6–10 are not uncommon in some areas); several hundred have been observed at good feeding grounds.

**ADULT**

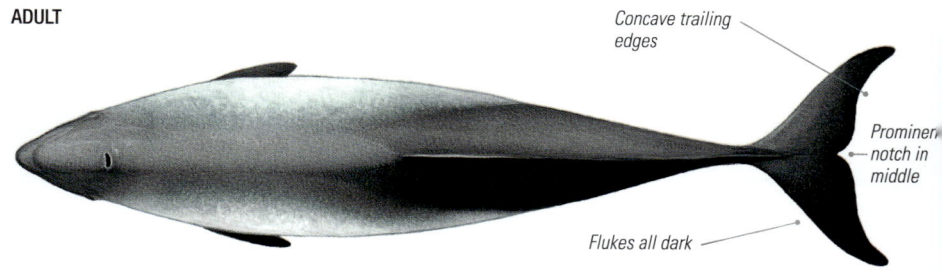

*Concave trailing edges*
*Prominent notch in middle*
*Flukes all dark*

**DIVE SEQUENCE** Surfaces with slow, forward-rolling motion (as if dorsal fin is mounted on revolving wheel and lifted briefly above surface, then withdrawn), and dives with little or no splash; when feeding (swimming fast and erratically) may produce distinctive spray (known as 'pop-splashing') – very different to 'rooster's tail' produced by Dall's porpoise. • **DEPTH** Typically 20–130m (maximum 410m). • **DURATION** Most dives *c.* 1 minute; maximum 6 minutes.
**BLOW** Indistinct, but on calm days a sharp, sneeze-like puffing sound may be audible.

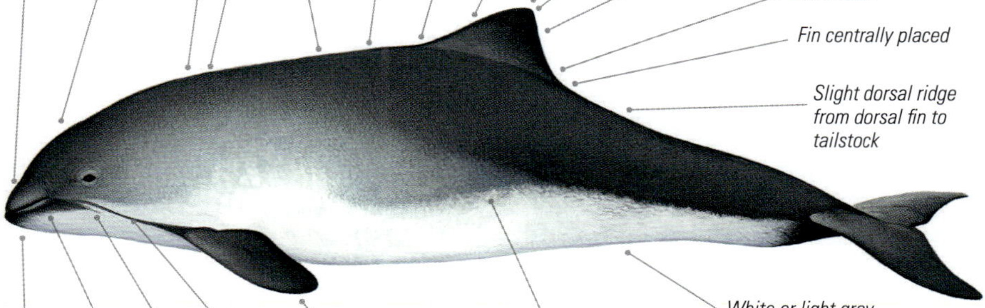

**ADULT**

- Overall impression of nondescript two-toned pigmentation
- Small, conical head (usually slightly larger in Atlantic than Pacific)
- Varying degree of bilateral asymmetry in pigmentation
- Many subtle differences in pigmentation between individuals
- Longer leading dorsal fin edge may have 12–19 small bumps (called tubercles or denticles), possibly serving hydrodynamic function
- Indistinct beak (usually slightly shorter in Atlantic than Pacific)
- Medium to dark gray upperside
- Robust body
- Low, triangular dorsal fin (appears large relative to visible portion of back when surfacing)
- Fin dark gray to black
- Concave trailing edge (variable)
- Broad base
- Fin centrally placed
- Slight dorsal ridge from dorsal fin to tailstock
- Dark gray or black 'lips'
- Straight mouthline slopes upward towards eye
- May be 1–3 dark streaks on chin (variable length, from 'lips' sometimes as far back as area between flippers)
- Small, dark flippers located in white area of body
- Dark gray stripe (variable width) from mouth to flipper (almost impossible to see in the field)
- Body color merges from dark to light through flecking or streaking, with various shades of gray
- White or light gray underside and chin

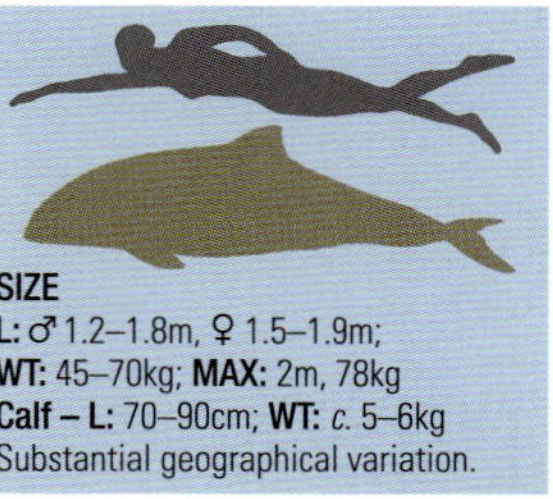

**SIZE**
**L:** ♂ 1.2–1.8m, ♀ 1.5–1.9m;
**WT:** 45–70kg; **MAX:** 2m, 78kg
**Calf – L:** 70–90cm; **WT:** c. 5–6kg
Substantial geographical variation.

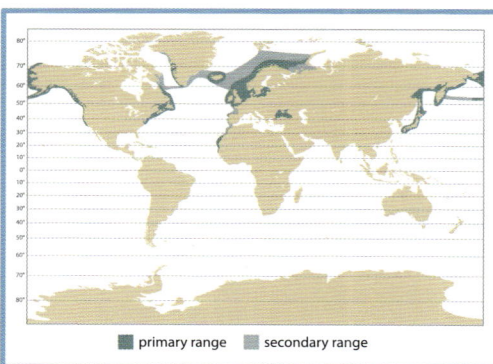

■ primary range   ■ secondary range

**AT A GLANCE** Cool temperate and sub-Arctic waters of the northern hemisphere • Small size and robust body • Dark back and low, triangular dorsal fin • No beak • Usually shy and undemonstrative • Usually alone or in small, loose groups • Slow, forward-rolling motion on surfacing

# VAQUITA
*Phocoena sinus*                                                                      Norris and McFarland, 1958

In imminent danger of extinction, the vaquita is the most endangered marine mammal in the world. The biggest threat for decades has been entanglement and accidental drowning in near-invisible gillnets; these are set for a variety of species, but the main concern recently has been those set illegally for a 2m-long sea bass-like fish, the totoaba (highly prized for its swim bladder, used in traditional Chinese medicine). Unless last-ditch conservation efforts are successful, it will not survive for much longer. Very little is known about its life and habits.

**IUCN status** Critically Endangered (2022).
**Population** *c.* 10 in 2024. 567 (1997); 245 (2008); 59 (2015); 30 (2016); fewer than 20 (2019). Calculated to have sustained a population of 5,000 for more than 250,000 years, until relatively recently when fishing began in the area. Decreasing.
**Classification** Odontoceti, family Phocoenidae.
**Taxonomy** No recognized forms or subspecies.
**Other names** Gulf of California harbor (or harbour) porpoise, cochito ('little pig'), Gulf porpoise, desert porpoise; 'vaquita' means 'little' cow'.
**DISTRIBUTION** Extreme northern end of the Gulf of California (Sea of Cortez), western Mexico (mainly north of 30°45'N and west of 114°20'W). Prefers shallow, murky, sediment-laden offshore waters with strong tidal mixing. In modern times, mostly in water 10–30m deep (rarely deeper than 40m). This is the most restricted distribution of any cetacean, and there is no evidence to suggest that it has retracted significantly in historical times. The entire range is less than 300km$^2$. Most recent sightings have been from near San Felipe, Roca Consag (a 90m-high granite outcrop 27km east-north-east of San Felipe), and El Golfo de Santa Clara (now predominantly off San Felipe).
**BEHAVIOR** Shy and retiring, typically surfacing away from vessels. Tends to avoid large motorized boats, but may occasionally approach quiet drifting boats. Does not bow-ride, and aerial displays such as breaching are unknown. Most sightings are fleeting and once only.
**FOOD AND FEEDING** Some 21 small (mainly bottom-feeding) fish species known; also takes two squid species. Feeding techniques unknown.
**TEETH** Upper jaw 32–44; lower jaw 34–40. Teeth are spatulate, as in all porpoises.
**GROUP SIZE AND STRUCTURE** Most sightings are of 1–3 individuals together, but up to 10 were observed in short-lived, loose aggregations; small groups often consisted of several mother–calf pairs.

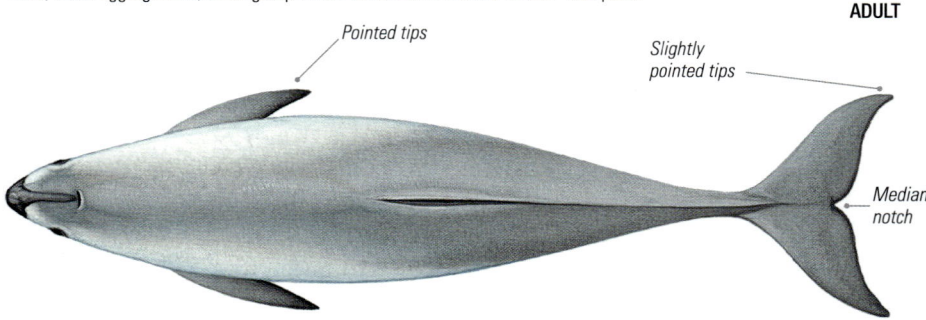

ADULT

**DIVE SEQUENCE** Surfaces slowly and inconspicuously (almost impossible to see in anything but flat-calm conditions), making slow, arching roll; often lifts head just above surface briefly; typically surfaces 3–5 times, followed by longer dive of 1–3 minutes. • **DEPTH** Shallow (rarely in water deeper than 40m). • **DURATION** Maximum at least 3 minutes.

**ADULT**

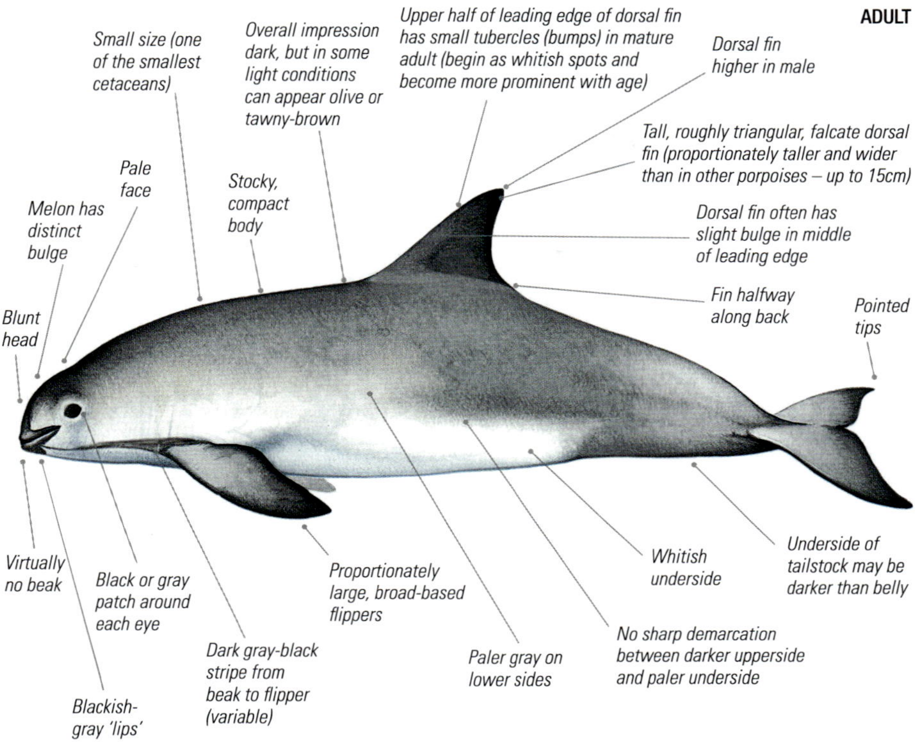

Small size (one of the smallest cetaceans)
Overall impression dark, but in some light conditions can appear olive or tawny-brown
Upper half of leading edge of dorsal fin has small tubercles (bumps) in mature adult (begin as whitish spots and become more prominent with age)
Dorsal fin higher in male
Tall, roughly triangular, falcate dorsal fin (proportionately taller and wider than in other porpoises – up to 15cm)
Pale face
Melon has distinct bulge
Stocky, compact body
Dorsal fin often has slight bulge in middle of leading edge
Blunt head
Fin halfway along back
Pointed tips
Virtually no beak
Black or gray patch around each eye
Proportionately large, broad-based flippers
Whitish underside
Underside of tailstock may be darker than belly
No sharp demarcation between darker upperside and paler underside
Dark gray-black stripe from beak to flipper (variable)
Paler gray on lower sides
Blackish-gray 'lips'

**CALF**

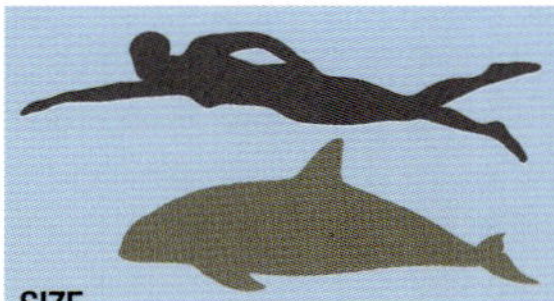

**SIZE**
**L:** ♂ 1.25–1.45m, ♀ 1.35–1.5m;
**WT:** 30–48kg; **MAX:** 1.5m, 55kg
**Calf – L:** 70–78cm; **WT:** 7.5kg

Tends to be darker than adult

Tijuana
MEXICO
UNITED STATES
MEXICO
Ensenada
BAJA CALIFORNIA
El Golfo de Santa Clara
Puerto Peñasco
San Felipe
Roca Consag
Puertecitos
SONORA
Gulf of California
PACIFIC OCEAN

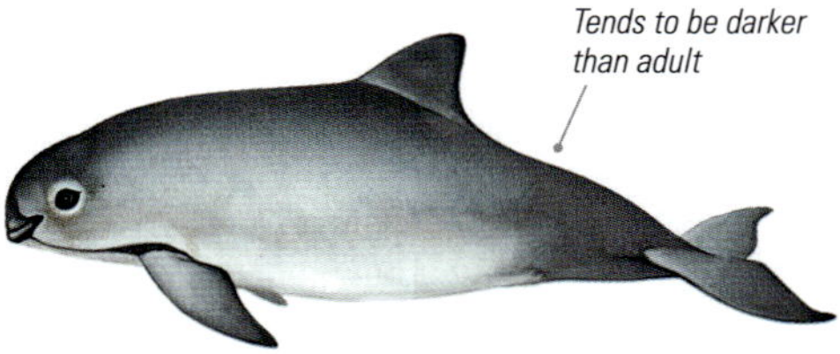

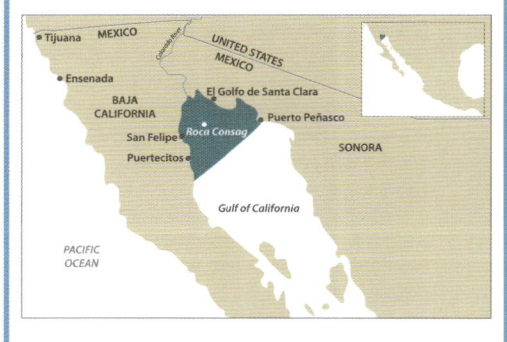

**AT A GLANCE** Extreme northern end of Gulf of California (Sea of Cortez) • Very small size • Appears all gray or gray-brown in good light • Prominent dorsal fin • Lack of prominent beak • Dark 'lips' and eye patch • Group size usually 1–3 • Typically surfaces slowly and inconspicuously and may lift head up briefly

# CARING FOR WHALES, DOLPHINS AND PORPOISES

Human impact has now reached every square kilometer of the Earth's oceans. In particular, commercial whaling and other forms of hunting, entanglement in fishing nets and myriad other conflicts with fisheries, overfishing, pollution, habitat degradation and disturbance, underwater noise, ingestion of marine debris, ship strikes and climate change are among the main threats faced by whales, dolphins and porpoises worldwide.

The IUCN Red List gives no fewer than 24 species – a quarter of all cetaceans – a threatened rating of Critically Endangered, Endangered or Vulnerable. But that reveals only part of the picture, because we simply do not know enough about most species to judge with certainty whether or not they are in trouble. At the same time, a frightening number have all but disappeared from many of their former haunts.

To the best of our knowledge one cetacean has become extinct in modern times: the Yangtze River dolphin, or baiji, from China (despite IUCN's precautionary Critically Endangered (Possibly Extinct) classification). The vaquita, a tiny porpoise from western Mexico, could be the next to go; probably fewer than 10 survivors cling on against all the odds. And several other species may not be far behind.

We nearly lost many of the larger whale species – if we had not stopped killing them at the 11th hour, we could have driven gray whales, blue whales, right whales, bowhead whales and several other species to extinction.

It is estimated that nearly 2.9 million great whales were killed during 1900–1999 alone (North Atlantic: 276,442; North Pacific: 563,696; and southern hemisphere: 2,053,956). By the time the worst of the slaughter was over, we were left merely with the tattered remains – in many cases, barely 5–10 per cent of their original populations. Some may never recover. There are still only about 370 North Atlantic right whales, for example, and recent population trends and ongoing threats are so dire that the species could become functionally (reproductively) extinct within 20 years.

Protecting cetaceans is no easy task: they are mobile and ignore political boundaries, and they face quite complex, insidious and cumulative threats. These days the odds are stacked against them and, for some, without more help, the future is undoubtedly bleak.

## HOW TO HELP

**Volunteer** Most conservation groups could not survive without the invaluable help of dedicated volunteers. If you have spare time, there are many ways to lend support – perhaps helping your favorite charity answer letters from enthusiastic children, training as a marine mammal medic who helps rescue stranded, injured or lost whales, dolphins or porpoises, or helping to clear a beach of rubbish. Or, if you are a photographer, for example, donate pictures to save the cost of buying them commercially.

**Raise urgently needed funds** Every conservation group is short of funds. With more money they could do more good work. You can raise funds to go into the general pot or for one particular species, place or project that you feel passionately about. Alternatively, why not consider adopting a whale or dolphin through various excellent schemes, or make regular donations by direct debit?

**Campaign** You could support the conservation campaigns of your favorite charities. They might invite you to write a letter or sign a petition, or you could take action on any other issue close to your heart.

**Raise awareness** The more people who are aware of conservation issues – and ultimately care about them – the better. It takes only one individual to affect the way hundreds of people see the world. You could: write letters to the national press about environmental stories, write articles for your local newspaper, offer yourself for interviews on local radio, or give talks to local clubs and organizations.

**Live a green life** There are many relatively easy ways to reduce your impact on the world's oceans. For example, ensure that any fish you eat comes from a sustainable fishery with minimal bycatch, reduce the amount of plastic you use, resist buying overpackaged goods and avoid supporting companies involved in harmful practices.

**Get involved with research** You could identify opportunities to help directly with research work. One way is to help researchers by sending suitable photographs for their photo-identification catalogs (enabling them to recognize, and follow, individual whales). Fieldwork is expensive, and the larger the catalogs, the more information they can provide. A brilliant place to submit ID photographs is happywhale.com. Images are forwarded to relevant catalogs, and photographers are informed of any matches made. Check how the species needs to be photographed – the underside of its tail, for example, or one particular side of its body – and do not forget to include your contact details, sighting date and location (ideally, the exact coordinates), vessel name and any other relevant information.

Whatever you are able to do, it all helps.

# GLOSSARY

**abyssal plain** Ocean floor beyond the continental shelf.

**ambergris** Dark grayish waxy substance formed in the intestines of sperm whales, once widely used in perfumes.

**amphipod** Small shrimp-like crustacean – a food source for some whales.

**anchor patch** Variable gray or white anchor- or W-shaped patch on the chest of some smaller toothed whales.

**Antarctic Convergence** Natural oceanic boundary in the Antarctic, where cold, less saline waters from the south sink below warmer, more saline waters from the north; considered the northern limit of 'biological Antarctica'. Otherwise known as the 'Polar Front'.

**balaenopterid** Member of the baleen whale family Balaenopteridae, otherwise known as 'rorqual'.

**baleen plate** Dense comb-like structure hanging down from the upper jaws of most large whales (baleen whales, Mysticeti); formerly known as 'whalebone'. Hundreds of baleen plates are packed tightly together to form a giant sieve for filter-feeding small prey.

**baleen whale** Member of the Mysticeti; a predominantly large whale with baleen plates instead of teeth.

**bathypelagic** Inhabiting the deepwater portion of the open ocean, 1,000–4,000m below the surface.

**beak** Elongated snout of many cetaceans; anterior portion of the skull that includes both upper and lower jaws.

**benthic** Living in, on or just above the ocean floor.

**blackfish** Colloquial term for six superficially similar members of the dolphin family (killer whale, false killer whale, pygmy killer whale, melon-headed whale, short-finned pilot whale and long-finned pilot whale).

**blaze** Light streaking, usually gray or white, on the side of a cetacean's body, usually starting below the dorsal fin and pointing up into the cape.

**blow** Refers to the act of breathing – the explosive exhalation followed immediately by an inhalation – and to the visible misty cloud of water droplets formed when a whale breathes; also known as a 'spout'.

**blowhole** Respiratory opening, or nostril, on the top of the head; baleen whales have two, toothed whales one.

**blubber** Layer of fatty tissue between the skin and underlying muscle of most marine mammals; important for insulation.

**bow-riding** Swimming or 'riding' on the pressure wave created in front of a vessel or large whale.

**breaching** Leaping completely (or almost so) out of the water. Officially, if more than 40 per cent of the whale's body leaves the water, it is a 'breach'; otherwise it is a 'lunge'.

**bubble-netting** Feeding technique used by humpback whales, producing 'fishing nets' by blowing bubbles underwater.

**bycatch** Animals caught accidentally or incidentally during fishing operations (when not the target species).

**callosity** Area of roughened, keratinised tissue on the head of a right whale, inhabited by whale lice.

**caudal fin** Tail fin.

**caudal peduncle** See 'tailstock'.

**cephalopod** Member of a group of benthic or swimming molluscs, including squid, cuttlefish and octopuses.

**cetacean** Any mammalian member of the Cetacea, a group of aquatic mammals that includes all whales, dolphins and porpoises.

**chevron** V- or U-shaped light-colored marking on the back or sides of a cetacean.

**continental shelf** Area of the seafloor around the edge of a continent; it slopes gently from the coastline to a drop-off point called the 'shelf break' or 'shelf edge'; from there the seafloor drops steeply, via the 'continental slope', to the ocean bottom.

**continental slope** Stretch of the ocean floor that drops steeply between the 'shelf break' and the 'abyssal plain'.

**cookiecutter shark** Small shark (up to 50cm) normally found in sub-tropical and tropical waters that takes bites out of marine mammals; the resulting wounds appear as round or oval craters (each about the size of an ice cream scoop).

**copepod** Minute shrimp-like crustacean (usually planktonic) that occurs in great abundance in the sea and is an important food source for some whales.

**crustacean** Member of a group of nearly 70,000 invertebrates (animals without backbones), including lobsters, crabs, shrimps and barnacles. Mostly aquatic, crustaceans are an important food source for many marine animals.

**delphinid** Member of the oceanic dolphin family Delphinidae.

**deep scattering layer (DSL)** A dense layer up to 200m thick of huge numbers of small fish, squid, crustaceans and plankton, in open oceans worldwide, which migrate up and down the water column within a single 24-hour period.

**demersal** Lives near the seafloor.

**diatom** Microscopic single-celled algae, abundant in marine and freshwater environments; often form a film that coats cetacean bodies, producing a yellowish, brownish or even greenish tinge.

**dorsal** Pertaining to the upper surface or the back (or the upper surface of any body part).

**dorsal cape** Distinct dark region on the backs of some toothed whales, dolphins and porpoises, mostly in front of the dorsal fin (sometimes stretching behind it).

**dorsal fin** Raised structure on the back of most cetaceans; not supported by bone.

**dorsal ridge** Ridge on the back as well as, or instead of, a dorsal fin; may also refer to ridges on the top of the rostrum in many baleen whales.

**driftnet** Fishing net that hangs in the water vertically, virtually unseen and undetectable, and is carried freely with the ocean currents and winds; a gillnet that is not anchored. Notorious for catching everything in its path, from seabirds and turtles to whales and dolphins.

**drive fishery** Technique used to capture dolphins and other small toothed whales, usually using speedboats to herd them into bays or shallow water, where they are killed.

**eastern boundary current** Found on the eastern side of oceanic basins, adjacent to the western coasts of continents, and flowing from high latitudes towards the tropics; relatively cold, shallow, broad and slow-flowing. Usually with more nutrient-rich upwellings than western boundary currents.

**echolocation** Process of sending out high-frequency sounds and interpreting the returning echoes to build up a 'sound picture', as in sonar; used by many cetaceans to orientate, navigate and find food.

**ecotype** Term that recognizes scientific uncertainty with regard to systematics and speciation in killer whales (and, to a lesser degree, some other species); it recognizes a work in progress, avoiding the immediate need to declare subspecies or species.

**El Niño** Complex global weather pattern marked by rising sea temperatures in the central and eastern tropical Pacific Ocean.

**epipelagic** Living within 200m of the surface in pelagic waters.

**euphausiid** Any member of the order Euphausiacea (86 known species of shrimp-like creatures called krill).

**extralimital** Occurrence outside the normal range.

**falcate** Sickle-shaped or back-curved; often used to describe the shape of a dorsal fin with a concave rear margin.

**flipper** Variably shaped, flattened, paddle-like forelimb of a cetacean (also known as the 'pectoral fin' or 'pec fin').

**flipper-slapping** Lying on the back or side, raising one or both flippers out of the water and then slapping it/them onto the surface. Also known as 'flipper-flopping', 'flippering', 'pectoral-slapping' or 'pec-slapping'.

**fluke (noun)** Horizontally flattened tail of a cetacean; two flukes comprise a cetacean's tail.

**fluke (verb), fluking** Natural extension of a deep or sounding dive – the whale bends its body towards the seabed and, as it rolls forward and down, the flukes rise above the surface; larger, more rotund whales fluke regularly; slimmer whales rarely or never fluke.

**flukeprint** A circular swirl of smooth water that resembles a sheen of oil, made by the downward movement of the tail creating a vortex, and left on the surface after a whale has dived.

**head-slap** When a whale lunges partially out of the water and forcefully slaps its throat onto the surface with a large splash; also known as a 'chin-slap'.

**home range** Area that an animal patrols regularly.

**intraspecific** Between individuals of the same species ('interspecific' means between individuals of different species).

**isobath** Imaginary line on a chart connecting all the points with the same depth underwater (rather like an underwater contour line).

**IWC (International Whaling Commission)**; in 1946, whalers adopted the International Convention for the Regulation of Whaling for the 'orderly development of the whaling industry'; the IWC was established as the convention's decision-making body and has attempted to regulate whaling (and, more recently, conserve whales) ever since.

**krill** About 86 species of small shrimp-like crustaceans (0.8–6cm long), comprising much of the ocean's zooplankton; a major food source for many large whales. Also known as a 'euphausiid'.

**lamprey** Primitive jawless, eel-like fish with a permanently open mouth bearing many teeth. Of 43 species, 18 are parasitic (boring into the flesh of cetaceans and other animals and sucking their blood); mainly in temperate waters.

**La Niña** Complex global weather pattern marked by falling sea temperatures in the central and eastern tropical Pacific Ocean; often follows an 'El Niño'.

**lobtailing** Lifting the tail clear of the water then slapping it down onto the surface, usually repeatedly and often forcibly; also known as 'tail-lobbing' or 'fluke-slapping'.

**logging** Lying at or just below the surface, inactive and usually horizontally, to rest.

**longline** Very long (sometimes tens of kilometers) fishing line, armed with multiple baited hooks (sometimes thousands) on shorter branch lines, set to catch large pelagic fish. Responsible for significant bycatch of dolphins and other species.

**lunging** Officially, if less than 40 per cent of the whale's body leaves the water, it is termed a 'lunge', otherwise it is a 'breach'; also known as a 'half-breach' or 'belly-flop'.

**mandible** Entire lower jaw.

**melon** Bulging fatty tissue forming the 'forehead' of toothed cetaceans, thought to focus and modulate sounds for echolocation.

**mesopelagic** Inhabiting the intermediate depths of the open ocean, typically at 200–1,000m.

**mesoplodont** Beaked whale belonging to the genus *Mesoplodon*.

**mysid** Small shrimp-like crustacean in the order Mysida (*c.* 1,200 species); mostly benthic.

**Mysticeti** One of two major groups of cetaceans, containing all the toothless or baleen whales (known as mysticetes).

**oceanic** Open sea environment beyond the edge of the continental shelf.

**Odontoceti** One of two major groups of cetaceans, containing all the toothed whales, dolphins and porpoises (known as odontocetes).

**offshore** Well away from the coast.

**pectoral fin** See 'flipper'.

**peduncle** See 'tailstock'.

**pelagic** Inhabiting offshore waters of the open ocean

**photo-identification (photo-ID)** Technique for studying cetaceans using photographs as a permanent record of identifiable individuals.

beyond the continental shelf; usually used to describe animals and plants living in upper portions of the water column; neither near to the coast, nor close to the bottom.

**plankton** Passively floating, or weakly swimming, plant (phytoplankton) and animal (zooplankton) life that occurs in swarms, usually near the surface of open waters.

**pod** Coordinated group of closely related killer whales or any group of socially affiliated, medium-sized toothed whales.

**polynya** Russian word for 'open water surrounded by ice'.

**porpoising** When members of the dolphin family (and, less commonly, some other cetaceans) travel at high speed and make low, arcing leaps clear of the water when they take a breath, before re-entering headfirst; sometimes called 'running'.

**purse-seine fishing** Vertical curtain of netting set around a shoal of fish, then gathered at the bottom and drawn in to form a 'purse' to prevent the fish from escaping; responsible for killing more dolphins in the past 50 years than any other human activity (though new regulations have reduced bycatch substantially).

**rake marks** Scarring produced by teeth during intraspecific fighting (usually between males) or from attacks by killer whales.

**remora** Type of fish that has modified its dorsal fin into a sucker (thus the alternative names 'whalesucker' or 'suckerfish') to attach onto large marine animals such as whales and dolphins.

**rorqual** Baleen whale in the family Balaenopteridae, characterized by a variable number of expandable pleats or grooves that run longitudinally from the chin towards the navel; the name comes from the Norwegian word rørkval, meaning 'the whale with pleats' (after the grooves); otherwise known as a 'balaenopterid'.

**rostrum** Beak-like projection at the front of a cetacean's head; also used specifically to describe the upper jaw.

**saddle patch** Light-colored, more-or-less saddle-shaped marking that straddles the back behind the dorsal fin of some cetaceans.

**school** Term for a coordinated group, normally used in association with dolphins, that swims and socializes together; often used synonymously with 'herd'.

**seamount** Underwater mountain, typically rising more than 1,000m above the surrounding deep-sea floor; usually an extinct volcano, with the summit some way beneath the surface; attracts an abundance of marine life.

**sexually dimorphic** When males and females of the same species differ in size and/or appearance.

**shelf break** Drop-off point at the edge of the continental shelf (from there, the seafloor drops steeply, via the 'continental slope', to the ocean bottom); also called the 'shelf edge'.

**skim-feeding** Technique used by some baleen whales that involves swimming slowly with the mouth open and constantly filtering food from the water, typically along or just below the surface.

**sounding dive** Deep (and usually longer) dive after a series of shallow dives; also known as the 'terminal dive'.

**splashguard** Elevated fleshy ridge immediately in front of the blowholes of a baleen whale, which helps to prevent water from pouring in when the blowholes are open.

**spout** See 'blow'.

**spyhopping** When a whale raises its head vertically out of the water, usually exposing the eyes to the air, before sinking smoothly below the surface without much splash. Also known as a 'head rise' or an 'eye-out'.

**stranding** The act of coming ashore, intentionally or accidentally, alive or dead.

**submarine canyon** Underwater canyon; deep, narrow, steep-sided valley cut into the seabed.

**tail breaching** Throwing the rear portion of the body, including the flukes, high out of the water and sideways across the surface, creating a huge splash; otherwise known as a 'peduncle throw' or 'peduncle slap'.

**tail-slapping** Like 'lobtailing', but normally associated with smaller cetaceans.

**tailstock** Muscular region of the tail between the dorsal fin and the flukes; also called the 'caudal peduncle' or 'peduncle'.

**throat grooves** V-shaped grooves (deep folds in the skin and blubber) on the throat, characteristic of beaked whales and gray whales.

**throat pleats** Longitudinal parallel furrows or grooves on the underside of many baleen whales (backward from the chin) that allow the throat to expand when engulfing huge quantities of water to capture prey; also known as 'ventral pleats'.

**toothed whale** See 'Odontoceti'.

**tubercle** Circular raised protuberance, or bump, found on some cetaceans (usually along the edges of pectoral and dorsal fins, but also on a humpback whale's head).

**upwelling** Process by which ocean water rises from the depths, forced up by currents, winds or density gradients; this brings nutrients to the surface.

**wake-riding** Swimming in the frothy wake of a boat or ship.

**water column** Anywhere between the surface and the seafloor.

**western boundary current** Found on the western side of oceanic basins, adjacent to the eastern coasts of continents, and flowing from the tropics to high latitudes; relatively warm, deep, narrow and fast-flowing; tends to have fewer nutrient-rich upwellings than eastern boundary currents, making it less productive.

**whalebone** See 'baleen plate'.

**whale louse** An amphipod crustacean (not an insect) adapted to living on the skin of cetaceans.

# NORTH AMERICAN SPECIES CHECKLIST

## MYSTICETI (BALEEN WHALES)

### Right and bowhead whales (family Balaenidae)
- North Atlantic right whale (*Eubalaena glacialis*)
- North Pacific right whale (*Eubalaena japonica*)
- Bowhead whale (*Balaena mysticetus*)

### Gray whale (family Eschrichtiidae)
- Gray whale (*Eschrichtius robustus*)

### Rorquals (family Balaenopteridae)
- Blue whale (*Balaenoptera musculus*)
- Fin whale (*Balaenoptera physalus*)
- Sei whale (*Balaenoptera borealis*)
- Bryde's whale (*Balaenoptera edeni*)
- Rice's whale (*Balaenoptera ricei*)
- Common minke whale (*Balaenoptera acutorostrata*)
- Humpback whale (*Megaptera novaeangliae*)

*Common minke whale*

## ODONTOCETI (TOOTHED WHALES)

### Sperm whale (family Physeteridae)
- Sperm whale (*Physeter macrocephalus*)

*Sperm whale*

### Pygmy and dwarf sperm whales (family Kogiidae)
- Pygmy sperm whale (*Kogia breviceps*)
- Dwarf sperm whale (*Kogia sima*)

*Beluga with diatoms*

*Narwhal*

### Narwhal and beluga (family Monodontidae)
- Narwhal (*Monodon monoceros*)
- Beluga (*Delphinapterus leucas*)

### Beaked whales (family Ziphiidae)
- Baird's beaked whale (*Berardius bairdii*)
- Sato's beaked whale (*Berardius minimus*)
- Cuvier's beaked whale (*Ziphius cavirostris*)
- Northern bottlenose whale (*Hyperoodon ampullatus*)
- Longman's beaked whale (*Indopacetus pacificus*)
- Perrin's beaked whale (*Mesoplodon perrini*)
- Pygmy beaked whale (*Mesoplodon peruvianus*)
- Ginkgo-toothed beaked whale (*Mesoplodon ginkgodens*)
- Hubbs' beaked whale (*Mesoplodon carlhubbsi*)
- Blainville's beaked whale (*Mesoplodon densirostris*)
- Sowerby's beaked whale (*Mesoplodon bidens*)
- True's beaked whale (*Mesoplodon mirus*)
- Stejneger's beaked whale (*Mesoplodon stejnegeri*)
- Gervais' beaked whale (*Mesoplodon europaeus*)

*Baird's beaked whale*

*Killer whale, resident*

*False killer whale*

**Marine dolphins (family Delphinidae)**

- Killer whale or orca (*Orcinus orca*)
- Short-finned pilot whale (*Globicephala macrorhynchus*)
- Long-finned pilot whale (*Globicephala melas*)
- False killer whale (*Pseudorca crassidens*)
- Pygmy killer whale (*Feresa attenuata*)
- Melon-headed whale (*Peponocephala electra*)
- Risso's dolphin (*Grampus griseus*)
- Fraser's dolphin (*Lagenodelphis hosei*)
- Atlantic white-sided dolphin (*Leucopleurus acutus*)
- Pacific white-sided dolphin (*Aethalodelphis obliquidens*)
- White-beaked dolphin (*Lagenorhynchus albirostris*)
- Northern right whale dolphin (*Lissodelphis borealis*)
- Rough-toothed dolphin (*Steno bredanensis*)
- Common bottlenose dolphin (*Tursiops truncatus*)
- Tamanend's bottlenose dolphin (*Tursiops erebennus*)
- Pantropical spotted dolphin (*Stenella attenuata*)
- Atlantic spotted dolphin (*Stenella frontalis*)
- Spinner dolphin (*Stenella longirostris*)
- Clymene dolphin (*Stenella clymene*)
- Striped dolphin (*Stenella coeruleoalba*)
- Common dolphin (*Delphinus delphis*)

*Risso's dolphin*

*Rough-toothed dolphin*

**Porpoises (family Phocoenidae)**

- Dall's porpoise (*Phocoenoides dalli*)
- Harbor porpoise (*Phocoena phocoena*)
- Vaquita (*Phocoena sinus*)

*Dall's porpoise,* dalli-*type*

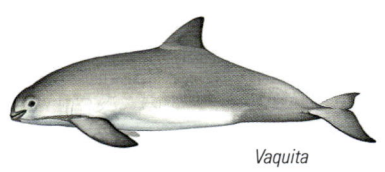
*Vaquita*

*White-beaked dolphin*

# SOURCES AND RESOURCES

I am very grateful to the innumerable scientists who have spent countless hours, days, months and years at sea, and in the laboratory, progressing our knowledge of cetaceans. I have read many thousands of their scientific papers during the preparation of this field guide, which represents the combined work of all these remarkable people. A comprehensive catalog of so many references would more than double the size of the book so, unfortunately, it is impossible to list them all. But here is a small but varied selection of good websites, books and journals that might be of interest.

## WEBSITES

American Cetacean Society – acsonline.org
European Cetacean Society – europeancetaceansociety.eu
Happy Whale – happywhale.com
International Whaling Commission (IWC) – iwc.int
IUCN Red List – iucnredlist.org
IUCN – Species Survival Commission (SSC) Cetacean Specialist Group – iucn-csg.org
Marine Conservation Society – mcsuk.org
NOAA Fisheries – fisheries.noaa.gov/whales
ORCA – orcaweb.org.uk
Porpoise Conservation Society – porpoise.org
Sea Watch Foundation – seawatchfoundation.org.uk
Society for Marine Mammalogy – marinemammalscience.org (Committee on Taxonomy 2024 List of Marine Mammal Species and Subspecies)
Whale and Dolphin Conservation – us.whales.org

## BOOKS

Baird, R. W. 2016. *The Lives of Hawai'i's Dolphins and Whales: Natural History and Conservation*. University of Hawai'i Press.

Berta, A., J. L. Sumich and K. M. Kovacs. 2015 (Third Edition). *Marine Mammals: Evolutionary Biology*. Academic Press.

Bortolotti, D. 2009. *Wild Blue: A Natural History of the World's Largest Animal*. Thomas Allen Publishers.

Brakes, P. and M. P. Simmonds. 2011. *Whales and Dolphins: Cognition, Culture, Conservation and Human Perceptions*. Earthscan.

Carwardine, M. 2017. *Mark Carwardine's Guide to Whale Watching in North America: USA, Canada, Mexico*. Bloomsbury.

Carwardine, M. 2020. *Handbook of Whales, Dolphins and Porpoises*. Bloomsbury.

Carwardine, M. 2022. *Field Guide to Whales, Dolphins and Porpoises*. Bloomsbury.

Darling, J. 2009. *Humpbacks: Unveiling the Mysteries*. Granville Island Publishing.

Ellis, R. 2011. *The Great Sperm Whale: A Natural History of the Ocean's Most Magnificent and Mysterious Creature*. University Press of Kansas.

Ellis, R. and J. G. Mead. 2017. *Beaked Whales: A Complete Guide to their Biology and Conservation*. Johns Hopkins University Press.

Fitzhugh, W. W. and M. T. Nweeia (eds). 2017. *Narwhal: Revealing an Arctic Legend*. IPI Press & Arctic Studies Center, National Museum of Natural History, Smithsonian Institution.

Ford, J. K. B. 2014. *Marine Mammals of British Columbia*. Royal BC Museum.

George, J. C. and J. G. M. Thewissen 2020. *The Bowhead Whale* Balaena mysticetus. Academic Press.

Heide-Jørgensen, M. P. and K. Laidre. 2006. *Greenland's Winter Whales*. Greenland Institute of Natural Resources.

Hoyt, E. 2011 (Second Edition). *Marine Protected Areas for Whales, Dolphins and Porpoises: A World Handbook for Cetacean Habitat Conservation and Planning*. Earthscan.

Jefferson, T. A., M. A. Webber and R. J. Pitman. 2015 (Second Edition). *Marine Mammals of the World: A Comprehensive Guide to their Identification*. Academic Press.

Jefferson, T. A. 2025. *Ridgway and Harrison's Handbook of Marine Mammals Volume 1: Coastal Dolphins and Porpoises*. Academic Press.

Kraus, S. D. and R. M. Rolland (eds). 2007. *The Urban Whale: North Atlantic Right Whales at the Crossroads*. Harvard University Press.

Laist, D. W. 2017. *North Atlantic Right Whales: From Hunted Leviathan to Conservation Icon*. John Hopkins University Press.

McLeish, T. 2013. *Narwhals: Arctic Whales in a Melting World*. University of Washington Press.

Mustill, T. 2023. *How to Speak Whale*. William Collins.

Reynolds, J. E. III, R. S. Wells and S. D. Eide. 2000. *The Bottlenose Dolphin: Biology and Conservation*. University Press of Florida.

Ridgway, S. H. and R. Harrison (eds). 1985. *Handbook of Marine Mammals, Vol. 3: The Sirenians and Baleen Whales*. Academic Press.

Ridgway, S. H. and R. Harrison (eds). 1989. *Handbook of Marine Mammals, Vol. 4: River Dolphins and the Larger Toothed Whales*. Academic Press.

Ridgway, S. H. and R. Harrison (eds). 1994. *Handbook of Marine Mammals, Vol. 5: The First Book of Dolphins.* Academic Press.

Ridgway, S. H. and R. Harrison (eds). 1999. *Handbook of Marine Mammals, Vol. 6: The Second Book of Dolphins and the Porpoises.* Academic Press.

Sumich, J. 2014. *E. robustus: Biology and Human History of Gray Whales.* Whale Cover Marine Education.

Swartz, S. L. 2014. *Lagoon Time: A Guide to Gray Whales and the Natural History of San Ignacio Lagoon.* The Ocean Foundation.

Turvey, S. 2008. *Witness to Extinction: how we failed to save the Yangtze River Dolphin.* Oxford University Press.

Whitehead, H. 2003. *Sperm Whales: Social Evolution in the Ocean.* University of Chicago Press.

Wilson, D. E. and R. A. Mittermeier. 2014. *Handbook of the Mammals of the World, Vol. 4. Sea Mammals.* Lynx Edicions.

Würsig, B., J. G. M. Thewissen and K. M. Kovacs (eds). 2018. *Encyclopedia of Marine Mammals.* Academic Press.

## JOURNALS

Aquatic Mammals – aquaticmammalsjournal.org

Endangered Species Research – int-res.com/journals/esr/esr-home

Frontiers in Marine Science – frontiersin.org/journals/marine-science

International Whaling Commission Journal of Cetacean Research and Management – journal.iwc.int

Journal of Mammalogy – academic.oup.com/jmammal

Marine Mammal Science – marinemammalscience.org/journal

Nature Scientific Reports – nature.com/srep

Polar Biology – springer.com/journal/300

Royal Society Open Science – royalsocietypublishing.org/journal/rsos

# ARTISTS' BIOGRAPHIES

**Martin Camm**, from Bedfordshire, UK, is one of the world's most renowned illustrators of aquatic life, specializing in cetaceans. He has contributed to hundreds of books, magazines and journals, and his work is widely published by many organizations, including the United Nations, the BBC, Greenpeace, International Fund for Animal Welfare (IFAW), Whale and Dolphin Conservation (WDC), and The Wildlife Trusts.

**Toni Llobet**, from Catalonia, Spain, thinks of himself more as a naturalist than an artist. He is rigorous in both disciplines, as his intricate work shows. He has worked as a wildlife illustrator for some 20 years, and his work includes massive projects such as illustrating the prestigious *Handbook of the Mammals of the World* (published by Lynx Edicions).

**Rebecca Robinson**, from Tasmania, Australia, graduated with a BSc in Zoology. In a bid to marry her passion for art and nature, she completed a Bachelor of Fine Arts degree (with Honors) in Wildlife Illustration. She then worked in the design department at the Marine Science Institute at the University of California Santa Barbara, and has been a freelance natural science illustrator ever since.

# IMAGE CREDITS

With the exception of the artworks listed below, all artworks in this book remain © Martin Camm (www.markcarwardine.com)

© Rebecca Robinson (www.markcarwardine.com): all dive sequences and whale blows.

© Toni Llobet: p18, p19, p71 (top two), p164, p165 (bottom), p175 (bottom); flukes on p8, p20, p21, p34, p39, p43, p47, p53, p57, p76, p81, p125.

© Toni Llobet from: Wilson, D. E. & Mittermeier, R. A. eds. (2014). *Handbook of the Mammals of the World. Vol. 4. Sea Mammals.* Lynx Edicions, Barcelona: p8 (second from top), p59, p60 (bottom), p90, p91 (top), p151 (top), p152 (middle), p165 (top).

© Jack Ashton: p15 (bottom), p16, p17 (lice).

# ACKNOWLEDGEMENTS

Many wonderful people have helped with this book – answering my endless questions, providing as yet unpublished information and commenting on early tentative drafts – and it couldn't have been written without them. Their generous contributions have helped me to make fewer mistakes than I would have made on my own and I'd like to thank them all for their time, enthusiasm, generosity and support. It is very much appreciated. All these whale biologists – friends and colleagues – are listed separately.

Special thanks go to the brilliant wildlife artists Martin Camm, Toni Llobet and Rebecca Robinson, who worked tirelessly on so many illustrations. Martin and I have collaborated on more whale books than either of us can remember, over more years than we dare admit, and it's always a great pleasure. And it has been wonderful working with Toni and Rebecca for the first time. A very big thank you to Rachel Ashton, my outstandingly efficient manager, for her never-ending patience, perseverance, encouragement, enthusiasm and general brilliance. I couldn't do it without you, Rachel. Анна Астафуова very kindly helped with research and, with her outstanding detective work, discovered all sorts of previously unpublished snippets of information. My literary agent, Doreen Montgomery, was a huge support, as always over the past quarter of a century; very sadly, she passed away while I was writing, and I will miss her terribly. Doreen's daughter, Caroline Montgomery, kindly took the reins and carries on the Montgomery tradition with aplomb. I am particularly indebted to Jim Martin at Bloomsbury (without whom this book wouldn't have happened at all) for his unwavering belief in the whole idea and for his genuine love and passion for the natural world; even when the project took many years longer than we had all anticipated, and became three books instead of one, he never stopped smiling and (at least seemingly) took it all in his stride. Thanks for everything, Jim. Thank you, also, to Alice Ward (Commissioning Editor, Bloomsbury Wildlife), who smoothed the way so calmly during production of the original *Handbook of Whales, Dolphins and Porpoises*; it was a pleasure working with you, Alice. The wonderful Jenny Campbell (Senior Editor, Bloomsbury Wildlife) took over for this updated and abbreviated version – the *Field Guide to Whales, Dolphins and Porpoises* – and was a fantastically understanding and encouraging support and help throughout; thank you for everything, Jenny. I was very lucky, too, to have James Lowen as the magnificently thorough copy-editor and the brilliant Nigel Redman as proof-reader. And we were very fortunate to have the incredibly talented Julie Dando as designer – an outstanding professional who went above and beyond the call of duty; thank you – I really appreciate it.

I'd like to thank my dear friends and family, who have enthused with me when things have been going well and cheered me on when I've been flagging: in particular, Peter Bassett and John Ruthven, for those welcome interludes over coffee in the Clifton Lido; John Craven, Nick Middleton and Marc Riley for putting up with my relentless chatter about field guides, whales, dolphins, porpoises, distribution maps, illustrations, scientific papers, etc., etc.; and Roz Kidman Cox and Michaela Strachan, for their knowing sympathy and good counsel. I'd also like to give my wonderful, kind, encouraging parents, David and Betty, a special mention: without their never-ending support and encouragement, life would have been so different. My brother Adam, sister-in-law Vanessa, nieces Jessica and Zoe, Beryl, Al and Jude, Florence and Miller have all kindly put up with my mind-bendingly long hours and preoccupation with what became known as 'the-project-that-never-seems-to-end'. Last, but by no means least, a huge heartfelt thank you for everything to my co-conspirator in life, Debra Taylor, who always makes things better.

At the end of the day, despite a phenomenal effort to make this field guide as accurate and complete as possible, I take full responsibility for any mistakes, oversights and inconsistencies that may have crept in. I'd welcome any thoughts, comments and suggestions – via www.markcarwardine.com – to incorporate in new, improved future editions. Thank you very much.

## WITH SPECIAL THANKS TO:
Robert L. Pitman
Charles Anderson

## ... AND ALL THE SCIENTISTS WHO KINDLY ADVISED ON INDIVIDUAL SPECIES
Àlex Aguilar
Wojtek Bachara
Robin W. Baird
Isabel Beasley
Chiara Giulia Bertulli
Arne Bjørge
Nancy Black
Moira Brown
Salvatore Cerchio
William Cioffi
Phillip J. Clapham
Diane Claridge
Rochelle Constantine
Barbara E. Curry
Merel Dalebout
Jim Darling
Natalia Dellabianca
David M. Donnelly
Simon H. Elwen
Ruth Esteban
James Fair
Ivan D. Fetudin
Andrew Foote
R. Ewan Fordyce
Ari Friedlaender
Sonja Heinrich
Denise Herzing
Sascha Hooker
Erich Hoyt
Miguel Iñíguez
Maria Iversen
Thomas A. Jefferson
Eve Jourdain
Catherine Kemper
Iain Kerr
Jeremy Kiszka
Kristin Laidre
Jack Lawson
Rob Lott
Donald McAlpine
Colin D. MacLeod
Amanda Madro
Anisul Islam Mahmud
Silvia S. Monteiro
Hilary Moors-Murphy
Dirk R. Neumann
Stephanie A. Norman
Giuseppe Notarbartolo di Sciara
Gregory O'Corry-Crowe
William F. Perrin
Cindy Peter
Róisín Pinfield
Andrew J. Read
Victoria Rowntree
Filipa Samarra
Jarrod A. Santora
Marcos César de Oliveira Santos
Richard Sears
Keiko Sekiguchi
Tammy L. Silva
Tiu Similä
Ravindra Kumar Sinha
Elisabeth Slooten
Kate Rose-Ann Sprogis
Steven Swartz
Jessica K. D. Taylor
Outi Tervo
Kirsten Thompson
Paul Thompson
Fernando Trujillo
Grigory A. Tsidulko
Samuel Turvey
Koen Van Waerebeek
Caroline R. Weir
Hal Whitehead
Tonya Wimmer
Bernd Würsig

# INDEX

## A

*Aethalodelphis obliquidens* 158
Afro-Iberian harbor porpoise 192
Antillean beaked whale 122
Antarctic blue whale 48
Arabian Sea humpback whale 72
arch-beaked whale 110
Arctic right whale 40
Arctic whale 40
Atlantic harbor porpoise 192
Atlantic right whale 32
Atlantic spinner dolphin 182
Atlantic spotted dolphin 174
Atlantic white-sided dolphin 156

## B

Baird's beaked whale 92
*Balaena mysticetus* 40
*Balaenoptera acutorostrata* 68
*Balaenoptera borealis* 58
*Balaenoptera edeni* 62
*Balaenoptera musculus* 48
*Balaenoptera physalus* 54
*Balaenoptera ricei* 66
beluga 90
beluga whale 90
*Berardius bairdii* 92
*Berardius minimus* 94
Bering Sea beaked whale 120
black Baird's beaked whale 94
blackfish 124, 138, 142, 144, 146
Black Sea bottlenose dolphin 166
Black Sea common dolphin 186
Black Sea harbor porpoise 192
Blainville's beaked whale 112
Blue Whale 48
Bornean dolphin 154
bottlehead 100
bottle-nosed dolphin. 166
bottlenose dolphin 166
bowhead whale 40
bridled dolphin 172
Bryde's whale 62

Burrunan bottlenose dolphin 166

## C

caaing whale 142
cachalot 78
California beaked whale 104
California gray whale 44
Central American spinner dolphin 176
Chilean blue whale 48
clymene dolphin 182
coalfish whale 58
cochito 194
common bottlenose dolphin 166
common bottlenose whale 100
common dolphin 186
common minke whale 68
common porpoise 186, 192
common rorqual 54
Costa Rican spinner dolphin 176
crisscross dolphin 186
Cuvier's beaked whale 96

## D

Dall's porpoise 190
*Delphinapterus leucas* 90
*Delphinus delphis* 186
dense-beaked whale 112
desert porpoise 194
devilfish 44
dwarf minke whale 68
dwarf sperm whale 84
dwarf spinner dolphin 176

## E

Eastern Pacific long-beaked common dolphin 186
eastern spinner dolphin 176
Eastern tropical Pacific bottlenose dolphin 166
Eden's whale 62
electra dolphin 148
*Eschrichtius robustus* 44
*Eubalaena glacialis* 32

*Eubalaena japonica* 36
euphrosyne dolphin 184
European beaked whale 122

## F

false killer whale 144
*Feresa attenuata* 146
finback 54
finfish 54
finner 54
fin whale 54
flathead 100
four-toothed whale 92
Fraser's dolphin 154

## G

Gervais' beaked whale 122
giant bottlenose whale 92
ginkgo-toothed beaked whale 108
ginkgo-toothed whale 108
*Globicephala macrorhynchus* 138
*Globicephala melas* 142
goose-beaked whale 96
goosebeak whale 96
grampus 124, 150
*Grampus griseus* 150
grayback 44
Gray's spinner 176
gray whale 44
Greenland right whale 40
Greenland whale 40
grey whale 44
Gulf of California harbor porpoise 194
Gulf of Mexico Bryde's whale 66
Gulf of Mexico whale 66
Gulf porpoise 194
Gulf Stream beaked whale 122

## H

harbor porpoise 192
harbour porpoise 192
hardhead 44
Hawaiian blackfish 148

Hawaiian spinner dolphin  176
helmet dolphin  182
herring hog  192
herring whale  54
Hubbs' beaked whale  110
hump-backed whale  72
humpback whale  72
*Hyperoodon ampullatus*  100

I

*Indopacetus pacificus*  102
Indo-Pacific beaked whale  102
Indo-Pacific common dolphin  186

J

Japanese beaked whale  108
jumper  156, 160

K

karasu  94
killer whale  124
  Bigg's  128
  fish-eating  127
  Icelandic summer-spawning herring-feeders  130
  large type B (pack ice killer whale)  134
  North-east Atlantic mackerel-feeders  131
  North-west Atlantic  132
  Norwegian spring-spawning herring-feeders  131
  offshore  129
  resident  127
  small type B (Gerlache killer whale)  135
  Strait of Gibraltar bluefin tuna-feeders  132
  transient  128
  type A (Antarctic killer whale)  133
  type C (Ross Sea killer whale)  136
  type D (sub-Antarctic killer whale)  137
  west coast community  131

*Kogia breviceps*  82
*Kogia sima*  84
kuro-tsuchi  94

L

lag  156, 158, 160
*Lagenodelphis hosei*  154
*Lagenorhynchus acutus*  156
*Lagenorhynchus albirostris*  160
*Lagenorhynchus obliquidens*  158
Lahille's bottlenose dolphin  166
least rorqual  68
lesser beaked whale  106
lesser cachalot  82
lesser finback  68
lesser fin whale  58
lesser rorqual  68
lesser sperm whale  82
*Leucopleurus acutus*  156
lisso  162
*Lissodelphis borealis*  162
little blackfish  148
little finner  68
little piked whale  68
long-beaked common dolphin  186
long-finned pilot whale  142
Longman's beaked whale  102
longsnout  176
long-snouted spinner dolphin  176

M

*Megaptera novaeangliae*  72
melon-headed whale  148
*Mesoplodon bidens*  116
*Mesoplodon carlhubbsi*  110
*Mesoplodon densirostris*  112
*Mesoplodon eueu*  119
*Mesoplodon europaeus*  122
*Mesoplodon ginkgodens*  108
*Mesoplodon mirus*  118
*Mesoplodon perrini*  104
*Mesoplodon peruvianus*  106
*Mesoplodon stejnegeri*  120
*Monodon monoceros*  86
mud-digger,  44

mussel-digger  44

N

narwhal  86
narwhale  86
North Atlantic beaked whale  116
North Atlantic fin whale  54
North Atlantic humpback whale  72
North Atlantic long-finned pilot whale  142
North Atlantic minke whale  68
North Atlantic right whale  32
northern blue whale  48
northern bottle-nosed whale  100
northern bottlenose whale  100
northern fourtooth whale  92
northern Indian Ocean blue whale  48
northern minke whale  68
northern right whale  32, 36
northern right whale dolphin  162
northern rorqual  58
northern sei whale  58
North Pacific beaked whale  120
North Pacific bottlenose whale  92
North Pacific fin whale  54
North Pacific humpback whale  72
North Pacific minke whale  68
North Pacific right whale  36
North Sea beaked whale  116

O

orca  124
*Orcinus o. ater*  124
*Orcinus orca*  124
*Orcinus o. rectipinnus*  124
Owen's pygmy whale  84

P

Pacific gray whale  44
Pacific harbor porpoise  192
Pacific right whale  36
Pacific white-sided dolphin  158
Pantropical spotted dolphin  172
pep  148
*Peponocephala electra*  148

Perrin's beaked whale  104
Peruvian beaked whale  106
*Phocoena phocoena*  192
*Phocoena sinus*  194
*Phocoenoides dalli*  190
*Physeter macrocephalus*  78
pikehead  68
pollack whale  58
pothead  138, 142
pseudorca  144
*Pseudorca crassidens*  144
puffer  192
puffing pig  192
pygmy beaked whale  106
pygmy blue whale  48
pygmy fin whale  54
pygmy killer whale  146
pygmy sperm whale  82

# R

Ramari's beaked whale  119
razorback  54
Rice's whale  66
Risso's dolphin  150
rollover  176
rough-toothed dolphin  164

# S

sabre-toothed beaked whale  120
saddleback dolphin  186
Sarawak dolphin  154
sardine whale  58
Sato's beaked whale  94
scrag whale  44
sea canary  90
sea unicorn  86
sei whale  58
Senegal dolphin  182
sharp-headed finner  68
short-beaked common dolphin  186
short-finned pilot whale  138
short-headed cachalot  82
short-headed sperm whale  82
short-snouted spinner dolphin  182
Sibbald's rorqual  48

sittang  62
slopehead  164
snub-nosed cachalot  84
southern fin whale  54
southern humpback whale  72
southern long-finned pilot whale  142
southern sei whale  58
Sowerby's beaked whale  116
sperm whale  78
spinner  176
spinner dolphin  176
spotted dolphin  174
spotter  172, 174
spray porpoise  190
springer  156, 160
squidhound  160
steephead  100
Stejneger's beaked whale  120
*Stenella attenuata*  172
*Stenella clymene*  182
*Stenella coeruleoalba*  184
*Stenella frontalis*  174
*Stenella longirostris*  176
*Steno bredanensis*  164
streaker  184
striped dolphin  184
sulphur-bottom  48
sulphur-bottomed whale  48

# T

Tamanend's bottlenose dolphin  166, 170
tropical beaked whale  112
tropical bottlenose whale  102
tropical whale  62
'true' blue whale  48
true porpoise  190
True's beaked whale  118
True's porpoise  190
*Tursiops erebennus*  166, 170
*Tursiops truncatus*  166

# U

unicorn whale  86

# V

vaquita  194

# W

Western Pacific harbor porpoise  192
white-beaked dolphin  160
white-bellied spinner  176
whitebelly  184
whitebelly spinner  176
white-flank porpoise  190
white-side  156
white whale  90
wonderful beaked whale  118

# Z

*Ziphius cavirostris*  96